Stoke-on-Trent Libraries
Approved for Sale

HORACE BARKS
REFERENCE LIBRARY

STOKE-ON-TRENT

ADVANCES IN CERAMICS • VOLUME 13

NEW DEVELOPMENTS IN MONOLITHIC REFRACTORIES

Volume 1   Grain Boundary Phenomena in Electronic Ceramics
Volume 2   Physics of Fiber Optics
Volume 3   Science and Technology of Zirconia
Volume 4   Nucleation and Crystallization in Glasses
Volume 5   Materials Processing in Space
Volume 6   Character of Grain Boundaries
Volume 7   Additives and Interfaces in Electronic Ceramics
Volume 8   Nuclear Waste Management
Volume 9   Forming of Ceramics
Volume 10 Structure and Properties of MgO and Al_2O_3 Ceramics
Volume 11 Processing for Improved Productivity
Volume 12 Science and Technology of Zirconia II

ADVANCES IN CERAMICS • VOLUME 13

NEW DEVELOPMENTS IN MONOLITHIC REFRACTORIES

Edited by
Robert E. Fisher
Plibrico Co.
Chicago, Illinois

The American Ceramic Society, Inc.
Columbus, Ohio

Proceedings of an international symposium on new developments in monolithic refractories held at the 86th Annual Meeting of the American Ceramic Society, April 29–May 3, 1984, Pittsburgh, Pennsylvania.

Library of Congress Cataloging in Publication Data

International Symposium on New developments in Monolithic Refractories (1984 : Pittsburgh, Pa.)
   New developments in monolithic refractories.

   (Advances in ceramics ; v. 13)
   "Proceedings of an International Symposium on New Developments in Monolithic Refractories held at the 86th Annual Meeting of the American Ceramic Society, April 29–May 3, 1984, Pittsburgh, Pennsylvania"--Copr. p.
   Includes index.
   1. Refractory Materials--Congresses. I. Fisher, Robert E. 1940-   . II. American Ceramic Society. Meeting (86th : 1984 : Pittsburgh, Pa.) III. Title. IV. Series.
TA418.26.T65   1985   666'.72   85-6136
ISBN 0-916094-66-9
Coden: ADCEDE

©1985 by The American Ceramic Society, Inc. All rights reserved.

No part of this book may be reproduced, stored in a retrieval system, or transmitted in any form or by any means, electronic, mechanical, photocopying, microfilming, recording, or otherwise, without written permission from the publisher.

Printed in the United States of America.

Preface

The International Symposium on New Developments in Monolithic Refractories was held on April 30 and May 1, 1984 in Pittsburgh, Pennsylvania in conjunction with the 86th Annual Meeting of the American Ceramic Society (ACerS). The symposium was jointly sponsored by Committee 547 on Refractory Concrete of the American Concrete Institute and the ACerS. The present volume, number 13 of the series *Advances in Ceramics* published by the ACerS, comprises the symposium proceedings.

The organizing committee consisted of:
Robert E. Fisher, Plibrico Co., U.S.
William E. Boyd, Kaiser Refractories, U.S.
John L. Evans, British Steel Corp., U.K.
Timothy J. Fowler, Monsanto Corp., U.S.
Joseph Kopanda, Alcoa, U.S.
Wolfgang Kronert, Technical Institute, Aachen, Germany
Earl Seward, Lehigh Portland Cement Co., U.S.
Kiyoshi Sugita, Nippon Steel Corp., Japan
Raymond W. Talley, General Refractories, U.S.
Alfred F. Woolley, Dravo Engineers, Inc., U.S.

The symposium provided a state-of-the-art picture of this rapidly evolving area of refractories technology. Interest in the subject was evidenced by the fact that 200 or more attended the technical sessions in Pittsburgh. Refractories technologists from more than 30 countries requested information. The international nature of the syposium was evident from the seven nationalities of the authors.

The editor is especially grateful to Messrs. Boyd & Evans from the organizing committee, who also acted as associate editors on the manuscripts. Dr. David Lankard (Chairman of ACI 547) and Dr. Jess Brown (a member of ACI 547) also deserve thanks for their efforts in editing the manuscripts.

R. E. Fisher
Plibrico Co.

Contents

SECTION I: THE EVOLUTION IN WORLDWIDE MONOLITHIC REFRACTORIES USAGE

Recent Progress in Monolithic Refractories Usage in the Japanese Steel Industry 1
    Y. Shinohara, H. Yaoi, and K. Sugita

Recent Progress in the Use of Monolithic Refractories in Europe 21
    W. Kronert

Evolution of Monolithic Refractory Technology in the United States 46
    D. R. Lankard

SECTION II: PROCEDURES FOR TESTING THE PROPERTIES OF MONOLITHICS

Creep of Refractories: Mathematical Modeling 69
    D. J. Bray

Test Methods for Monolithic Materials 81
    G. C. Padgett and F. T. Palin

Compressive Stress/Strain Measurement of Monolithic Refractories at Elevated Temperatures 97
    W. R. Alder and J. S. Masaryk

Aggregate Distribution Effects on the Mechanical Properties and Thermal Shock Behavior of Model Monolithic Refractory Systems 110
    J. Homeny and R. C. Bradt

The Heat Evolution Test for Setting Time of Cements and Castables 131
    C. H. Fentiman, C. M. George, and R. G. J. Montgomery

SECTION III: STATE-OF-THE-ART ACTIVITIES IN INSTALLATION AND BAKEOUT OF MONOLITHICS

Introduction of Automatic Gunning Machines for Tundish Linings 139
    T. Morimoto, K. Ogasahara, A. Matsuo, and S. Miyagawa

Properties and Service Experience of Organic Fiber-Containing Monoliths 149
    T. R. Kleeb and J. A. Caprio

The Development of Dry Refractory Technology in the United States 161
    J. L. Turner, Jr. and D. M. Myers

A Review of International Experiences in Plastic Gunning ... 165
L. P. Krietz, G. Wilson, D. Hofmann, and M. Tsukino

Viscosity and Gunning of Basic Specialties 175
W. Siegl

Dryouts and Heatups of Refractory Monoliths 192
N. W. Severin

SECTION IV: LOW-CEMENT CASTABLES

Effect of Microsilica on Physical Properties and Mineralogical Composition of Refractory Concretes 201
B. Monsen, A. Seltveit, B. Sandberg, and S. Bentsen

Vibrated Castables with a Thixotropic Behavior 211
R. Stieling, H.-J. Kunkel, and U. Martin

High-Technology Castables 219
E. P. Weaver, R. W. Talley, and A. J. Engel

High-Performance Castables for Severe Applications 230
C. Richmond and C. E. Chaille

Progress of Additives in Monolithic Refractories 245
Y. Naruse, S. Fujimoto, S. Kiwaki, and M. Mishima

Low-Moisture Castables: Properties and Applications 257
S. Banerjee, R. V. Kilgore, and D. A. Knowlton

A New Generation of Low-Cement Castables 274
B. Clavaud, J. P. Kiehl, and J. P. Radal

Calcium Aluminate Cements for Emerging Castable Technology 285
G. MacZura, J. E. Kopanda, and F. J. Rohr

SECTION V: MONOLITHICS FOR BLAST FURNACE USAGE

Designing a Casthouse for Preformed Shapes 305
R. A. Howe, J. W. Kelley, and T. A. Dannemiller

Application of Dry-Forming Method to Blast Furnace Troughs 313
S. Nishizawa and A. Kondo

The Use of Monolithic Refractories in Blast Furnaces 323
L. Krietz, R. Woodhead, S. Chadhuri, and A. Egami

Wear Mechanisms in Alumina-Silicon Carbide-Carbon Blast Furnace Trough Refractories 331
S. B. Bonsall and D. K. Henry

Progress in Casting Trough Materials and Installing Techniques for Large Blast Furnaces 341
Y. Toritani, T. Yamane, S. Yamasaki, I. Nishijima, T. Kawakami, and Y. Kadota

Comparison of Monolithic Refractories for Blast Furnace Troughs and Runners 355
    C. M. Jones

SECTION VI: MISCELLANEOUS NEW APPLICATIONS FOR MONOLITHICS

Reactions of Alumina-Rich Ramming Mixes with Lignite Ashes in Reducing Atmospheres at High Temperatures 365
    P. Dietrichs and W. Kronert

The Properties and Applications of Dolomite Ramming Mixes 388
    J. W. Stendera

SiC Monolithics for Waste Incinerators: Experiences, Problems, and Possible Improvements 395
    G. S. Dhupia, W. Kronert, and E. Goerenz

Development of Spinel-Based Specialties: Mortars to Monoliths 411
    A. Cisar, W. W. Henslee, and G. W. Strother

Section I
The Evolution in Worldwide Monolithic Refractories Usage

Recent Progress in Monolithic Refractories Usage in the Japanese Steel Industry 1
    Y. Shinohara, H. Yaoi, and K. Sugita

Recent Progress in the Use of Monolithic Refractories in Europe 21
    W. Kronert

Evolution of Monolithic Refractory Technology in the United States 46
    D. R. Lankard

Recent Progress in Monolithic Refractories Usage in the Japanese Steel Industry

Yasuaki Shinohara, Hideo Yaoi, and Kiyoshi Sugita

Nippon Steel Corp.
6-3 Otemachi, 2-Chome, Chiyodaku
Tokyo, Japan

The use of monolithic refractories in the Japanese steel industry is approaching 50% of the total refractories used in this industry. In the technology for using monolithic refractories, material selection techniques and lining techniques are closely connected. Typical methods of lining with monolithic refractories include casting, vibration, and ramming, and these refractories are applied to a wide variety of equipment, including iron troughs, ladles, and tundishes. Further technical studies must be made on (1) drying and heating techniques, (2) highly corrosion-resistant materials, and (3) techniques for controlling the quality of refractory linings.

About 30 years have passed since monolithic refractories were introduced into Japan, and today monolithic refractories are indispensable to steel manufacturing technology.

At first they were used mainly for the lining of slab reheating furnaces and soaking pits. Influenced by the introduction of the sand-slinger method for lining ladles in the early 1970s, such various lining methods as casting, ramming, and vibration-forming were developed, with the result that monolithic refractories came to be widely used to line molten metal troughs and vessels as well.

Repair techniques were also developed for furnace life extension and refractory cost reduction; monolithic refractories play an important role in this area, also.

With these circumstances for a backdrop, the use of monolithic refractories by the Japanese steel industry has continued to increase and now comprises nearly 50% of refractories consumed by the industry (Fig. 1).[1]

This paper reports mainly on technology for using monolithic refractories in the Japanese steel industry's major equipment. Future subjects for study, and prospects for the future regarding these materials are also described.

Monolithic Refractories for Ironmaking

Iron Trough Refractories

Technical innovations for iron trough refractories began to develop rapidly in the late 1970s. Although ramming had been the traditional method used to line iron troughs, vibration-forming, casting, and other new methods also came to be used to raise the lining efficiency, save lining labor, and extend the service lives of iron runners. The fact that these new lining

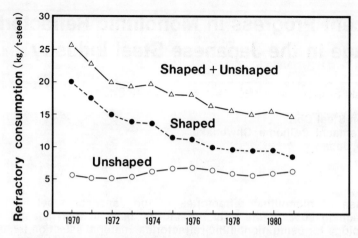

Fig. 1. Refractory consumption in the Japanese steel industry.

methods can effectively perform intermediate repairs is significant both technically and economically.

The advent of these new lining methods was made possible by the development not only of new lining devices but also new material techniques. The use of ideas and principles of powder technology and rheology (and particularly the utilization of submicrometer particles) made an important contribution to the development of new lining materials.

Vibration-Forming Method: The vibration-forming (VF) process and the Shinagawa vibration process (SVP) method are the methods presently used in Japan.[2,3] The VF process, which was developed in the early 1970s, provided a basis for developing subsequent processes. This process utilizes the thixotropic properties of low-moisture, refractory particle mixes that become fluid when mechanically vibrated and return to the hardened state upon standing. Figure 2 shows the repair of a main iron trough using this process. The

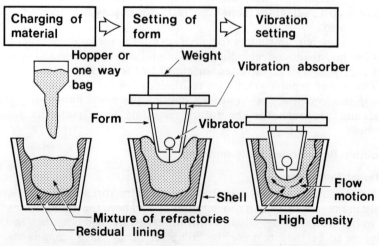

Fig. 2. Trough relining procedure using the VF process.

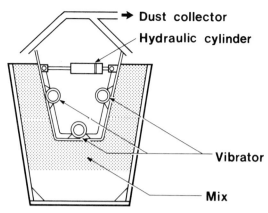

Fig. 3. SVP vibration process for trough relining.

refractory material is charged into the iron trough, and the form is settled into the material and vibrated to reline the trough.

Figure 3 shows a schematic view of the SVP method, a dry-type vibration method developed in the late 1970s. While the VF process uses a mix containing about 5% moisture, the SVP method uses a mix which contains no liquid binder (except 0.5% oil for dust prevention). This method produces a lining that does not need drying and is free from the problem of explosion

Table I. Properties of Trough Lining Materials

	VF Process	SVP Method
Chemical composition (%)		
Al_2O_3	77	76
SiO_2	3	2
SiC + C	17	15
Water content (%)	5.0	0
Apparent porosity (%)	110°C/24 h 17.2 1400°C/ 1 h 22.2	180°C/18 h 26.1 1500°C/ 3 h 25.0
Bulk density (g/cm³)	110°C/24 h 2.92 1400°C/ 1 h 2.82	180°C/18 h 2.72 1500°C/ 3 h 2.71
Modulus of rupture (kg/cm²)	110°C/24 h 51 1400°C/ 1 h 10	180°C/18 h 55 1500°C/ 3 h 35

Table II. Performance Result of Main-Trough Linings

	VF Process	SVP Method
Life (day)	25~35	30~32
Hot metal throughput (tons)	50 000~100 000	60 000~70 000

even on rapid heating. It is a lining method particularly suitable for stationary troughs.

Tables I and II show examples of the properties and the performances of iron trough refractories in which the vibration-forming methods were used.

Casting Method: The first trough refractory casting method, the N-CAST process, is based on the mechanization and systematization of the element operations in the conventional forming of refractory concretes.[4]

The refractory material is mixed in a large, continuous mixer, transferred to the site by feed pump, and cast into a preplaced form. Figure 4 is a diagram of these operations. The greatest advantage of this method is that relining can be performed without having to demolish the residual lining, thereby greatly saving refractory material.

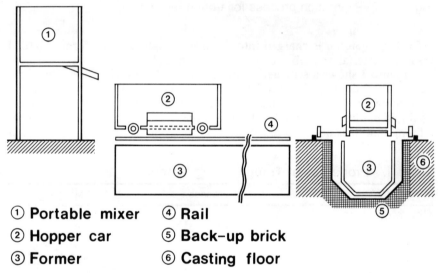

① **Portable mixer**   ④ **Rail**
② **Hopper car**   ⑤ **Back-up brick**
③ **Former**   ⑥ **Casting floor**

Fig. 4. N-CAST process for trough relining.

Figure 5 shows an example of the reduction in the refractory consumption at large blast furnaces achieved by the use of this method.[5]

Utilizing the colloidal property of clay, both fluidization and solidification characteristics are imparted to the refractories used in the N-CAST process by the addition of small quantities of binders. Table III shows an example of the quality of trough refractories used in the N-CAST process.

Although casting methods other than the N-CAST process have been developed and commercialized, they are not much different from the N-CAST process in principle and technique.[3] All these casting methods have spread rapidly over Japan and are now used by almost all Japanese steelmakers.

Refractories for Hot Metal Transfer Equipment

Hot metal transfer equipment consists of hot metal ladles and mixer cars. The sand-slinger method was used to line hot metal ladles in some cases

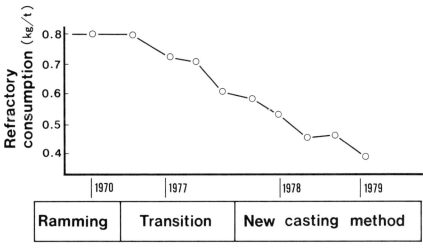

Fig. 5. Typical trend in main trough-lining consumption.

in the past, but at present this is rarely used for this purpose.[6] There are two reasons for this: (1) hot metal ladles are being superseded by torpedo ladles, and (2) the method of lining hot metal ladles was influenced by the change in the method of lining molten steel ladles (as described later) because the ladles lined using the sand-slinger method were used as both hot metal ladles and molten steel ladles. In torpedo cars, monolithic refractories are used mainly for lining the mouth and for intermediate relining of the cars. High-alumina, low-cement, castable refractories are mostly used as monolithic refractories for lining the mouths of mixer cars.

Table III. Properties of Trough Lining Material (N-CAST Process)

Chemical composition (%)	SiO_2	3– 4
	Al_2O_3	67–68
	SiC	17–18
	F·C	5– 6
Water content (%)		9.0
Bulk density (g/cm^3)	110 °C/24 h	2.63
	1450 °C/ 2 h	2.62
Apparent porosity (%)	110 °C/24 h	21.8
	1450 °C/ 2 h	25.3
Modulus of rupture (kg/cm^2)	110 °C/24 h	12.7
	1450 °C/ 2 h	36.9
Crushing strength (kg/cm^2)	110 °C/24 h	150.0
	1450 °C/ 2 h	209.0
Reheat linear change (%)	1450 °C/ 2 h	0.0

Monolithic Refractories for Steelmaking

Blast Oven Furnaces (BOF)

Although monolithic refractories are applied to some BOF linings in the United States, the application has not yet been commercialized in Japan. Firebrick is used for BOF brickwork in this country.

In the late 1950s when BOF was introduced into Japan, the furnace life was 100–400 heats, and the refractory consumption was 6–12 kg/ton steel. In the 1970s, furnace lives of over 5 000 heats were recorded in succession, and even the record of 10 000 heats was broken due to the subsequent progress in operation techniques and refractory techniques.[7] The progress of gunning repair and other repairing techniques greatly contributed to prolonging BOF life, although dynamic furnace control operation, slag control, and other operational techniques also made considerable contributions to extending lining life.

The BOF gunning repair method was introduced in Japan in the late 1960s. At first the wet-type gunning repair method was used, but in the late 1970s it was replaced by the more durable dry-type gunning repair method.

Silicate binders were first used for the gunning mix. Phosphate binders developed in the early 1970s showed much better results, and at present these are used in most cases, except for the repair of stainless steel refining furnaces. Table IV shows the quality of representative gunning mixes.

Table IV. Properties of Gunning Mixes (BOF)

	Phosphate Bond	Silicate Bond	Phosphate Bond + Carbon Bond
Chemical Composition (%)			
MgO	80	93	76
CaO	11	1	10
P	2		2
C			5
Max. grain size (m/m)	3	3	3
Hot modulus of rupture (kg/cm^2) 1400°C/15 min	4	4	14

In the 1980s, a new repairing technique, flame gunning repair, is being commercialized[8-10] (Fig. 6). Flame gunning repair for BOFs developed or introduced in Japan are classified into three types by the kind of the fuel used: gaseous fuel, liquid fuel, and solid fuel. With all three fuels, refractory particles are melted or half-melted in a 2000°–2800°C flame obtained by burning fuel with oxygen, and jetting the molten or half-molten particles onto the part to be repaired to obtain a dense and highly corrosion-resistant lining. The flame gunning method is far more effective than conventional gunning repair.

Table V shows the features of the various flame gunning methods commercially used in Japan for repairing BOFs. Figure 6 shows the gaseous-fuel-type BOF lava flame device.

Table V. Flame Gunning Repair for BOF

	A	B	C
Fuel	Coke	Kerosene	Propane
Capacity (ton/h)	9~36	3	1.5
Flame temperature (°C)	1800~2000	2400~2500	2300~2400
Gunning mix	$MgO\text{-}SiO_2$	$MgO\text{-}SiO_2$	$MgO\text{-}SiO_2$
Energy consumption (kcal/ton·mix)	2700×10^3	6500×10^3	3500×10^3
Durability SUS	max. 2 ch	max. 9 ch	max. 8 ch
Plain steel	max. 11 ch	max. 23 ch	max. 18 ch

Although BOF gunning mix consumption is decreasing due to the change in BOF operating conditions (such as the advent of highly durable magnesia carbon brick), the share of gunning mixes in the total consumption of refractories for BOFs is still large at about 30%.

Ladles

The use of monolithic refractories for ladles in Japan began with the introduction of the sand-slinger method in the early 1970s. The sand-slinger method was introduced from Europe to cope with the short supply of ladle brick in Japan in those days, as well as to meet the requirements for mechanization of ladle relining to save labor. These needs served as a stimulus to the development of the ramming, vibration-forming, and casting methods.

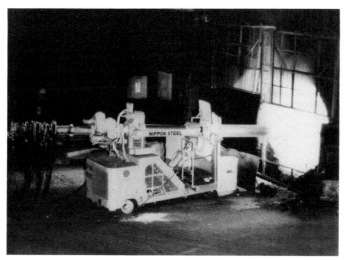

Fig. 6. Flame gunning for BOF.

Table VI. Monolithic Refractory Lining Methods for Ladles in the Japanese Steel Industry

Method	Industry	
Slinger	1 Company	1 Work
Casting	3 Companies	5 Works
Ramming	3 Companies	3 Works
Vibration	1 Company	1 Work

Table VI shows the present status of the usage of monolithic refractories for ladles by Japanese steel companies. The number of ladles lined with monolithic refractories has now stopped increasing, partly because of investment control and labor supply stabilization resulting from a changed economic situation, but mainly because of such changed operational conditions as the rise in the BOF tapping temperature, the increase in the secondary refining ratio and continuous casting ratio, and the trend toward high-purity steels. These changes in operational conditions require higher performance than ever before in ladle linings, and the present refractories cannot always meet the requirements adequately.

To increase the use of monolithic refractories for ladle lining, it is necessary to develop both high-corrosion-resistant materials (such as basic refractories) and new lining methods. If the development of these materials cannot be advanced, the use of monolithic refractories for ladles will continue at its present level.

Slinger Method: The slinger method of ladle relining was first commercialized in West Germany in the late 1960s and spread to other European steelmaking countries.[11] In the early 1970s, Japan was introduced to this method. In this method, the ladle is relined by slinging a refractory between the permanent lining and a form set up in the ladle.

Table VII. Properties of Slinging Mass

	A	B	C
Chemical composition (%)			
SiO_2	90.5	65.9	19.2
Al_2O_3	5.4	7.6	2.7
ZrO_2		23.5	
MgO			71.2
Water content (%)	7~8.5	5~7	
Apparent porosity at 1500°C/2 h (%)	15.6	20.6	23.0
Thermal expansion at 1000°C (%)	1.48	1.30	1.17

The main refractory used in this method is Belgian natural sand in Europe. In Japan, synthetic sand is mainly used, since no natural sand of this kind is available. High-silicate sand and zircon sand were used depending on service conditions, and basic sand was also tried.[12] Table VII shows the properties of these refractories.

As the packing density of sand-slung linings is lower than that of brick, the service life is generally only 80-90% of that of brick-lined ladles. However, there is one report on a ladle lined by the slinger method and used for a period longer than brick-lined ladles.[13]

The slinger method is advantageous in that it takes only a short time to line ladles and has a large labor-saving effect; however, it has its drawbacks: The packing density is low, and care must be taken for proper drying and heating after relining. It is also a weakness of this method that the lining thickness must be balanced according to the quality of the refractory, since the use of a fixed-shape form does not permit variations in the lining thickness.

The slinger method in Japan was at its peak in the late 1970s, when three steel companies used it. Thereafter, the use of this method decreased, primarily because of the reduction in lining life due to the growing severity of ladle service conditions and the development of new techniques for relining ladles with monolithic refractories. At present, it is used at only one steelworks, so it may be safely said that the era of the slinger method in Japan has ended.

Ramming Method: The ramming method of relining ladles is advantageous in that the physical properties of the installed linings are similar to those of brick and also that jointless linings (which cannot be expected from brickwork) are obtained. Ramming is performed in many parts of the United States and Europe. The ramming method developed in Japan is called the LSM process.[14]

Figure 7 shows an outline of the ladle stamping method (LSM) process.

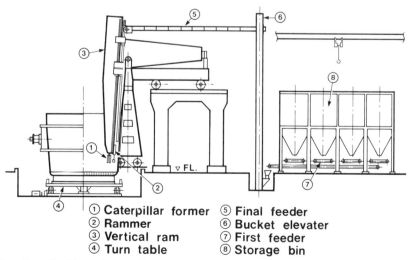

① Caterpillar former   ⑤ Final feeder
② Rammer   ⑥ Bucket elevater
③ Vertical ram   ⑦ First feeder
④ Turn table   ⑧ Storage bin

Fig. 7. Outline of the LSM process.

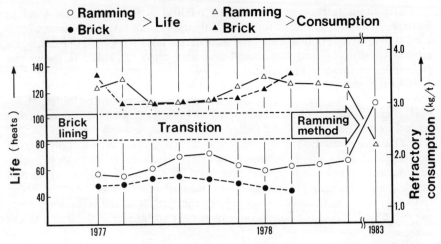

Fig. 8. Lining life and refractory consumption of ladle (ramming method).

The greatest feature of this process is the nonuse of forms in relining; a caterpillar-type, rotating body is used instead. While the refractory is being fed into the space between the caterpillar and the permanent lining, one or more rams are operated to automatically line the ladle. Figure 8 compares a brick-lined, 320-ton ladle with a 320-ton ladle lined by the LSM process in terms of lining life and refractory consumption. This figure shows that the latter ladle is superior in both lining life and refractory saving. While the lining efficiency for the 320-ton ladles was 1.7 ton/man/7 h for bricklaying, that for the LSM process was 4.0 ton/man/7 h.[15]

The refractories used in this process are mainly zircon-roseki materials; special basic materials are sometimes used for the upper and lower slag lines. Table VIII shows the properties of these refractories.

Ramming-process linings so produced are dense and consequently slow to wear, but it takes a longer time than the slinger method to reline ladles, the

Table VIII. Properties For Ramming Mixes

	Semi + Zircon (A)	Semi + Zircon (B)	Special Basic
Chemical composition (%)			
SiO_2	61	54	11
Al_2O_3	9	12	37
ZrO_2	27	30	
MgO			10
Cr_2O_3			24
Apparent porosity (%)	14.5	15.4	17.0
Bulk density (g/cm³)	2.82	2.87	3.15
Crushing strength (kg/cm²)	125.0	96.0	350.0

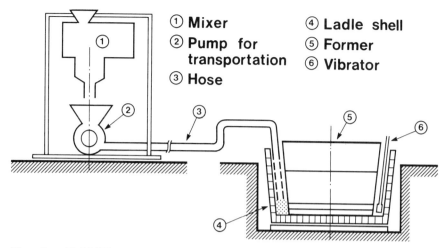

Fig. 9. N-CAST process for ladle lining.

lining thickness can hardly be changed at different parts of the ladle, and additional relining cannot be performed.

Casting Method: Japan's earliest casting method for relining ladles is the N-CAST process established in the late 1970s.[16] The principle of this process is the same as the N-CAST process for troughs. The material to which fluidization and solidification characteristics are imparted in advance is cast between the form and the permanent lining (Fig. 9).

The greatest advantage of this process is that, in addition to being able to perform overall repair of ladle linings, it can restore partly damaged linings to their original state by additional relining (Fig. 10). The lining life can be extended indefinitely by repeating this additional relining. There have been cases in which the lining life exceeded 1000 heats (Fig. 11).[17] The reduction in refractory consumption is substantial—up to 6-8 relinings, but the reduction

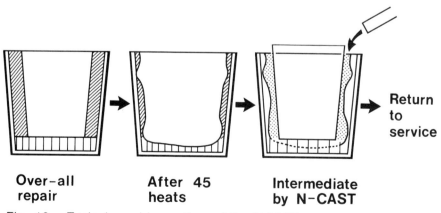

Fig. 10. Typical repairing pattern of the N-CAST process.

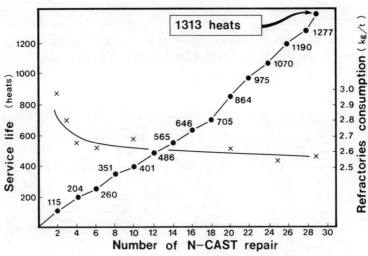

Fig. 11. Performance results of ladle lining (N-CAST process).

after that is small. Therefore, extending the lining life beyond 300–500 heats is of little use in terms of refractory costs. Zircon-roseki materials are mainly used in this process (Table IX). Moreover, this process has the following

Table IX. Properties of Casting Mixes

Items	A	B	C
Main component	ZrO_2-SiO_2	ZrO_2-SiO_2	SiO_2
Chemical analysis (%)			
SiO_2	58	51	93
Al_2O_3	8	6	4
ZrO_2	30	38	
Grain size (%)			
>1 mm	43	33	43
<0.074 mm	32	31	30
Water addition (wt%)	7	7	8~5
Bulk density after drying	2.72	2.78	2.17
1400°C/3 h	2.40	2.64	1.85
Apparent porosity (%)			
after drying	17	20	18
1400°C/3 h	22	21	24
Crushing strength (kg/cm^2)			
after drying	3	129	28
1400°C/3 h	192	390	236
Linear shrinkage after drying	−0.26	−0.19	−0.10
1400°C/3 h	+3.35	+2.12	+5.20

additional advantages: The labor-saving effect is great, the installation investment is small, and no large space is needed for equipment installation.

However, some problems attributable to the added moisture are inevitable in the casting method. The material used in this process requires a moisture content of 7-8% minimum to secure the fluidity necessary for casting. This value is higher than the moisture content of the materials used in any other lining method mentioned above. Therefore, the porosity of the linings is relatively high, and precautions are necessary to control the drying and heating operations. These disadvantages must be eliminated in the future development of this method of casting refractory materials.

Vibration-Forming Method: The only vibration-forming method commercialized in Japan for lining ladles is the VF process.[18] Its principle is the same as that of the VF process for lining troughs (Fig. 12). The materials used in this process are mainly zircon-roseki materials, and the moisture addition is 5-6% lower than for the cast refractory (Table X).

This process has the advantages of low moisture addition to the material, dense and highly corrosion-resistant linings, and the possibility of additional relining as in the N-CAST process. Its disadvantage is high equipment cost. As the service conditions for ladles become more severe in the future, the use of this process and the ramming process will probably increase.

Tundish

Monolithic refractories are used in continuous-caster tundishes for two purposes: to line the tundishes and to coat the surface of linings. Roseki brick is generally used to line tundishes because of its low price. The tundish lining seldom wears as it is covered with coating material or insulation board, but partial or total relining must be performed from time to time because the brick cracks, or the lining becomes loose due to the repetition of heating and cooling, or the removal of adhering metal becomes necessary.

Monolithic refractories are used to line tundishes to save labor needed for relining and to reduce refractory cost by preventing the loosening of the

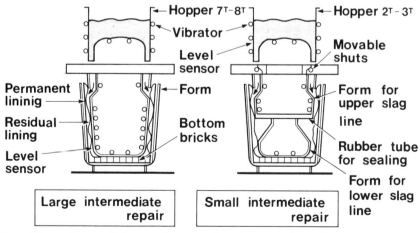

Fig. 12. Intermediate repair procedure for ladle by VF process.

Table X. Properties of Vibration-Forming Mixes

		A	B
Chemical composition (%)	SiO_2	54.78	49.5
	Al_2O_3	4.04	2.90
	ZrO_2	39.50	46.11
Permanent linear change (%)	1400°C/ 2 h	+1.63	+1.98
Modulus of rupture (kg/cm^2)	110°C/24 h	53.8	59.8
	1400°C/ 2 h	34.4	35.7
Crushing strength (kg/cm^2)	110°C/24 h	123	114
	1400°C/ 2 h	221	178
Apparent porosity (%)	110°C/24 h	14.8	6.0
Bulk density (gr/cm^3)	110°C/24 h	3.00	3.17
Apparent porosity (%)	110°C/ 2 h	9.7	21.1
Bulk density (gr/cm^3)	110°C/ 2 h	2.78	2.94
Water content (%)		4.5	4.3

lining. The following methods are mainly used: the vibration-forming method, the casting method, the block method, or the ramming method.

The lining surface is generally covered with coating material or insulation board to facilitate the removal of metal and scum from the lining surface and to prevent the contamination of molten steel with such adhering materials on the tundish surface at the time of the subsequent cast. In Japan, coating materials are used more widely than insulation board for these purposes.

Vibration-Forming Method: The only vibration-forming method applied to tundishes in Japan again is the VF process. Its principle is the same as that of the VF process for lining ladles and troughs. The form used is of the settling type.

Fireclay-zircon (e.g., SiO_2: 48%; ZrO_2: 33%; Al_2O_3: 17%) is used in this process. It has a life 2.0–2.5 times longer than brick (brick: 33–47 heats, VF: 75–89 heats), and refractory consumption is about one-half (brick: 0.92–1.16 kg/ton steel, VF: 0.48–0.57 kg/ton steel). Also, the lining efficiency is about 3.5 times higher than for brick linings.

Casting Method: The N-CAST process and processes similar to it are used in Japan to line and coat tundishes by casting.[19]

Zircon-roseki materials (e.g., SiO_2: 51%; ZrO_2: 40%; Al_2O_3: 7%) are mainly used in these processes. Four Japanese steelmakers use the casting method at four steelworks.

According to one report, the refractory consumption is 0.65 of brick lining.[20]

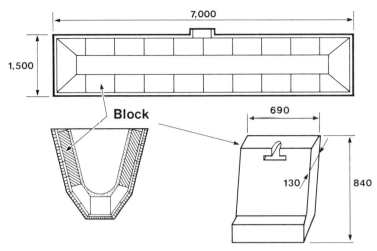

Fig. 13. Block lining method for tundish.

Block Method: This method uses large blocks to line tundishes. It was developed to decrease the number of joints in a lining and to reduce labor needed for relining tundishes.

Figure 13 gives an outline of this method. In this example, the tundish sidewalls are built using large, preformed blocks, each about 150 kg.

Preformed blocks of fireclay or high-alumina castable refractories are used to reduce refractory cost by about 30% and raise the bricklaying efficiency by about 20%.[21] In other cases, preformed blocks of ramming material are used.

Ramming Method: Since it is relatively easy to use the ramming method for lining tundishes, this method is already commercialized mainly for lining the bottom of small tundishes. However, since this method has not yet been fully mechanized, ways to improve its mechanization must be developed.

Coating Method: As mentioned previously, the tundish lining surface is covered with insulation board or coating material when the molten steel is received, and the relative merits and demerits of these covering means are

Table XI. Comparison of Insulation Boards and Coating Materials (Tundish)

	Insulation Boards		Coating Materials
Erosion resistance	Worse	<	Better
Energy savings	Better	>	Worse
Steel quality	Worse	<	Better
Workability	Worse	≤	Better
Cost	Worse	<	Better

qualitatively shown in Table XI. In Japan, coating material is used in more cases than insulation board.

At first, high-alumina coating materials (Al_2O_3: 50–60%) were mainly used, but at present magnesia coating materials are mainly used because material containing less SiO_2 is required to ensure product quality.

Insulation board is superior to coating material in its insulation property, so the development and commercialization of insulation coating material is also being promoted.[22]

Table XII shows the quality characteristics of various coating materials. Although the coating of tundish lining surfaces was formerly performed manually, gunning a coating was recently developed for labor reduction and is already commercialized at several steelworks.[23]

Table XII. Properties of Coating Materials for Tundish

	A	B	C
Chemical analysis (%)			
SiO_2	5	7	32
Al_2O_3	1	13	61
MgO	91	78	
Bulk density (gr/cm³)			
110°C/24 h	2.29	1.66	2.24
1500°C/ 3 h	2.36	1.64	2.15

Others

In addition to the above equipment, there are other important items of steelmaking equipment for which monolithic refractories are used, such as tubes of degassing systems, the lances for hot metal pretreatment, and molten steel secondary refining. These are omitted from this paper.

Monolithic Refractories for Rolling

The history of castable refractories and air-set-type plastic refractories in the United States dates back to the 1930s. They were introduced into Japan in the late 1940s as materials for lining boilers. It was in the late 1960s that monolithic refractories were first used regularly to line rolling mill furnaces in the Japanese steel industry. The application started with soaking pit covers and expanded to the sidewalls and roofs of soaking pits and slab reheating furnaces. In the 1970s, monolithic refractories began to be used for lining all parts of furnaces except for hearths.

Monolithic refractories for rolling mill furnaces have improved in both material quality and application. The clay-bond castable refractories developed in the mid-1970s combine the high workability of castable refractories and the high durability of plastic refractories.[24] They are presently used widely in Japan to line the sidewalls and roofs of rolling mill furnaces.

Also, the low-cement, castable refractory technique introduced from France in 1977 influenced many Japanese castable refractory techniques and led to the development of various castable refractories and related techniques. This technique is based on the dispersion and coagulation of submicrometer particles and is used to line skid pipes of reheating furnaces and the gas injection lances in the steelmaking area. Favorable results are shown in these applications because of the high strength and high corrosion resistance peculiar to these refractories.

Plastic refractories which can be applied by gunning were recently developed and have been commercialized. Table XIII shows the characteristics of the above new materials. Several techniques for the application of these materials are already established to save labor and shorten work schedules.

Table XIII. Properties of Advanced Monolithic Refractories

	Clay-Bond, Castable Ref.	Low-Cement, Castable Ref.	Plastic Ref. For Gunning
Chemical analysis (%)			
SiO_2	60	65	42
Al_2O_3	35	32	52
CaO		1	
Bulk density (%)			
110°C/24 h	2.31	2.45	2.13
1500°C/ 3 h	2.22	2.40	
Modulus of rupture			
110°C/24 h	55	85	40
1500°C/ 3 h	315	200	87

Several examples of castable refractory application techniques are high-efficiency applications using large-capacity mixers and feed pumps, and prefabrication of sidewall and roof linings.[25]

Examples of plastic refractory application techniques include lining using small blocks,[26] the gunning technique mentioned above, as well as the prefabrication method which can be used as in the case of castable refractories.

Techniques for applying monolithic refractories to rolling mill furnaces are established, but the present surrounding conditions, including the adoption of the continuous casting-direct rolling (CC-DR) process, call for the extension of furnace lives and the shortening of relining time. The advent of new materials and new methods is looked forward to by the steel industry.

Future Prospects and Subjects for Study

Although the ratio of use of monolithic refractories by the Japanese steel industry appears to have stopped increasing recently, it will increase further in the long run due to energy-saving, labor-saving, and resource-conservation needs. Although predictions for its future growth are not in complete agreement, the share of monolithic refractories in the total consumption of refractories by the industry 10 years from now may be estimated at 55-65%.

The advantages of the use of monolithic refractories may be summarized as follows.

Energy Savings

Table XIV shows a sample calculation of the total energy consumed for the refractories used by a steelmaker.[27] The energy needed for manufacturing refractories accounts for 20% of the total energy, and the rest is accounted for mostly by the energy for firing them. Although a large energy-saving effect can be expected from making refractories free from the necessity of firing, the energy for preparing raw materials must be considered. The preparation of raw materials for monolithic refractories tends to require greater energy than the preparation of those for fireclay. Therefore, due consideration must be given to overall energy savings.

Table XIV. Energy Consumption of Refractories in Production and in Application at Steelworks ($\times 10^3$ kcal/tons·Ref.)

Raw Material	Production	Transport	Installing	Drying	Total
4500 (62%)	1400 (20%)	220 (3%)	80 (1%)	950 (13%)	7150 (100%)

Labor Savings

The adoption of monolithic refractories will save labor in both the manufacture and application of refractories. Japanese refractory makers' firebricks manufacturing efficiency is about 10 ton/man/month, while their monolithic refractories manufacturing efficiency is 50–100 ton/man/month. A major problem involved in this manufacturing efficiency is the size of the manufacturing lots, particularly as the kinds of Japanese monolithic refractories are increasing in recent years.

Mechanization of the application of monolithic refractories is easy because of the nature of the refractories. In the case of the ladle application mentioned above, the lining efficiency is 1.5–2.5 times that for brick.[13]

Resource Conservation

The greatest feature of a monolithic refractories lining is that additional repair (relining) is possible. Demolition of refractories theoretically can be reduced to zero when relining. Figure 14 shows an example of a material balance for refractories at an integrated steelworks for 1981 vs 1974.[28] (1981 information is indicated by solid lines, and 1974 data are given by dotted lines.) As this figure shows, the quantity of demolished refractories decreases with increased use of monolithic refractories.

Although the consumption of monolithic refractories is expected to increase further, there are several problems to be solved regarding their manufacture.

Drying and Heating Techniques

Most monolithic refractories contain 3–4% moisture at the time of application, and some of them contain even more than 10% moisture. If their drying and heating conditions are not appropriate, troubles occur, such as bursting due to the sudden generation of steam inside the lining, and the deterioration of the structure due to the shift of the binders as a result of the

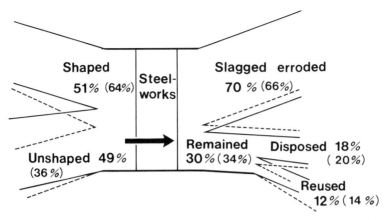

Fig. 14. A material balance for refractories at an integrated steelworks (1981 vs 1974).

shift of the moisture. For this reason, heating equipment and operations tend to become more complicated than those for brick lining. As a means of solving this problem, the microwave drying technique was developed and put to commercial use for ladle lining.[29]

High-Corrosion-Resistant Monolithic Refractories

Service conditions for refractories are becoming more and more severe (mainly in the steelmaking area) due to the adoption of hot-metal pretreatment, the diversification of secondary refining, and the increase in the continuous casting ratio. This circumstance impedes the application of monolithic refractories in the steelmaking area, particularly to ladles. To break through this situation, it is essential to develop highly corrosion-resistant monolithic refractories (including basic materials) and techniques for applying these materials. The development of such new materials and techniques will make it possible to use monolithic refractories for lining mixer cars, BOFs, degassing systems, etc., which at present cannot be lined with these refractories.

Techniques for Controlling the Quality of Linings

To a greater extent than for firebricks, the application quality of monolithic refractories influences the quality of the linings produced. Although the lining operation for monolithic refractories is largely mechanized, the present techniques cannot completely obviate human error. Under such circumstances, the techniques for deciding whether or not monolithic refractory linings can adequately withstand service are not sufficient. As monolithic refractories are expected to be used in more important parts of equipment in the future, the development of quality control methods or test methods for installed linings must be studied.

Summary

Monolithic refractory techniques used by the Japanese steel industry are outlined above mainly regarding application techniques. Prospects for the future of these materials are not always bright, and there remain many problems to be solved. However, these difficulties with monolithic refractories will be overcome and their position in the industry established with the advent of new application systems and new materials developed by the cooperation between refractories manufacturers and users.

References

[1] M. Saigusa, Shiraishi Memorial Seminar. July 1983.
[2] T. Ochiai et al., *Tetsu-to-Hagane,* **61** [12] 9 (1975).
[3] M. Nishi et al., *Taikabutsu Overseas,* **1** [1] 163 (1981).
[4] H. Tanaka et al., Iron and Steel Inst. Japan, Refractories Joint Committee Report No. 21-2-10 (1975).
[5] K. Sugita et al., Autumn Meeting of the Refractory Materials Section of the British Ceramic Society. Middlesborough, October 1979.
[6] K. Goto et al., *Seitetsu Kenkyu* (Nippon Steel Technical Report) No. 283; p. 71 1975.
[7] K. Sugita, *Tetsu-to-Hagane,* **65** [9] 1462 (1979).
[8] T. Hagiwara, *Tetsu-to-Hagane,* **68** [11] 232 (1982).
[9] T. Morimoto et al., 1st International Conference on Refractories. Tokyo, November 1983.
[10] T. Kurahashi et al., 1st International Conference on Refractories. Tokyo, November 1983.
[11] H. Kleechult, *Stahl und Eisen,* **89** [16] 859 (1969).
[12] O. Terada et al., *Taikabutsu,* **32** [268] 273 (1980).
[13] M. Tate, *Taikabutsu,* **26** [202] 515 (1974).
[14] H. Shibata, *Inter. Ceram.,* **127,** 304 (1978).
[15] Nippon Steel Corp. Oita Works; Iron and Steel Inst. Japan, Refractories Joint Committee Report, 1983.
[16] H. Tanaka et al., *Taikabutsu,* **30** [243] 215 (1978).
[17] T. Yamaguchi et al., *Taikabutsu,* **31** [261] 540 (1979).
[18] H. Shibata; ILAFA-ALAFA Congress. Lima, November 1980.
[19] H. Shibata et al., *Taikabutsu,* **31** [260] 468 (1979).
[20] K. Shimada et al., *Tetsu-to-Hagane,* **68** [4] S182 (1982).
[21] H. Katayama et al., *Taikabutsu,* **31** [256] 231 (1979).
[22] K. Isomura; AIME Steelmaking Conference. Pittsburgh, March 1984.
[23] H. Shibata, Ceramic Data Book '80; p. 137, 1980.
[24] H. Hujie et al., *Taikabutsu,* **27** [209] 249 (1975).
[25] M. Handa, *Taikabutsu,* **26** [198] 324 (1974).
[26] A. Ohba, Seitetsu Kenkyu Nippon Steel Technical Report No. 283; p. 100, 1975.
[27] K. Sugita et al., XXIVth International Colloquium on Refractories. Aachen, September 1981.
[28] K. Hiragushi, Shiraishi Memorial Seminar. July 1983.
[29] T. Ochiai et al., *Taikabutsu Overseas,* **1** [2] 92 (1981).

Recent Progress in the Use of Monolithic Refractories in Europe

W. KRÖNERT

Institute of Ceramics, Glass, Cement and Refractories
Technical University of Aachen
Federal Republic of Germany

Introduction

Initially, some data are presented concerning production and shipment of unshaped materials of the European Community (EC) and West Germany. Only a few unofficial figures are available for the production of monolithics in the EC. A rough estimate is given in Table I. From 1974 to 1982, a 20% decrease in production from 1.75 to 1.40 million tons is evident. The reasons for this trend, which is similar to the situation in West Germany, follow. The statistical figures for West Germany are compiled by the German Refractory Association. In German plants, records are kept on all products sent to domestic or foreign customers. The following figures represent about 90% of the local German refractory production, since not all of the German refractory manufacturers are also members of the Refractory Association. Data shown in the following tables start with 1974. From this year on, ISO 1927 definitions were used for the different types of monolithics. In 1984, ISO 1927 was revised, and an approved version was introduced that allows for more detailed statistics (Table II). Tables III and IV show the development of sales in tons and deutsche marks (DM) for brick and monolithics from 1974 to 1983. The high SiO_2 mixes are listed separately. Table III demonstrates clearly the dramatic decline in the production of brick and high-silica monolithics. The other refractory mixes, jointing materials, and coatings remained essentially constant during the last decade, so that the percentage of

Table I. Estimate for the Production of Monolithics in the Countries of the EC.

Production per year in 1000 metric tons	
1974	1.750
1978	1.550
1982	1.400

Table II. Raw Materials for Monolithic Refractories: Classifications Made According to the New Draft of ISO R 1927.

Class	Name	Limiting values of main oxides
I	alumina-rich products	$Al_2O_3 \geq 45\%$
II	schamotte products	$10\% \leq Al_2O_3 < 45\%$
III	acidic products	$85\% \leq SiO_2, Al_2O_3 < 10\%$
IV	basic products: magnesia, chromite, forsterite, dolomite, other alkali earth oxides	
V	special products: carbon; carbides, nitrides, zirconium silicate etc. and their mixtures	
VI	mixed products: mixtures of classes I - V with one main component from class I - IV	
VII	special mixed products: mixtures of classes I - V with one main component from class V	

Raw materials for monolithic refractories
Classification of the monolithic refractories
according to ISO R 1927 (new draft)

Table III. Development of Sales for Brick and Monolithics (in tons).

type of unshaped material (monolithics)	1974	1978	1980	1981	1982	1983
	shipment to customers per year in 1000 metric tons					
high-SiO_2 mixes	441	361	355	304	271	262
other refractory mixes, jointing materials and coatings	297	276	294	294	274	283
refractory bricks	1350	932	965	885	758	742
total	2088	1569	1614	1483	1303	1287
all unshaped materials (monolithics)	35,3%	40,6%	40,2%	40,3%	41,8%	42,3%
refractory bricks	64,7%	59,4%	59,8%	59,7%	58,2%	57,7%

Statistics of Refractory Association in Bonn/FRG

the total monolithics production shows an upward trend. The shipment of unshaped material reached 42.3% in 1983. A similar development in the production of monolithics is found in Japan (Fig. 1), where a 40% share of the total refractory production of unshaped refractory products is projected. Due to the steel crisis and the improvement of refractory know-how, the consumption of refractories has declined in the iron and steel industries as well as in all other consumer industries.

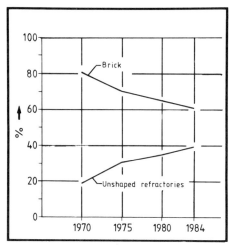

Fig. 1. Production of brick and unshaped refractories in Japan (% of total).

This decrease cannot be recognized in the sales figures because the inflation rate and trend toward high-performance products make up for the loss in tonnage (Table IV). The share of monolithics in the total refractory production amounts to 22% in sales in 1983, compared with 16.7% in 1974. This increase of the monolithic share both in tonnage and in DM can be attributed to the rising importance of unshaped materials that have replaced brick for many applications. Tables V and VI show the different types of monolithics, according to ISO 1927, in tons and sales over the past 10 years. Table V shows a steep decrease for high-SiO_2 mixes, moldable mixes, and jointing materials. A less prominent regression is noticeable in ramming mixes and insulating castables. The production regarding dense castables is on the increase, whereas basic monolithics have hardly changed.

Mixes that are extensively used in blast furnace runners that contain carbides, nitrides, carbon, zircon, etc. exhibited a remarkable increase. The importance of dense castables can be attributed to installation by vibration and gunning techniques, which are economical and time-saving.

The decreased sales of insulating mixes can be attributed to several reasons. Despite the fact that saving energy has become increasingly important after two oil crises, the use of insulating castables has declined. The main reason for this development is the appearance of new and sophisticated insulation materials such as modules based on refractory fibers of different kinds.

Table IV. Development of Sales for Brick and Monolithics (in DM).

type of unshaped material (monolithics)	1974	1978	1980	1981	1982	1983
	shipment to customers per year in million DM					
high-SiO_2 mixes	28	32	36	34	34	34
other refractory mixes, jointing materials and coatings	152	174	216	228	232	240
refractory bricks	899	826	989	992	929	971
total	1079	1032	1241	1254	1195	1245
all unshaped materials (monolithics)	16,7%	20,0%	20,3%	20,9%	22,3%	22,0%
refractory bricks	83,3%	80,0%	79,7%	79,1%	77,7%	78,0%

Statistics of Refractory Association in Bonn/FRG

Another reason for the decrease lies in the poor economic situation of the petrochemical industry which always has been an important customer for insulation castables.

The figures in Table V may be misleading regarding the use of basic monolithics during the past decade, especially in the iron and steel industry, since dolomitic material is not completely included in the statistics and greater amounts of magnesitic mixes are imported from such other countries

Table V. Development of Sales of the Different Types of ISO 1927 Monolithics (in tons).

type of unshaped material (monolithics)	ISO Class	1974	1976	1978	1980	1982	1983
		shipment to customers per year in 1000 metric tons					
high-SiO_2 mixes	III	441,3	386,4	360,8	354,7	270,8	262,5
mouldable mixes with Al_2O_3-SiO_2 base	I, II	37,7	30,1	24,8	26,0	20,0	18,2
ramming mixes with Al_2O_3-SiO_2 base	I, II	28,9	29,2	23,7	31,1	24,2	22,3
concretes (castables and gunning mixes), insulating mixes	I, II	80,6 / 21,0	81,3 / 20,7	80,8 / 20,3	85,7 / 17,3	83,5 / 16,1	95,0 / 16,7
mixes containing magnesia, chrome ore, forsterite,(dolomite)	IV, IV	41,2	49,6	51,1	45,5	48,9	49,1
mixes containing carbides, nitrides, carbon, zircon and others	V, VII	8,1	10,0	18,8	26,0	26,7	27,9
jointing materials and coatings	I-VII	54,5	49,6	37,5	42,8	37,2	37,6
other mixes not classified,	–	21,2	14,4	15,6	15,0	13,2	11,7
prefabricated shapes	–	4,3	4,0	3,9	4,6	4,5	3,8

Statistics of Refractory Association in Bonn/FRG

Table VI. Development of Sales of the Different Types of ISO 1927 Monolithics (in DM).

type of unshaped material (monolithics)	ISO Class	1974	1976	1978	1980	1982	1983
		shipment to customers per year in million DM					
high-SiO_2 mixes	III	28,3	29,7	31,7	35,6	33,8	34,3
mouldable mixes with Al_2O_3-SiO_2 base	I, II	21,3	19,0	16,8	20,6	17,9	15,0
ramming mixes with Al_2O_3-SiO_2 base	I, II	21,6	24,6	20,6	27,8	24,8	22,9
concretes (castables and gunning mixes)	I, II	56,1	64,6	65,3	79,0	89,7	98,8
mixes containing magnesia, chrome ore, forsterite, (dolomite)	IV, IV	18,0	25,9	27,3	28,8	32,1	34,2
mixes containing carbides, nitrides, carbon, zircon and others	V, VII	11,0	12,7	21,8	31,2	35,3	36,9
jointing materials and coatings	I–VII	13,1	14,7	13,4	17,4	18,5	19,3
other mixes not classified	–	5,6	4,2	3,4	4,2	4,4	4,7
prefabricated shapes	–	5,1	4,9	5,5	7,3	8,9	8,0
Statistics of Refractory Association in Bonn/FRG							

as Austria, Ireland, Czechoslovakia, and Sweden. The consumption of ceramically bonded mortars has decreased due to the decline in the use of brick. The use of chemically bonded mortars, however, has remained at a constant level, mainly because they have advantages over the customary ceramic bond-forming mortar.

Prefabricated monolithic shapes, mostly in the form of vibrated refractory concretes, have shown no increase in use for the past decade. It is anticipated that prefabricated shapes will be used more extensively in the future because of their easy production and installation. Table VI again shows the sales figures of the different types of monolithics. The trend toward high-performance materials is not clearly visible due to the inflation rate.

Table VII shows the tendencies of ISO classes I–III for Al_2O_3-SiO_2 monolithics without jointing materials and coatings. The statistics indicate that monolithics with a high alumina content are preferred. Generally, an upward trend toward high-performance materials is noticed. Furthermore,

Table VII. Development of Sales of Al_2O_3-SiO_2 Monolithics (in tons).

type of unshaped materials (monolithics) without jointing materials and coatings	1974	1976	1978	1980	1982	1983
	shipment to customers per year in 1000 metric tons					
mixes with Al_2O_3-content >45% ISO class I	76	77	70	83	78	81
mixes with content of 10% < Al_2O_3 < 45% ISO class II	92	85	80	77	66	71
mixes with content of $SiO_2 \geq 85\%$ Al_2O_3 < 10% ISO class III	441	386	361	355	271	262
Statistics of the Refractory Association in Bonn/FRG						

monolithics that can be installed easily by machines will be of greater importance. In the future, these time-saving, new, improved techniques will replace the present installation methods.

New Material Developments

Recently, an international colloquium was held at Aachen in the field of blast furnace troughs and runners; therefore, it is not necessary to go into details.[1] The colloquium's findings generally indicated that today high-grade materials consisting of alumina or alumina-rich grog, combined with clay-bonded SiC, carbon, and/or silicon nitride, are widely used in runners and troughs. There are moldings and ramming mixes, castables, and precast shapes. Attempts are being made to use gunning techniques as well. Of special importance for new lining methods are newly developed vibration and casting materials with particular rheological properties.

The introduction of low-cement castables and cement-free castables has been one of the most impressive developments in the field of monolithics in recent years. To understand why these refractory materials are superior in some cases to normally cement-bonded castables, it is necessary to compare some of their properties.

The composition of any monolithic is basically similar to that of the corresponding brick type: Both are composed of a grog and a bond. However, whereas the grog requires refractoriness and structural and volume stability up to the maximum use temperature, the bond has different properties. Although the chemical nature of grog and bond is normally similar in brick and moldings, the chemical composition of the cement in castables is normally different from that of the grog, which in most cases leads to a decrease in refractoriness. The cement composition is of lower mechanical stability and may react with the grog to form new phases, depending on the temperature. Another important point is the water content of brick and monolithics in the green state. The addition of water to the mixture for brickmaking is as low as 0.5-3%. The brick is shaped by high pressure, and the ceramic bond is developed by firing at high temperatures. The porosity of brick can be kept very low, depending on the manufacturing process (e.g., isostatic pressing). However, castables need 8-15% added water to allow the hydration reaction and for final shaping by casting or vibration. The resulting porosity of the fired refractory concrete is around 25%.

Disadvantages of Conventional Refractory Concretes

Conventional concretes contain 15-30% alumina cement. This amount is necessary to achieve satisfactory strength at room and medium temperatures. The 8-15% water added during processing is used as follows: (1) 0-15% of the water is taken up by the porosity of the grains (grog); (2) 6-10% of the water is needed for the hydraulic bond; and (3) 2-6% of the water causes the concrete to flow, which allows it to become dense.

Compared with a brick based on the same raw materials, dense refractory concrete has the following disadvantages: higher porosity, high CaO contents, high Fe_2O_3 content in fused alumina cement, and low hot strength.

The high concentrations of fluxes (CaO, Fe_2O_3) have a particularly detrimental effect on corrosion resistance and hot strength. Hot strengths of refractory concretes must be differentiated between medium (250°-800°C)

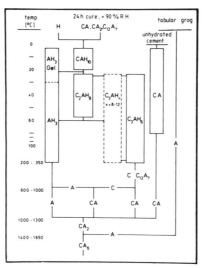

Fig. 2. Hydration and dehydration of CA-corundum concretes according to Givan et al. (Ref. 2).

and high temperatures (>1100°C). The hydrate phases CAH_{10}, C_2AH_8, C_3AH_6, AH_3, and AH are largely converted between 250°–350°C to CA, CA_2, and $\alpha - Al_2O_3$, the conversion being completed by 800°C (Fig. 2).[2] These phase transformations lead to a reduction in the mechanical strength.

The amount and temperature range of strength reduction depend on the cement used and the grain-size distribution. The lowest hot-strength values lie between 250° and 700°C (Fig. 3). A ceramic bond is formed between 900° and 1000°C, which results in increased strength. The development of a liquid

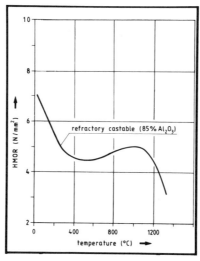

Fig. 3. HMOR of 85% Al_2O_3 refractory castable.

phase in turn leads to a decrease in hot strength. For conventional concretes, one cannot go below a CaO content of 3–4%. This is possible only for concretes with low cement contents.

Concretes with low cement contents were first mentioned in a French patent in 1969.[3] Reduction of the cement content to 5–8% without a detrimental effect on the strength was achieved through the addition of a fine-grained refractory material and a deflocculant, promoting a homogeneous distribution of the cement and fine-grained addition and reducing the amount of mixing water needed. Through the homogeneous distribution of cement, the hydraulic bond can be fully utilized. The first concretes with less cement had lower porosity, higher corrosion resistance, and very good hot strengths at medium and high temperatures. However, their sensitivity to quick heating had to be taken into account, since the water was set free in a very narrow temperature range.

Another patent in 1976 reduced the cement content of such concretes to 1%.[4] In addition to the deflocculant, a very high packing density is achieved through the use of two special, fine grain sizes. One material should have a grain size between 1 and 100 μm, and the second between 0.01 and 0.10 μm. The first should be inert with respect to hydration; the second should not form gels with water. The mixture of both these materials in a 1:100 ratio causes the finer particles to fill the spaces between the coarser grains, which in turn reduces the water consumption, and thus the porosity of the concrete, and increases the strength of materials with low cement contents.

Microsilica[5] (average size 0.15 μm) has recently come into use as the finer material. This amorphous SiO_2 is formed through the vapor phase as a by-product in a certain electrometallurgical process (Fig. 4). Refractory concretes produced with the help of such materials have high mechanical strength, high erosion resistance, good thermal-shock behavior, and low shrinkage.

Fig. 4. Microsilica (average size 0.15 μm).

These concretes have a pseudozeolithic bond. Since they do not contain the hydrate phases of conventional concretes, the water in the bond is not freed within a narrow temperature region, but rather is freed slowly between 150° and 450°C. Therefore, even with refractory concretes containing only

10% porosity, there is no fear of explosion if the recommended heatup schedule is not followed strictly (Fig. 5). To remove the water more easily, some producers add about 2% organic, low-melting fibers to the castable (Figs. 6-8). Thus, pores are formed at rather low temperatures for a more rapid water transport.

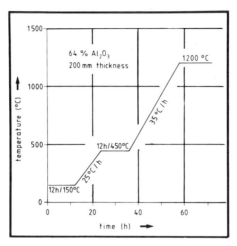

Fig. 5. Heatup scheme of a low-cement, refractory castable.

The castables are processed by vibration compacting, since their low water contents preclude casting. Hardening begins about 4 h after the mixing and is practically completed after 48 h. The mold can be removed after 12-24 h, and the concrete further hardens in the air. Low-cement castables have the following advantages over conventional refractory concretes:

(1) Low CaO content: 2% for fireclay, low-cement castables compared with 8% for conventional concrete; 1% in corundum, low-cement castables compared with 3-5% for conventional concrete (Fig. 9).

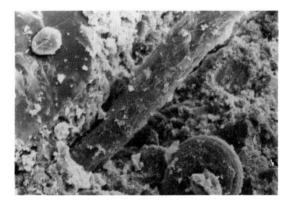

Fig. 6. Refractory concrete with organic fibers before heat treatment.

Fig. 7. Refractory concrete with organic fibers after 6 h at 120°C.

Fig. 8. Refractory concrete with organic fibers after 24 h at 150°C.

(2) Low porosity (8-14%) after firing to 1000°C compared with 25-30% for conventional concrete.

(3) Higher hot strength: at 1000°C, about 20 MPa compared with 5 MPa for conventional concrete.

(4) High abrasion resistance.

(5) Improved thermal-shock resistance.

(6) Higher corrosion resistance.

A critical point is the water addition: These improved properties are achieved only if the water is added exactly as recommended by the producer.

Figure 10 shows the various possibilities of processing under practical conditions based on the consistency experiment (according to 26. PRE recommendation). With the given retardation period of 15 s, 2 quantities of mixing water are possible, i.e., 10/L and 13L/100 kg. However, with a water dosage of 3L/100 kg, the segregation starts during vibration, which is another example of the strong influence of specimen preparation on the results of technical measurements.[6] Estimated usage of the three types of

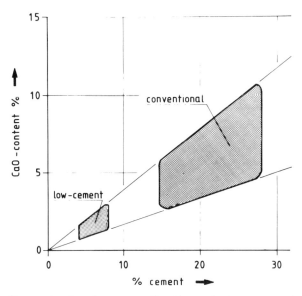

Fig. 9. CaO content of conventional and low-cement refractory castables.

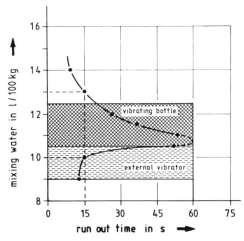

Fig. 10. Consistency test of a dense, refractory concrete according to 26. PRE (R = 2.0 g/cm³).

castables is: usual refractory concrete—70–80%; low-cement concrete—10%; and high-strength concrete (ULC)—10–20%.

Cement-Free Concretes

A further development is cement-free refractory concrete, which uses chemical binders such as phosphoric acid, phosphates, water glass, magnesium chloride (sorel bond), and inorganic polymers. In some cases, a setting agent such as magnesium oxide must also be added. However, the chemical makeup of such refractory concretes with a chemical bond and setting accelerators is quite complex, and these concretes are very sensitive to an incorrect dosage of water.

Utilization Areas of Low-Cement Concretes

Iron and Steel Industries: Utilization in soaking pits, in which the upper part is greatly strained by high thermal shock and mechanical shock is advantageous. A fireclay, low-concrete castable can be processed at the site or used as prefabricated material. Another potential application is in the lining of walking beams or stationary beams in walking-beam furnaces. Burner blocks in walking-beam furnaces (refractory concrete of low cement content with about 60% Al_2O_3) can be improved because of the greater thermal shock resistance. In rotary hearth kilns, low-cement concretes can be used in lining the hearth, where thermal shocks and high pressures occur. Burner blocks in bogie-hearth furnaces are another area of application.

Aluminum Industry: Anode production, electrolyte kilns, transfer ladles, and linings of holding and melting furnaces are prime areas for low-cement concrete use.

Cement Industry: Nose rings in the rotary kiln, where strain is generated from periodic changes in temperature from 500° to 1350°C could also benefit from the use of low-cement concrete.

Petrochemical Industry: Special concretes of low cement content with Cr_2O_3 is another application area, as well as for coal gasifiers and carbon-black reactors.

Steel Fiber-Reinforced Refractory Concrete

It is well known that the tensile strength of refractory concretes can be increased considerably through addition of steel fibers. This, however, leads to a number of problems. The improved tensile strength and increased resistance to thermal shock, i.e., prevention of premature crack formation and spalling, are offset by increased costs, more difficult installation, and reduced refractoriness. However, experience has shown that steel fiber-reinforced concretes can be used at temperatures used for pig iron or molten steel as long as the operation is not continuous.

To date, steel fiber-reinforced concretes have been employed successfully in the following areas: waste gas sliders, pig iron trumpet assemblies, the discharge-end area in rotary cement kilns, ladle roof covers, lining of carrier cars of bogie-hearth furnaces, and in installations for steel and pig iron desulfurization.

During an investigation on the physical properties of steel fiber-reinforced concretes, the thermal conductivity was also studied. It was initially assumed that the refractory concrete would show an increase in thermal conductivity, since the thermal conductivity of steel fibers is much higher than that of refractory concrete. The measurements (Fig. 11), however, show the opposite effect, at least at low temperatures. The explanation for this effect is that, due to a slightly increased water content, the total porosity is greater than that of unreinforced concrete and the individual fibers partly undergo an incomplete bond with the concrete after firing. The steel fibers thus do not build a network within the concrete which would lead to higher thermal conductivity. As a result of the experiments conducted with steel fiber-reinforced concretes, the following can be stated:

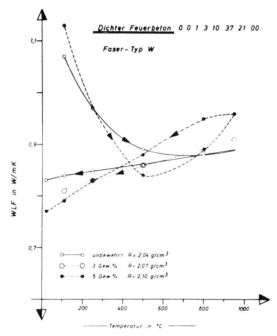

Fig. 11. Thermal conductivity according to DIN 51046 of a refractory concrete with and without fibers.

(1) The preparation of refractory concretes requires an increased effort with regard to mixing and the intensity of mixing, as well as with regard to safety precautions (hands and eyes).

(2) The shrinkage is reduced with increasing fiber content and can also be converted to expansion.

(3) The crushing strength is only partly improved. The highest improvement is registered for fireclay concrete with low bulk density. For concretes with a high Al_2O_3 content and high bulk density, the crushing strengths are partly decreased.

(4) The fracture energy is increased. Differences due to the amounts added were not registered.

(5) The thermal shock resistance is increased through the addition of steel fibers—up to double the quenching cycles, depending on the amount added.

(6) The thermal conductivity is either not increased at all or shows only a small increase.

Significance of these Results

(1) Steel fiber reinforcement should not be used where no thermal shock takes place and the temperatures lie above the oxidation temperatures of the

steel fibers. The high Fe_2O_3 content then acts as a flux and leads to premature destruction of the concrete.

(2) In case of thermal shock and high mechanical loads, the life of the refractory concrete can be increased through the addition of steel fibers. In this case, it is immaterial whether the temperatures on the hot face achieve the oxidation temperatures of the fibers. This is confirmed by practical experience; for example, the life of pig iron trumpet assemblies could be doubled through the addition of 2 wt% fibers. In this case, the limitations of steel fibers were also evident, since an increase in fiber content led to a reduction of lining life to values for nonreinforced concretes. Another area for the use of steel fiber-reinforced concretes is the rotary cement kiln, where the discharge-end zones in particular undergo strong thermal shock and high mechanical load.

(3) According to our experience, 2 wt% fibers are sufficient with regard to processing and cost increases. The latter aspect is particularly important, since the addition of steel fibers should not lead to an undue increase in the cost of the material.

(4) With regard to fiber length, 25 mm is satisfactory. Longer fibers improve neither the fracture energy nor the strength, and mixing is made more difficult.

Fig. 12. Stages of CO destruction of a high-alumina castable after 200 h at 500°C according to ASTM C 288-78.

These results show that not only the type of fiber but also the type of refractory concrete used have an effect on the properties. Therefore, in the course of planning, an investigation should be conducted on the refractory concrete plus steel fiber envisaged.

Testing of Monolithics

A few remarks should be made concerning the testing of monolithics. One point of interest is the stability of mixes and castables against CO disintegration. Here, laboratory methods have been developed to simulate CO attack. The trend has been to use mixtures with water vapor and other gases rather than pure CO gas in testing (Fig. 12). Another problem is the exact knowledge of the thermal conductivity, since there is no simple, reliable testing technique. Values shown in data sheets, especially for insulating materials, are often too low, due to an inappropriate testing method (Figs. 13 and 14).[6] For castables, too, the water content, as it relates to the state of preheating or prefiring, has a major influence on thermal conductivity (Fig. 15).[6]

In Germany, an agreement was contracted some years ago between the refractory producers and the iron and steel industry to introduce a numerical code for all monolithics.[7] This 12-digit code designates the class of the monolithic (designation), the delivery state, the type of bonding, the processing technique, the main component of the grog, the percentage of the main component, the mass requirement, and the maximum use temperature. There are difficulties, of course, in giving the exact maximum use temperature, as it depends on the running conditions of the furnace. Thus, it was proposed that a classification temperature of the monolithic be designated.

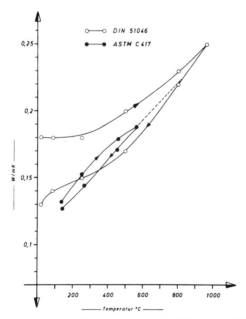

Fig. 13. Thermal conductivity according to DIN and ASTM for an insulating refractory concrete (R = 0.5 g/cm³).

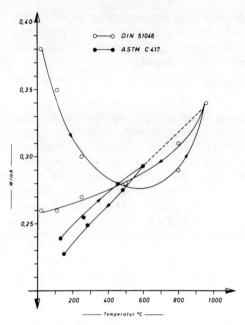

Fig. 14. Thermal conductivity according to DIN and ASTM for an insulating refractory concrete ($R = 0.5$ g/cm³).

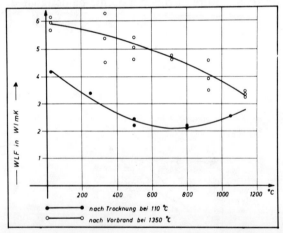

Fig. 15. Thermal conductivity of a 90% Al_2O_3, corundum-based, ramming mix according to the hot-wire method (DIN 51046).

In the past, monolithics were classified according to their total linear shrinkage after firing for 5 h. At the classification temperature, shrinkage should not exceed 1.5% for castables, 2% for ramming mixes, and 3% for moldables. This classification method is unreliable, as the shrinkage depends, among other things, on the conditions of the test-specimen production.

Therefore, a further test will be conducted in which the refractoriness under load is measured. Only if T_2 is above the classification temperature mentioned before will this temperature be supposed correct (Fig. 16.)[8,9] The proposed new method to find the correct classification temperature is thus a combination of both testing procedures: total linear shrinkage after firing, and refractoriness under load.

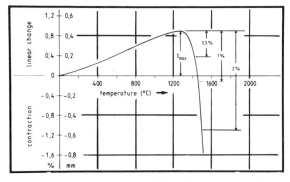

Fig. 16. Refractoriness under load (DIN 51 053, part 1).

Finally, some examples for the use of monolithics are given. Figure 17 shows a hex-mesh installation for units where a severe erosion attack can occur, for instance, in gas transfer pipes.[6] The hex mesh, made from high-alloyed steel, is welded to the pipe wall. Two types of castables are used: an insulation concrete, and a low-shrinkage, dense material which guarantees high abrasion resistance.

Figure 18 shows the inner wall of a tire incinerator during installation. This is an example of a plastic gunning application. A higher densification of the plastic and a lower shrinkage are obtained by this kind of lining technique (Fig. 19).[6] Under the prevailing clean conditions, the rebound can be reused for gunning.

Fig. 17. Hex-mesh installation.

Fig. 18. Inner wall of a tire incinerator during installation.

Fig. 19. Lining with higher densification and lower shrinkage of the plastic.

Fig. 20. Window case of a bottom-mounted flair.

Figure 20 shows the window case of a bottom-mounted flair made by a moldable mix or by prefabricated shapes. The material is phosphate-bonded in both cases. These monolithics show a higher thermal shock resistance than bricks.[6]

Fig. 21. End-pusher furnace.

Figure 21 demonstrates the use of monolithics for the lining of a large, end-pusher furnace (27 m long and 7.6 m wide). The furnace roof is constructed completely with a moldable mix, whereas the skid pipes are prefabricated.[6]

Figures 22–24 show a walking-beam furnace, 26 m long and 14.6 m wide. The roof and the burner front walls are installed completely with prefabricated shapes of a chemically bonded mix containing 65% Al_2O_3. The bottom of the hearth is made of high-prefired shapes.[6]

Fig. 22. Walking-beam furnace.

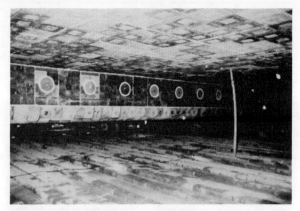

Fig. 23. Walking-beam furnace.

Fig. 24. Walking-beam furnace.

Fig. 25. Construction of a coke-oven door with prefabricated shapes.

The next three figures (Figs. 25-27) demonstrate the construction of a coke-oven door with prefabricated shapes of a castable.

The blocks are assembled to the steel support by clamping bolts; after attachment, the holes in the shapes are casted.

Fig. 26. Construction of a coke-oven door with prefabricated shapes.

Fig. 27. Construction of a coke-oven door with prefabricated shapes.

Figures 28 and 29 show the inner wall of a rotary ring furnace with gunned insulation, prepared for the gunning of moldables. The anchors still have a protective covering (cardboard capsules) to prevent the insulation gunning mix from filling the anchor grooves. In this case, the gunning machine had a capacity of 4 tons/h. The total length of the wall is about 145 m. The hearth lining consisted of a prefabricated, chemical-ceramic-bonded, vibration castable, with a minimum cold crushing strength of 80 N/mm^2, pretempered at 400°C.[10]

Fig. 28.   Inner wall of a rotary ring furnace.

Fig. 29.   Inner wall of a rotary ring furnace after completion.

Figures 30 and 31 show the gunning of an aluminum double-inductor furnace. The lining was installed with a 55% acid-bonded, plastic gunning mix (phosphate-bonded), with a gunning capacity of 1.5 tons/h.[10] The gun-

Fig. 30. Gunning of an aluminum, double-inductor furnace.

Fig. 31. Gunning of an aluminum, double-inductor furnace.

ning of a 250-ton capacity steel ladle, with the help of a recently developed, horizontal manipulator-gunning machine (telescope system) that uses a chemically bonded, magnesite-based, gunning material is shown in Fig. 32. Gunning took place primarily in the hearth and slag regions. The gunning machine operates with a rotating nozzle, so that the gunning angle is variable. Rebound during hot gunning is about 10–15%. The material is pumped dry up to the nozzle and then mixed with water.[10]

The last three figures (Figs. 33–35) show the gunning of the roof of a reheat walking-beam furnace, with roof burners of a plastic material. A 60% Al_2O_3, air-bonded plastic is used, and setting is through atmospheric drying. Rebound is 20%, which was then reused, so that the effective rebound was only 4%.[10]

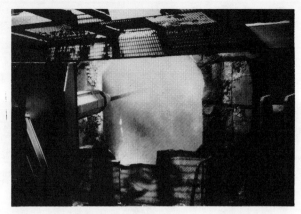

Fig. 32. Gunning of a steel ladle.

Fig. 33. Gunning of the roof of a reheat walking-beam furnace.

Fig. 34. Gunning of the roof of a reheat walking-beam furnace.

Fig. 35. Gunning of the roof of a reheat walking-beam furnace.

Finally, it should be mentioned that, in Europe, a rotor gunning machine has been used recently which can add the exact predetermined amount of needles during gunning of castables and plastic mixes. The needles fall regularly into the rotor and are mixed very uniformly.[10]

References

[1] Proceedings of the XXV International Colloquium on Refractories, Aachen, West Germany (1982).
[2] G.V. Givan, L.R.D. Mart, R.P. Heilich, and G. MacZura, *Am. Ceram. Soc. Bull.*, **54** [8] 710–13 (1975).
[3] Fr. Pat. No. 69.34405.
[4] Fr. Pat. No. 77.14717.
[5] B. Monsen, A. Seltveit, B. Sandberg, and S. Bentsen; this proceeding.
[6] R. Krebs; private communications.
[7] Stahl-Eisenwerkstoffblatt 916, 1st ed. VDEh, Düsseldorf, 1973.
[8] W. Krönert, *Keram. Z.*, **33** 702–08 (1981).
[9] P. Artelt, G. Heintges, and A. Majdic, *Stahl und Eisen*, **101** 1433–36 (1981).
[10] E. Struzik; private communications.

Evolution of Monolithic Refractory Technology in the United States

DAVID R. LANKARD

Lankard Materials Laboratory, Inc.
Columbus OH 43207

Like the wheel, monolithic refractories continue to be reinvented and also are seen on inspection to be simply an improvement on ancient practices. Archeological evidence shows that as long ago as 6000 B.C. the inhabitants of Western Asia created vitrified clay hearths through long use of pits dug in the earth to contain cooking fires. Later peoples transported and smeared wet clays on these hearths to provide an even more heat-resistant surface and to repair cracks and spalls.[1] Who can deny that this practice represents the first application of plastic refractories: allowing process heat to develop the ceramic bond in a refractory material formed on-site. Evidence exists, too, showing that the ancient Greeks and Romans used mortars and concretes in constructions where heat was involved in a process function, (e.g., ore roasters and lime kilns), anticipating by several thousand years our use of refractory concretes.

In our time, the need for monolithic, jointless refractories probably arose from manufacturing, construction, performance, and maintenance

Fig. 1. Spalling and delamination in a brick and mortar refractory construction.

problems associated with traditional brick-and-mortar refractory construction (see Fig. 1). In the United States, monolithic refractories first surfaced as distinct refractory materials around 1914.[1] The idea for monolithics had very likely occurred to at least a few workers prior to that time, to brick masons or brickmakers, perhaps, working with mortars (then called refractory cements), to workers in the portland cement concrete field, or to workers who used wet clays to patch furnace linings. In all of these cases, the refractory materials were molded or formed at room temperature into a desired shape or configuration prior to hardening through chemical reactions or exposure to heat.

The first plastic refractories were simple mixtures of plastic clay and crushed-brick grog or calcined clay furnished to the user in a wet, moldable form. The first refractory concretes were simple mixtures of hydraulic cements and aggregates available at the time. Despite some obvious differences, refractory concretes and plastic refractories have some important features in common. They are both composite materials composed of a continuous fine-grained binder (matrix) phase surrounding a particulate, discrete aggregate phase (see Fig. 2). Also, both types of monolithic refractories are, following the addition of water, amenable to forming or molding into desired shapes having few or no separations or joints (see Fig. 3).

From simple beginnings in the early 1900s, monolithic refractories have evolved over the succeeding 70 years or so into a versatile, widely used class of refractory construction materials that offer performance and cost-effectiveness on a par with, and in some cases, superior to prefired refractory bricks and shapes. During this evolutionary period, the basic concept of monolithic refractories has not changed significantly. They are still two-phase composites composed of a binder phase and an aggregate phase, and rendered initially moldable by the addition of liquids and/or the use of energy to make them flow.

The success of monolithics has not been due to dramatic changes in their concept, but to significant advances in the type and quality of the binders,

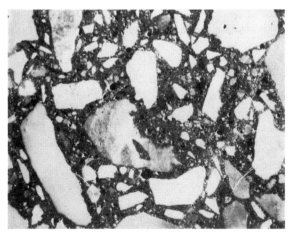

Fig. 2. Microstructure of a monolithic refractory showing the binder (continuous) and aggregate (discrete) phases.

Fig. 3. An example of monolithic refractory construction.

aggregates, and additives used in them and to innovations in their design and installation in refractory structures. The impetus and direction for the development of improved monolithic refractories was provided by the increasingly more severe demands (technical and economical) placed on refractories in general by all of the refractory-consuming industries. Therein lies the story of the successful evolution of monolithic refractories.

The Period 1900 to 1940

The most significant event of this era was, of course, the actual introduction of monolithic refractories in the United States on a commercial basis.

Most turn-of-the-century furnaces were constructed of refractory brick and mortar with the brickwork functioning in both a structural and a heat-containment role as depicted in Fig. 4. When furnace designers began to use

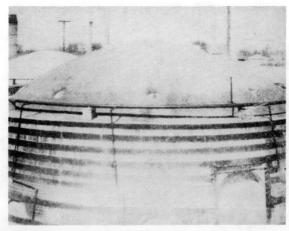

Fig. 4. Round kiln of the early 1900s in which brick performed both structural and refractory roles.

steel plate for furnace shells, however, the structural role of the refractory was eliminated or at least minimized. Attention was therefore focused on an inherent deficiency of brick and mortar construction—a multitude of joints. In 1912 and even as late as 1924, papers concerning recent developments in the refractories industry, e.g., in *Transactions of the American Ceramic Society* and the *Annual Report of the Society of Chemical Industries* failed to mention any potential offered by monolithic construction relative to this problem.[2,3] That is not to say, however, that the question was being ignored.

Plastic Refractories: Some time before 1914, W. A. L. Schaefer considered the possibility of installing a freshly made, fireclay-brick composition as a lining over a steel shell and developing the ceramic bond with the firing system of the furnace itself[1]; thus, the idea of the plastic refractory was rediscovered. In 1914, Schaefer established the Pliable Firebrick Company in Chicago to manufacture the new product and to assist users in its installation. The name Plibrico was adopted for the company in 1917.

Based on the idea that the joints in brickwork doom this construction to failure, claims made for Pliable Firebrick stated:

> It is firebrick in plastic, putty-like form. It is semi-baked and preshrunk. It is installed chunk by chunk with a steel hammer. It is then trimmed with a trowel and baked out with a slow fire giving a monolithic furnace lining without a single vulnerable joint. It is not applied as a coating over old firebrick walls. It is installed next to the outer wall after the old firebrick has been removed. It is resistant to 3100F. It is formulated of the finest refractory clays mined from our own mines.[4]

Plastic firebrick was initially supplied in sealed steel drums containing around 295 kg of material (0.14 m^3 at 2082 kg/m^3).

Workers cut the drums open to remove the material and rendered it into small pieces using a shovel as shown in Fig. 5. The pieces were then installed chunk-by-chunk by workers using a wooden or steel hammer, as shown in Fig. 6. Installation instructions noted, "The harder you pound, the better results you will get."[4]

Fig. 5. Plastic firebrick as supplied ca. 1918 in 295-kg units in sealed steel drums.

49

Fig. 6. Installation of plastic firebrick ca. 1918 using a steel hammer.

As shown in Fig. 7, pockets arranged in checkerboard fashion were left in the common brick setting to key and hold the plastic in place. Eight to ten hours of a "slow" flame were then recommended to remove the moisture.

By the early 1920s, other U.S. refractory manufacturers began to offer plastic refractories. Although there were at least 10 manufacturers of plastic refractories in 1930, Plibrico claimed at that time to "produce more than all other manufacturers combined (20 000 users)."[5]

The greatest initial use for plastic refractories in the United States was in boilers where they were used for linings, fireboxes, arches, bridge walls, curtain walls, burner ports, and blow-off piers. Subsequently it was claimed that plastic refractories could be used "wherever firebrick is used." Successful

Fig. 7. Early 1920s installation of a plastic refractory lining showing pockets left in common brick setting to key the plastic in place.

applications of plastics in forge furnaces, gas generators, heat-treating furnaces, oil stills, incinerators, coke ovens, melting furnaces, baking ovens, and in air- and water-cooled furnace walls were claimed by 1930.[5] Plastics thinned with water were also used as a stucco surfacing on old firebrick.

Some of the early claims for plastic refractories were improved spalling resistance and superior insulating value.[6] In 1926 the term "plastic fireclay refractory" first appeared in the open technical literature. A number of articles were published that year citing the advantages and convenience of plastics for quick repairs, replacing special shapes, and boiler applications.[7-9]

As a user of plastic refractories in 1926, Weightman offered some valuable advice regarding their formulation and use:

> Plastic refractories first appeared on the market as plastic fire brick. Since that time many such materials have appeared, some better and some not fit for use. There has been much of the "follow the leader" type of production in these materials with the result that few, if any, of such materials are as good as they might be. If the plastics are to be the best possible in service quality, more cooperation must be given the manufacturer by the user.
>
> When a setting is made with fire brick the greater part of the structure has been pre-fired to a high temperature, not more than 3 to 15% consisting of unfired mortar. With plastic refractories the entire structure consists of wet materials with a grog or fired portion of only 25 to 50%.[9]

Weightman focused attention on the possible adverse consequences of heating plastic refractories from one side only. He advised the user to design the plastic installation to avoid loadings that might cause buckling and to minimize problems with metallic reinforcements.

On strength and grain sizing, Weightman commented,

> The outstanding deficiency in plastic refractories is that of mechanical strength. When these are installed the material is pounded into forms with a mallet. The strength of the wall depends on how well this pounding is done. To simulate practice the plastic should be pounded into the molds.
>
> The dry strength of some materials tests as high as 500 pounds per square inch. These materials were made of well-graded grog and considerable plastic clay.
>
> Many plastic refractories are made of poorly graded materials. Occasional large pieces of ganister exert a tremendous pressure, spalling the refractory. The screen sizes used in the manufacture of plastic refractories should be such that the sample can be easily built into thin sections. Finer and more uniformly graded material will greatly improve the forming of plastic refractories and give greater mechanical strength.[9]

Finally, Weightman's comments on the price of plastics in 1926 deserve mention:

> Price is one factor that has retarded the more general use of plastic refractories. There appears to be no reason why such fancy prices should be asked for these materials. Some plastic refractories sell at $24 to $60 a ton and a few fancy ones at as much as $160 a ton. Even considering warehousing and extensive sales work, these prices are far out of line when it is considered that fired shapes of equal and better quality may be purchased for $20 a ton and less.
>
> In spite of these exorbitant prices, plastic refractories are increasing in use on account of the ease with which they may be handled, and the ever-

increasing cost of the furnace brick layers. Few, if any, plastic furnace linings exceed the life of a good firebrick installation, though often the operating labor can make a better plastic installation than to lay a firebrick setting. A reduction in the price to a more equitable level will greatly increase the use of plastic refractories.[9]

Significantly, it was also in 1926 that the U.S. government developed specifications for plastic refractories for use in boiler construction[10] and the use of sodium silicate as a binder was discussed.[7,9]

In a "recent developments" paper published in 1929, reference is made to plastic refractories containing sillimanite aggregate and to rammed-chrome ore-based hearths.[11]

By the early 1930s, manufacturers began to offer plastic refractories in the more convenient 45-kg, presliced form in common use today. During this time, the use of pneumatic ramming devices were first advocated for the installation of plastics. Also by 1930, the use of anchors for installing plastic refractory linings was in practice. Figure 8 shows an installation in the early 1930s using flexible steel anchors, although ceramic anchors were also available for use.

Fig. 8. Installation of a plastic refractory lining in the early 1930s using flexible steel anchors.

Despite an increasing interest in, and availability of, plastic refractories from 1914 to 1940, there were few technical publications on the subject (at least as evidenced by article titles dealing principally with these materials).

Refractory Concretes: The "discovery" of the cementing qualities of the calcium aluminates and of refractory concretes is usually attributed to Sainte-Claire Deville in France sometime before 1856.[12] He heated mixtures of alumina and lime and mixed this reaction product with corundum aggregate and water to produce crucibles suitable for use at very high temperatures. It was not until 1918, however, that the Lafarge Company in France commercially offered a calcium aluminate cement (produced from bauxite and lime) that at this time was touted as an alternative to portland cement, since it provided more rapid hardening and resistance to sulfate attack.

In the United States, Spackman in 1902 produced calcium aluminates by heating low-grade aluminous materials with limestone. In 1909, the Aluminates Patents Company was formed to exploit Spackman's patents. Spackman's calcium-aluminate material (named "Alca" cement) was marketed between 1910 and 1917, but was not advocated for use as a cement by itself but rather as a rapid hardening additive to natural cement and plaster materials.[13,14]

Prior to 1920, there are few references in the American technical literature to the manufacture and uses of calcium aluminate cements[15] and no references to refractory concretes, although at least one American Ceramic Society article described a refractory mortar based on sodium silicate mortar and waste abrasive material and recommended for the repair of kiln furniture.[16]

As a reflection of the increased interest in calcium aluminate cements in the United States, France, and Germany, the term "alumina cement" first appeared in the subject index of *Abstracts of the American Ceramic Society* in 1922. Not surprisingly, much of the literature in the early and mid-1920s on calcium aluminate cement dealt with its use as an alternative to portland cement and no mention was made of its potential in refractory applications.[17-22] There are, however, references during this time to patching materials based on sodium silicate binders combined with refractory aggregate.[23]*

In 1924, the Universal Atlas Cement Division of the U.S. Steel Corporation began manufacturing Lumnite calcium aluminate cement at Northampton, Pennsylvania using bauxite and limestone.[14,21] The development in the United States of a refractory concrete using calcium aluminate cement as a binder is usually ascribed to P.J. Kestner[14] who received a U.S. patent in 1926 for a mixture of calcium aluminate cement, refractory grog, and calcined bauxite. It is speculated that in the next few years experiments with field mixes of calcium aluminate cement and local refractory aggregates (typically crushed brick) got underway in the United States.[24] Robson suggests that "some remarkably ambitious refractory concrete work was carried out in the United States at that time (1924-1926) for furnace linings and even for blast furnace linings."[25]

It was around 1928 that the commercial production of bagged mixes of refractory aggregates and calcium aluminate cement for monolithic refractory concrete construction was started.[24]† By 1934, there were at least 25 of these proprietary mixes (called castables by the manufacturers) on the market. Originally offered in large 272-kg drums, by 1934 or so the 45-kg bag had come into favor (see Fig. 9).

The number of articles published in the open literature on refractory concretes from 1924 to 1940 strongly suggests that the advent of commercial castables stimulated a more widespread use of refractory concretes. However, field mixes remained in common use up to 1940.

---

*It is interesting that, before 1940, refractory mortars for setting brick were frequently referred to as "refractory cements," making it difficult to determine in the literature whether it is, in fact, mortars or true cements that are being referred to.

†At this time, calcium aluminate cement cost about $2.00 to $3.50 per bbl. (Ref. 26).

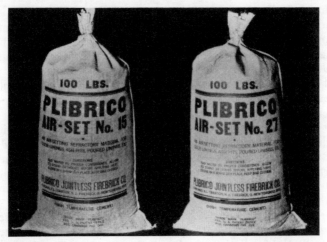

Fig. 9. Castable of the 1930s furnished in 46-kg quantities in burlap bags.

Another stimulus to the development of refractory concretes was the realization that plastics could not be used in some applications, such as boiler baffles and furnace doors, where temperatures did not get hot enough to "thoroughly bake the plastic material."[5]

During the early days of refractory concretes, mixing was commonly done by hand in a mortar box or wheelbarrow (although machine mixing was also used). Casting, slap-troweling, and hand-forming were the most common early forms of installation of refractory concretes, although some gunning was also done (the "cement gun" having been introduced around 1910).[24]

In 1938, Babcock and Wilcox Company in the United States began production of a purer form of calcium aluminate cement (for their own use) and shortly after that offered a 1649°C castable based on this cement and calcined clay aggregates.

Insulating castables were available as early as 1925, based on calcium aluminate cement and lightweight aggregates such as expanded shale, perlite, and vermiculite.[27,28]

The technical literature on refractory concretes through 1939 principally discusses applications for the "new" material, which included boiler applications, incinerators, take-off pipes (coke ovens), induction furnaces, ceramic kilns, furnace roof arches, furnace doors, ash pit linings and walls, and hearths in rotary oil burners, and for the production of such special shapes as burner blocks.

Binder, Aggregate, and Product Development

Prior to 1925, the principal aggregates available for use in monolithic refractory products were calcined clays and crushed forms of fired refractory brick and shapes.[29] Missouri diaspore was used as a source of high alumina in refractory products in the early 1920s but apparently was not widely used in monolithics. Early in 1925, kyanite began to be imported to the United States from India (Lapsa Bura) and quickly found its way into monolithic refractory products (although on a limited basis). Many of the aggregates first used in refractory concretes were those that had been used in portland cement concretes. They included normal-weight aggregates such as traprock, silica sand, and blast furnace slag, as well as lightweight aggregates such as diatomaceous earth, expanded clays and shales, expanded vermiculite, and expanded perlite. Other aggregates used in early monolithic products on a more limited basis included chrome ores, calcined bauxite, fused alumina, and even silicon carbide.[29,30]

Development work was started by Alcoa on tabular alumina in the late 1930s to provide a high-purity material for aircraft spark plug compositions.[31] Available in the 1940s, tabular alumina was not widely used in monolithic refractories until the 1950s, presumably due to its relatively high cost.

Most of the castable and field-mixed refractory concretes of this time (1920–1940) contained Lumnite as the source of calcium aluminate cement. However, as mentioned, Babcock and Wilcox Company first produced a high-purity, calcium aluminate cement in 1938 for use in their own products. The cement (termed C3) was developed by MIT researchers for Babcock and Wilcox using pure alumina and limestone as precursor materials. The cement contained approximately 70% Al_2O_3 and 30% CaO and was produced by firing relatively large, wedge-shaped blocks in a tunnel kiln at temperatures up to 1649°C. It was originally thought that the principal cementing phase formed was C_3A_5, but subsequent work revealed a mixture of monocalcium aluminate (CA) and CA_2.

The number of castable products available from U.S. refractory manufacturers prior to 1940 was quite limited. For example, in 1933 Plibrico offered 4 products: a 1482°C castable, a 815.5°C castable, a castable for the construction of boiler baffles, and a castable for oil burner and stoker installations. In 1938, Johns-Manville offered 3 castables: a "standard" version with 6-mesh, maximum grain size useful to 1315.5°C, a high-temperature version for use to 1538°C, and a lightweight version (1201 kg/m^3) useful to 1315.5°C.

The workhorse plastic of this era contained plastic clay as the binder and crushed firebrick or calcined clay as the aggregate. It offered use up to 1704°C in applications not involving contact with molten metals or slags. As with castables, the variety of plastics offered by manufacturers was very limited (one or two products) prior to 1940.

Summary

The period 1914 to 1940 can be characterized as a time of trial and experimentation for monolithic refractories. By 1940, monolithic refractories commanded only about 2 to 3% of the total refractory market, with manufacturers offering a relatively limited number of products in the form of castables, plastic refractories, and mortars.

The Period 1940 to 1960

This two-decade era saw large gains made in both the variety and quality of monolithic refractory products in the United States, particularly after World War II. Strong, refractory, abrasion-resistant aggregates became more widely available and major breakthroughs were made in the binder area with the introduction of intermediate and high-purity calcium aluminate cements and the development and application of phosphate bonding in refractories.

1940 to 1950

The status of plastics, castables, and mortars through the 1940s is reflected in the term "refractory specialties" used to identify them, implying that they were "special" products used in "special" applications and as such represented no threat to prefired refractory shapes. As might be expected, the war years of 1941 to 1945 stimulated increased use of monolithic refractories due to their relatively rapid rate of installation which speeded construction and frequently eliminated the need for complete rebuilds.

Refractory concretes were applied during the 1940s by casting, gunning, and slap-troweling techniques. Castables were typically supplied in moisture-proof paper bags, a change from the burlap bags of the 1930s. Field mixes were made with Lumnite cement containing vermiculite, crushed brick, olivine, traprock, or expanded shale. Applications for refractory concretes during this time included kiln car tops, furnace roofs, doors, covers, arches and hearths, boiler applications (combustion chambers, burner pits), precast shapes, linings for open hearth flues and ducts, and soaking pit sidewalls and covers. For these applications, the producers of Lumnite recommended an increased aggregate content in the concrete as service temperatures increased from 1093° to >1427°C (1 bag Lumnite to 0.11 m^3 aggregate at temperatures up to 1093°C; 1 bag Lumnite to 0.17 or 0.20 m^3 of aggregate at temperatures >1427°C). Machine mixing of refractory concretes became more common during the 1940s.

Plastic refractories in the 1940s were typically supplied in 45-kg units in sliced form, and installation with pneumatic rammers instead of hand-held hammers became common practice. Graphitic plastics were first introduced in 1941 (Kaiser Helspot). The use of graphitic materials in plastics further improved the resistance of these materials to wetting and penetrations of molten slags and metals.

Toward the end of the 1940s, refractory manufacturers were offering as many as 20 to 25 products in their "refractory specialty" lines, including heat- and air-setting mortars, heat-setting plastics, graphitic plastics (offered by three companies in 1950), and normal-weight and insulating refractory castables. At this time, refractory concretes began a rapid growth in petroleum- and petrochemical-related application areas (FCCUs and process furnaces), many applied by gunning.[32,33] Insulating castables replaced block insulation in FCCUs, largely solving the hot gas bypassing problem. By 1947, the gross value for plastic refractories in comparison to all refractories was about 3%, whereas the similar figure for refractory concretes was 1.7%.

By 1950, plastic refractories, refractory concretes, and, of course, mortars were well-established, despite their reference as "specialty products." In a 1950 publication of the Mexico Refractories Company, a glossary is given which includes the terms "plastic firebrick," "castable refractory," and "mortar refractory specialties," but not "ramming mixes." Even following two decades or so of use, refractory concretes were still viewed in the early 1950s as a relatively new and novel material.[34-36] Lack of understanding of the material is reflected in an opinion of that time that refractory concrete has "no appreciable drying shrinkage or after-contraction," is "stable under load at 1300° to 1600°C," and "does not spall."[34] One writer in 1950 lamented the lack of good guidelines for choosing refractory concretes for various applications.[35] The situation was well summed up in 1951 by West and Sutton, who said:

At the present time most producers and users of refractory castables have little knowledge of cements, their mineral composition, and hydration reactions. On the other hand most men in the cement industry have little understanding of the uses of their cements in the refractories industry. Much progress could be made by bringing these two groups together.[36]

As it turned out, West and Sutton appear to have had at least a few receptive readers.

1950 to 1960

The period 1950 to 1960 saw tremendous gains made in the monolithic refractory field. These gains were made on the bases of (1) the availability of new and improved binder phases, (2) the availability of new and improved refractory aggregates, (3) the innovative uses of additives, (4) improved mixing and installation procedures and equipment, and (5) improved co-operation between the manufacturers and users of monolithic refractories. It was during the 1950s that it first became obvious that monolithic refractories had the potential for competing head-to-head with many fired refractories.

Calcium Aluminate Cements: Low-purity Lumnite cement was the only domestic calcium aluminate cement that was available on the open market in the United States from its introduction in 1924 until the early 1950s. Then, based on development work started in 1951, Alcoa, in 1955, began marketing a high-purity calcium aluminate cement labeled CA-25.[37] CA-25 was produced from mixtures of pure alumina and limestone and contained very small quantities of silica and iron oxide. It was determined by Alcoa researchers that CA alone was not refractory enough to be used in 1649°–1760°C aluminosilicate-based castables. A satisfactory balance in strength and refractoriness was achieved in a cement based on a $CaO \cdot 2.5Al_2O_3$ composition. Filling the gap between high-purity CA-25 cement and low-purity Lumnite, Universal Atlas offered the intermediate-purity Refcon cement around 1961 following extensive development efforts in the mid- and late-1950s. Additionally, during the 1950s, a second low-purity cement was offered for sale in the United States (Lone Star/Lafarge's ciment Fondu).

Table I. Composition of Low-, Intermediate-, and High-Purity Calcium Aluminate Cements in the United States*

Type of Cement	Percent of Indicated Oxide			
	Al_2O_3	CaO	SiO_2	Fe_2O_3
Low-purity	36–47	35–42	3.5–9.0	7–16
Intermediate-purity	48–62	26–39	3.5–9.0	1–3
High-purity	70–80	18–26	0–0.5	0.1–0.2

*Ref. 24.

Table I shows the compositional ranges for low, intermediate, and high-purity cements (reflecting current values). A common and essential feature of all of these calcium aluminate cements is the presence of CA as the principal anhydrous aluminate phase. The rapid development of strength, characteristic of calcium aluminate cements, is due principally to this phase.

Initially, calcium aluminate cements were made by a full-melting process in rotary kilns, but techniques were developed later to produce clinker with virtually no liquid-phase sintering involved.[38]

Phosphate Bonding: During the late 1940s, the American Refractories Institute sponsored a program under the direction of W.D. Kingery at MIT to explore the potential for phosphate bonding as related to refractory technology. A summary of this work, reported in the early 1950s,[39,40] stimulated considerable interest in this subject.[41-45]

Ramtite Company was the first in the United States to market a phosphate-bonded, plastic refractory (90 Ram). Subsequently, phosphate-bonding systems (particularly those based on aluminum phosphates) were shown to provide high strengths with full bonding developing at temperatures as low as 343°C. Refractories based on aluminum phosphate bonding exhibited high strength, dimensional stability, and excellent erosion/abrasion resistance through temperatures up to 1871°C. Phosphate bonding of alumina and aluminosilicate refractories became common practice in monolithic refractories during the 1950s.

Aggregate Development: During the 1940s, the principal aggregates used in plastic refractories were crushed firebrick, calcined clays, chrome ore, sillimanite/kyanite, and, less frequently, calcined bauxite. Castables and field mixes of this period contained blast furnace slag, crushed firebrick or insulating brick, calcined flint clay, chrome ore, sillimanite/kyanite/mullite, calcined bauxite, calcined diaspore, fused alumina, and, less frequently, calcined magnesite.

Following World War II, trials were conducted with tabular alumina aggregates having Al_2O_3 contents of 99.5% (developed in 1936), and monolithic refractories containing these aggregates, as well as pure calcined aluminas, were offered on a commercial basis in the early 1950s. Ramtite was one of the first companies to offer ramming mixes based on tabular alumina aggregate.

Tabular alumina castables that contained high-purity calcium aluminate cement had an improved erosion resistance, good resistance to wetting and penetration by molten metals, good strengths in the intermediate-temperature range, and improved refractoriness relative to low- and intermediate-purity cements available at the time.[37] Most monolithic refractory producers offered such a product prior to 1960.

During the 1950s, high-MgO casting mixes (originating with Kaiser Permanente) were developed for open hearth and electric steel furnace bottoms, based on seawater-derived magnesia.[46] The casting of these materials provided cost savings and reduced downtime relative to rammed or brick bottoms.

Summary: By the end of the 1950s, some refractory manufacturers offered as many as 90 different "refractory specialty" products. By 1960, castables based on high-purity calcium aluminate cement and tabular alumina aggregates were quite common, claiming advantages in the areas of refractoriness, erosion and abrasion resistance, and hot load-bearing ability. Ramming mixes began to show up in manufacturers' product catalogs with alumina contents over 90% and useful at temperatures up to 1871°C. Ramming mixes based on chrome ore/periclase were developed for the copper industry. Graphitic and high-alumina plastics and ramming mixes (air-setting and phosphate-bonded) were widely offered.

By the end of the 1950s, plastics, ramming mixes, and castables were being touted for use in many severe applications formerly serviced only by fired brick. These uses included applications where refractories were in contact with molten metals and slags under erosive and abrasive conditions, such as alumina melting furnace hearths and lower sidewalls, iron and steel ladle linings, blast furnace troughs and runners, and nonferrous melting applications. The use of refractory concretes in petrochemical and refinery process vessels showed dramatic gains by 1960.

Perhaps the best indication of the growth of monolithic refractories from 1950 to 1960 is the fact that in 1960 they represented roughly 30% of the total value of refractories in the United States compared to less than 10% in 1950. Certainly, it was time for the "specialty" term to be dropped for these materials.

The Period 1960 to 1984

Writing in the early 1960s, William Boyd of Kaiser Refractories cited the growth of monolithic refractories through the 1940s and 1950s and predicted that the trend to increased refractory construction with monolithics would continue and that we would look back upon the 1960s as a decade for revolutionary changes in refractories and refractory applications.[47] He certainly knew what he was talking about; the trend continued not only through the 1960s but is evident today as well.

Boyd's prognosis was based on evidence reflecting both improved performance and increased user appeal of monolithics over the period 1942-1962. Improved performance in monolithics and installation advantages were cited as reasons for a trend from the use of monolithics principally in patching and repair situations to their use in complete new unit construction as original linings. Boyd particularly singled out the gunning of monolithics (both hot and cold; refractory concrete and plastic) as an important factor in the growth trend.

From 1960 to the present, there has been a steady increase in the use of monolithic refractories by all of the refractory-consuming industries at the expense of prefired refractories. At least one industry (refining and petrochemical) has gone almost exclusively to the use of monolithics. In 1960, the combined value of monolithics relative to all refractories was around 30%. By 1982, the combined value for monolithics was around 37%, 45% in the clay refractory category, and 30% in the nonclay category.[48]

A major trend of monolithic refractory development during this time is one of product specialization wherein specific products have been developed for specific applications and even for specific locations within a given type of process vessel. In 1960, major refractory manufacturers were offering 70 to 90 different products. By 1984, this number had at least doubled, with some manufacturers offering as many as 180 different monolithic refractory products.

The diversification of monolithic refractory products during this period did not come about principally as a result of the availability of new and improved aggregates and binders but rather through innovative changes in raw material grain sizing, through reduced binder contents, through innovative uses of additives and admixtures, through improved installation procedures and refractory designs, and through innovations in the form in which the monolithics are supplied (e.g., large extruded or cast blocks). Much of this product diversification has come about as a result of continually improving cooperation between the manufacturers and the users of refractories.

Currently the vast majority of refractory concretes (both gunning and cast) are used as prepackaged, proprietary castables rather than mixed in the field from standard mix designs. Refractory concretes are available which contain low-, intermediate-, or high-purity calcium aluminate cements (as well as portland cement or phosphate binders) combined with a wide variety of aggregates and additives (e.g., calcined clays, calcined bauxite/clay mixtures,‡ calcined bauxite, tabular alumina, fused alumina, calcined alumina, magnesia, chrome ore, zircon, kyanite, mullite, silicon carbide, fused silica, and such old standbys as crushed brick and expanded mineral products. Refractory concretes are available in dense, intermediate, and lightweight forms (3524 to 481 kg/m^3), in coarse- and fine-grained versions, and are installed by hand-forming, vibration casting, slap-troweling, gunning (hot and cold), gun-casting, and pumping. Improved performance in refractory concretes at high temperatures is being gained through the use of reduced cement and water contents[49] and through the use of steel fiber reinforcements.[50] Significant advances have been made in the curing and bakeout procedures for refractory concretes,[51] and additives for minimizing the risk of explosive spalling have been identified.[52,53] Large precast blocks of refractory concrete are currently being evaluated in a number of applications.[54]

Plastics and ramming mixes are currently available which contain any of the dense aggregates used in refractory concretes. Heat- and air-setting versions of these materials are available, as well as phosphate and organic-resin bonded products. Plastics are now available that can be installed by ramming, vibration, or gunning.[55] Automated equipment has been developed for installing rammed and gunned linings.[56] Large extruded blocks of plastic (placed with cranes) have been used successfully in a number of applications. Products that can be rammed or vibrated in place and are completely free of water are also currently available.[57]

Considering the advances made in the field of monolithic refractory technology, it is somewhat surprising that the technical literature on the subject during the period 1960 to 1984 is relatively sparse. This situation perhaps reflects a presumed need by manufacturers to avoid the dissemination of proprietary information. A majority of technical papers in this era deal with applications and practices for monolithic refractories, although in the late 1970s and early 1980s discussions are seen of rational design procedures using monolithic refractories[58-60] and the use of nondestructive testing methods for inspection of installed monolithics.[61,62]

Summary and Future

The evolution of the monolithic refractory industry (refractory concrete, plastic refractories, ramming refractories) in the United States has been traced from its beginnings around 1914 to the present, a period of some 70 years. During this time, monolithic refractories evolved from simple materials with a very limited range of applications into a versatile, widely used class of refractory construction materials offering performance and cost-effectiveness on a par with, and in some cases superior to, prefired refractory bricks and shapes.

Figure 10 shows the growth of monolithic refractories in the United

---

‡Mulcoa grains produced synthetically from mixtures of bauxite and a high-purity kaolin and containing 47, 60, or 70% Al_2O_3. First offered by CE Minerals (then Mulcoa) in 1970.

States since 1914, expressed as the percent of total refractory value. From 1914 until the late 1940s, monolithics represented less than 5% of the total refractory market. This was a period of introduction, trial, and experimentation for monolithic refractories. As shown in Fig. 11, the number of different monolithic refractory products offered by manufacturers was quite limited through the late 1940s. During this period (1914 to about 1948), monolithic refractories were not viewed as serious competition for traditional brick-and-mortar construction.

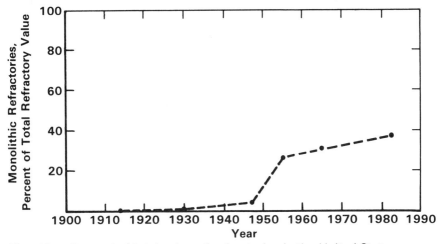

Fig. 10. Percent of total value of refractories in the United States represented by monolithic refractories for the period 1914 to 1984.

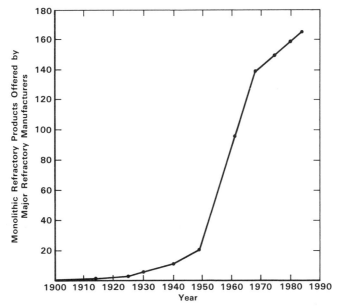

Fig. 11. Number of different monolithic refractory products offered by major U.S. refractory manufacturers for the period 1914 to 1984.

Following World War II, tremendous gains were made in the monolithic refractories field in the United States, reflected in both the market value of the materials (Fig. 10) and in the variety of products that were offered (Fig. 11). The period 1950 to 1960 saw the development of new and improved binders and aggregates, innovative uses of additives, improved mixing and installation procedures, an improved understanding of the nature of monolithic refractories (e.g., cement hydration reactions), and improved cooperation between the manufacturers and users of monolithic refractories. By 1960, it was obvious to many that monolithic refractories had the potential for competing head-to-head with many prefired refractories.

From 1960 to the present, a major trend of monolithic refractory development has been one of product specialization. Working together, manufac-

Table II. Milestones in the Evolution of Monolithic Refractories in the United States.

Date	Event
1914	W.A.L. Schaefer establishes the Pliable Firebrick Company in Chicago, Illinois (the first commercial plastic refractory)
1922	The term "alumina cement" first appears in the subject index of *Abstracts of the American Ceramic Society*
1924	Universal-Atlas begins manufacture of low-purity calcium aluminate cement (Lumnite) in Northampton, Pennsylvania
1924	First use of refractory concrete in the United States (field mixes)
1925	Insulating refractory concretes produced with Lumnite cement and expanded mineral aggregates
1926	The term "plastic fireclay refractory" first appears in the open technical literature
1928	First commercial production of proprietary castable products
Early 1930s	Plastic refractories offered in presliced, 45.36-kg (100-lb.) units
	Pneumatic ramming devices advocated for the installation of plastic refractories
	Metal and ceramic anchors used in construction of monolithic plastic refractory linings
	Refractory concretes applied using the "cement gun"
1938	Babcock and Wilcox Company begins production of a high-purity calcium aluminate cement for use in their own products

1941	Graphitic plastic introduced (Kaiser Helspot)
Early 1950s	Initial work on phosphate bonding begins stimulated by Kingery papers in the *Bulletin of the American Ceramic Society*
	First commercial phosphate-bonded plastic (Ramtite 90 Ram).
	Tabular alumina (Alcoa) is first used as an aggregate in monolithic ramming mixes (Ramtite)
	Magnesia casting mixes developed for open hearth and electric furnace bottoms (Kaiser Permanente)
	Hot gunning repair techniques developed
Mid-1950s	Phosphate-bonded plastics and ramming mixes become available
1955	High-purity calcium aluminate cement available on a commercial basis in the United States (Alcoa CA-25)
1956	Initial work reported on castables based on high-purity calcium aluminate cement and tabular alumina aggregates
1960	Monolithic refractories represent 30% of the total value of the refractories in the United States
Early 1960s	First use of large precast refractory concrete blocks in soaking pit walls
1961	Universal-Atlas Cement Company begins marketing of intermediate-purity calcium aluminate cement (Refcon)
Mid-1960s	Monolithic refractories resistant to CO disintegration developed
1965	Hotwork, Inc. is the first company in the United States exclusively devoted to dryouts and initial heat-ups of refractory installations
1970	Concept of steel fiber reinforcement for monolithic refractories first introduced (Battelle)
	Synthetic calcined aggregates containing 47, 60, or 70% Al_2O_3 become available (Mulcoa Corp.)
Early 1970s	Alumina-chromia monolithics introduced
1970s	Gun-casting of refractory concretes developed
	Development of phosphate-bonded, erosion- and abrasion-resistant castables
	Development of vibratable plastic refractories
Late 1970s	Initial use of low-cement-content castables in the United States
Early 1980s	Gunning of conventional plastic refractories

turers and users of refractories have developed specific monolithic refractory products for specific applications with the aim of providing cost-effective performance. Currently, some major refractory manufacturers offer as many as 180 different monolithic refractory products, many with well-defined, limited areas of application.

Table II lists the more important milestones in the evolution of monolithic refractories in the United States. Reference to this table and to Figs. 10 and 11 leads to the conclusion that, for monolithic refractories, the evolutionary process is far from complete. A simple review of the titles of the papers being presented at this Symposium indicates where current development trends lie for monolithic refractories and what the short-range future holds.

(1) Increased emphasis on developing meaningful tests for monolithic refractories.

(2) Continued innovation in equipment and procedures for installing and heating of monolithics (e.g., gunning plastics, automated ramming equipment).

(3) Continued innovation in material formulations to achieve improved installed properties (e.g., dry vibratables, explosive spall-resistant castables).

(4) Continued innovation in material formulations and in the use of additives or admixtures to achieve denser, stronger, abrasion-/erosion-resistant (at temperature) monolithics (e.g., water-reducing admixtures, low-cement-content castables, silica fume, steel fiber reinforcement).

(5) Continued interest in the use of large, precast blocks and shapes of refractory concretes and plastic refractories to reduce installation time and provide improved properties (e.g., prefabricated casthouse shapes).

(6) Continued tailoring of monolithic refractory products to meet specific service demands (e.g., refractories for blast furnace troughs, coal gasification equipment, incinerators).

(7) Continued development of monolithic refractories containing basic aggregates.

Almost 20 years ago, at a time when monolithic refractories commanded about 32% of the total refractory market in the United States, a lecturer speaking on the use of monolithics in steel plant applications said, "If they [monolithics] had been fit to take over the whole market, they should have done so by now."[63] Today, with monolithics still growing (40% of the U.S. market), the same is undoubtedly still true. However, it should be obvious to all concerned that in 1984 monolithic refractories should no longer be referred to as "specialties." They are, in fact, in a class by themselves and their future seems assured.

References

[1]"Technology of Monolithic Refractories," Plibrico Japan Company, Ltd., Tokyo, Japan, p. 598, 1984.

[2]F.T. Havard, "Recent Developments in the Refractories Industry," *Trans. Am. Ceram. Soc., XIV,* 480–88 (1912).

[3]W.J. Rees, "Refractories in 1924," *Annual Rept. Soc. Cham. Ind.,* **9,** 284–292 (1924).

[4]"Cutting Furnace Costs," Plibrico Jointless Firebrick Company, Chicago, IL, pp. 1–39 (1918).

[5]"Cutting Furnace Costs," Plibrico Jointless Firebrick Company, Chicago, IL, pp. 1–39 (1930).

[6]"Refractories and Furance Design," Plibrico Jointless Firebrick Company, Chicago, IL, pp. 1-31 (1923).
[7]"Plastic Refractories," *Brt. Clayworker,* **35** [14] (1926).
[8]H.E. Weightman, "Use of Plastic Refractories in Boiler Furnaces," *Power,* **63** [3] 90 (1926).
[9]H.E. Weightman, "Service Requirements For Plastic Refractories," *Am. Ceram. Soc. Bull.,* **5** [4] 210-13 (1926).
[10]"Specifications For Plastic Fireclay Refractories," *Brick and Clay Record,* **69** [6] 430 (1926).
[11]M.C. Booze, "Development in Manufacturing Uses and Applications of Refractories," *Fuels and Furnace,* **7** [4] 567-71 (1929).
[12]H. Sainte-Claire Deville, *Ann. Phys. Chim.,* **46** [3] 196 (1856).
[13]H.S. Spackman, *Proc. ASTM,* **10,** 315 (1910).
[14]T.D. Robson, p. 263 in High Alumina Cements and Concretes. Wiley & Sons, New York, 1962.
[15]E.J. Crane and E. Hockett, *Trans. Am. Ceram. Soc. Collec. Index,* 1-70, (1920).
[16]S. Rusoff, "Experiences in Mending Kiln Furniture," *Trans. Am. Ceram. Soc., XVII,* 200-203 (1915).
[17]H.S. Spackman, "Alumina Cement; It's Development, Use and Manufacture," *Eng. News-Rec.,* **88,** 831-34 (1922).
[18]"High Alumina Cements," *Engr. (London),* 114, 180, (1922).
[19]P.H. Bates, "The Cementing Qualities of the Calcium Aluminates," *Bur. Stand. Tech. Paper,* p. 197 (1922).
[20]H.S. Spackman, "What We May Expect To Do With Aluminate Cement," *Concrete,* **24** [3] 88-90 (1924).
[21]E. Eckel, "The Actual Uses of Alumina Cements," *Concrete,* **23,** 175-76 (1923).
[22]R.J. Anderson, "Aluminum and Bauxite," *Mineral Ind.,* **33,** 11-50 (1924).
[23]R.C. Gosreau, "Bonding High Temperature Refractories," *Chem. Met. Eng.,* **31,** 696-98 (1924).
[24]"Refractory Concrete," American Concrete Institute Committee Report ACI 547R-79, p. 224, 1979.
[25]T.D. Robson, "Refractory Concretes: Past, Present, & Future," in Refractory Concrete, American Concrete Institute Publication SP-57, pp. 1-10, 1978.
[26]E. Eckel, "Opportunities For Alumina Cement Manufacturing," *Mfrs. Rec.,* **97** [25] 59 (1930).
[27]N.J. Kent, "Lightweight Refractory Concrete," *Heat Treat. Forg.,* **7** (1935).
[28]"Insulcrete, Insulating Refractory Concrete," *Iron Age,* **9,** 33-35 (1925).
[29]"Chromcrete, New Neutral Base Castable Refractory," *Am. Gas J.,* **4,** 16 (1933).
[30]"Chrome Base Castable Shows Merit in Steel Heating Furnaces, Hearthcrete," *Steel,* **2,** 21-22 (1934).
[31]B.L. Bryson, "Tabular Alumina—A New Sophistication in Refractory Raw Materials," *Refrac. J.,* **11,** 6-9 (1971).
[32]W.B. Paul, Jr., "Monolithic Refractories in Fluid Catalytic Cracking Refinery Units," *Am. Ceram. Soc. Bull.,* **33** [4] 108-10 (1954).
[33]J.F. Wyant, and W.L. Bulkley, "Refractory Concrete For Refinery Vessel Linings," *Am. Ceram. Soc. Bull.,* **33** [8] 233-39 (1954).
[34]A.E. Williams, "Refractory Concrete," *The Chemical Age,* May 10, 1952; pp. 721-24.
[35]J.I. Cordwell, "Refractories Development: An International Review of Recent Work and Techniques," *Refrac. J.,* **7,** 264-69 (1950).
[36]W.J. Sutton and R.R. West, "Manufacture and Use of Fireclay Grog Refractories," *Am. Ceram. Soc. Bull.,* **30** [2] 35-40 (1951).
[37]W.H. Gitzen, L.D. Hart, and G. MacZura, "Properties of Some Calcium Aluminate Cement Compositions," *J. Am. Ceram. Soc.,* **40** [5] 158-67 (1957).
[38]"Calcium Aluminate Cements Produced By Solid State Reaction," *Ind. Heat.,* **11,** 53-55 (1973).
[39]W.D. Kingery, "Fundamental Study of Phosphate Bonding in Refractory, I, Literature Review, II, Cold-Setting Properties, III, Phosphate Adsorption by Clay and Bond Migration," *J. Am. Ceram. Soc.,* **33** [8] 239-50 (1950).
[40]W.D. Kingery, "IV, Mortars Bonded with Monoaluminum and Monomagnesium Phosphate," *J. Am. Ceram. Soc.,* **35** [3] 61-63 (1952).
[41]W.H. Gitzen, L.D. Hart, and G. MacZura, "Phosphate-Bonded Alumina Castables: Some Properties and Applications," *Am. Ceram. Soc. Bull.,* **35** [6] 217-23 (1956).
[42]G.J. Grott, "Phosphate-Bonded Refractories For Specialty Applications," *AIME Electric Furnace Steel Proc.,* **17,** 391 (1959).
[43]"Phosphate and Ceramic Bonded High-Alumina Refractories Developed for Aluminum Melting Furnaces," *Ind. Heat.,* **27,** 1694 (1960).
[44]P.A. Gilham-Dayton, "The Phosphate Bonding of Refractory Materials," *Trans. Brt. Ceram. Soc.,* **62** [11] 895 (1963).

[45]H.D. Sheets, J.J. Buloff, and W.H. Duckworth, "Phosphate Bonding of Refractory Compositions," *Brick and Clay Rec.*, 55-57 (1958).

[46]H.M. Kraner, et-al., "Casting Large Sections of Basic Refractories," *Am. Ceram. Soc. Bull.*, **39** [9] 456-59 (1960).

[47]W.E. Boyd, "Trends in Monolithic Refractories," *Ind. Heat.*, **10**, 1989-2007 (1964).

[48]"Refractories," Current Industrial Report Summary for 1982, *U.S. Dept. of Commerce*, November 1983, pp. 1-20.

[49]B. Clavard, J.P. Kiehl, and R.D. Schmidt-Whitley, "Fifteen Years of Low Cement Castables in Steelmaking," *Proceedings First International Conference on Refractories*, November 15-18, 1983, Tokyo, Japan. Published by Technical Association of Refractories, Japan, 1983; 589-606.

[50]D.R. Lankard, "Steel Fiber Reinforced Refractory Concrete," American Concrete Institute Publication SP-57 (Refractory Concrete) 1978; pp. 241-63.

[51]N.W. Severin, "Dryouts and Heatups (Bakeouts) of Refractory Monoliths;" for abstract see *Am. Ceram. Soc. Bull.*, **63** [3] 432 (1984).

[52]T. Kleeb and J. Caprio, "Properties and Service Experiences of Organic Fiber-Containing Monoliths;" for abstract see *Am. Ceram. Soc. Bull.*, **63** [3] 431 (1984).

[53]D.R. Lankard and L. Hackman, "Use of Admixtures in Refractory Concretes," *Am. Ceram. Soc. Bull.*, **62** [9] 1019-25 (1983).

[54]R.A. Howe, J.W. Kelley, and T.A. Dannemiller, "Prefabricated Casthouse Shapes—Advantages, Evolution & Design Considerations;" for abstract see *Am. Ceram. Soc. Bull.*, **63** [3] 432 (1984).

[55]G. Wilson, D. Hofmann, M. Tsukino, and L. Krietz, "A Review of International Experiences in Plastic Gunning;" for abstract see *Am. Ceram. Soc. Bull.*, **63** [3] 432 (1984).

[56]T. Morimoto, K. Ogasahara, A. Matsuo, and S. Miyagawa, "Introduction of Automatic Gunning Machine for Tundish Lining;" for abstract see *Am. Ceram. Soc. Bull.*, **63** [3] 431 (1984).

[57]D.M. Myers, and G.M. Turner, Jr., "The Development of Dry Refractory Technology in the United States;" for abstract see *Am. Ceram. Soc. Bull.*, **63** [3] 432 (1984).

[58]T.J. Fowler, "Rational Design With Monolithic Refractories," American Concrete Institute Publication SP-74, Monolithic Refractories, 1982; pp. 47-72.

[59]S.A. Bortz, R.F. Firestone, and M.J. Greaves, "Castable Refractory Design Requirements;" American Concrete Institute Publication SP-74, Monolithic Refractories, 1982; pp. 17-32.

[60]R.E. Farris, "Refractory Concrete—Use of Engineering Property Data For Design;" American Concrete Institute Publication SP-57, Refractory Concrete, 1978; pp. 151-66.

[61]W.A. Ellingson, "Advances in Non-Destructive Evaluation Methods For Inspection of Refractory Concretes;" American Concrete Institute Publication SP-74, Monolithic Refractories, 1982; pp. 33-56.

[62]M.S. Crowley, "Inspection of Refractory Concrete Linings;" for abstract see *Am. Ceram. Soc. Bull.*, **61** [8] 829 (1982).

[63]G.M. Workman, "Where Have All the Firebricks Gone?," *Refract. J.*, **6**, 196-203 (1965).

Section II
Procedures for Testing the Properties of Monolithics

Creep of Refractories: Mathematical Modeling 69
    D. J. Bray

Test Methods for Monolithic Materials 81
    G. C. Padgett and F. T. Palin

Compressive Stress/Strain Measurement of Monolithic Refractories at Elevated Temperatures 97
    W. R. Alder and J. S. Masaryk

Aggregate Distribution Effects on the Mechanical Properties and Thermal Shock Behavior of Model Monolithic Refractory Systems 110
    J. Homeny and R. C. Bradt

The Heat Evolution Test for Setting Time of Cements and Castables 131
    C. H. Fentiman, C. M. George, and R. G. J. Montgomery

Creep of Refractories: Mathematical Modeling

D. J. Bray

Aluminum Company of America
Chemical and Ceramics Division
Alcoa Center, PA 15069

Ceramic materials are of technical importance largely due to their high-temperature properties. A primary factor limiting structural usefulness of ceramic materials is high-temperature creep. While creep of metals has been studied extensively, and creep of pure oxides somewhat less, creep of refractories as a class of materials has been left relatively unstudied because of its complexity. Classical creep theory developed for metals and single-crystal oxides does not apply directly to refractories; however, empirical analyses can be made by using similar techniques to determine the general effect of certain variables on deformation behavior.[1]

Once data are generated, the subsequent analysis is often inadequate for understanding the deformation mechanisms taking place. Standard analyses generally end at comparing total strain under a particular set of conditions between materials of interest. If steady state strain rates are determined, it is usually not as a function of stress or temperature and often not for adequate time periods. Also, little importance is placed on the primary creep and associated mechanisms in ceramic materials.

In this study, a method is proposed for analyzing creep data of refractories using empirical mechanical analogs and nonlinear least-squares techniques to provide both a better analysis for refractory selection and to increase understanding of the creep behavior of refractories.

Creep Theory

Creep is defined as a thermally activated deformation process that occurs when materials are exposed to high temperatures ($>0.5T_{\text{melting point}}$) under some applied stress.[2-4] Most ceramic materials exhibit a characteristic sigmoidal curve when strain is plotted vs time under constant stress and temperature conditions (Fig. 1). This creep curve exhibits four distinct stages: (1) directly on loading, the specimen exhibits an instantaneous, elastic strain; (2) this is followed by the primary creep stage, where strain rate declines gradually as a function of time; (3) the secondary stage (steady-state) follows, where the strain rate is constant; and (4) a tertiary creep region sometimes follows, where the strain rate begins to accelerate and leads to catastrophic failure or creep rupture.[5]

The amount of strain or creep (ϵ) is a function of temperature (T), stress (σ), time (t), and structure (S): $\epsilon = f(T, \sigma, t, S)$. The structure term can include both macrostructures such as grain size and phase distribution, and microstructures such as crystal structure, dislocation configurations, etc.

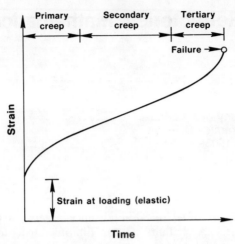

Fig. 1. Typical sigmoidal curve of strain vs time under constant pressure and temperature for ceramic materials.

Primary creep is generally short-time creep and represents a small percentage of the total deformation in most cases. Secondary creep represents the long-time deformation processes, and therefore, many analyses have been developed for this portion of the creep curve. Steady-state creep is of engineering importance, because it leads to irreversible or permanent strains.[6] From an absolute-rate theory approach, the steady-state strain rate, $d\epsilon/dt_{ss}$, will exhibit an Arrhenius dependence as a function of temperature.[5,7] Additionally, it has been found that $d\epsilon/dt_{ss}$ is proportional to stress to some power so that:

$$d\epsilon/dt_{ss} = \dot{\epsilon}_{ss} = A\sigma^\eta \exp\left(-\frac{\Delta H}{RT}\right) \qquad (1)$$

where $\dot{\epsilon}_{ss}$ = steady state strain rate; A = constant including structure term; σ = stress; η = stress exponent; ΔH = activation energy for creep; R = universal gas constant; and T = temperature, which is the general creep equation for materials.

Creep mechanisms are usually divided into two distinct groups. First there are "lattice mechanisms" that are entirely intragranular in nature. These mechanisms are equally likely to occur in polycrystalline materials and single crystals. Second, there are "boundary mechanisms," which depend on grain boundaries and are therefore relevant to polycrystalline materials. Boundary mechanisms are processes in which strains occur by the movement of grains relative to one another. Two types of boundaries exist: those containing a liquid or liquidlike second phase (Newtonian viscous character), and those with no second phase.[4,5] In refractory systems, both mechanisms are operative.

Creep of Refractories

Although the term "refractories" encompasses a wide range of materials, it is generally set apart from polycrystalline, fine-grained technical ceramics. Important commercial refractories are almost entirely polycrystalline, with the vast majority also being polyphase.[1,8] There are a few single-phase, polycrystalline refractories, but generally a refractory material consists of a coarse material bonded by a second phase. The coarse material is usually a well-developed crystalline aggregate, whereas the second phase is often poorly developed, with a lower degree of crystallinity or is, in many cases, completely amorphous. A poorly developed second phase generally plays a major role in the properties, including creep.[9-12] It is probable that boundary mechanisms would dominate the creep behavior of such materials.

By the very nature of their application, refractory materials are generally load-bearing structural materials. The many factors cited above contribute to deformation of refractories, and therefore analysis is difficult. Creep analysis is usually carried out on the steady-state portion of the curve using the classical creep equation. Model refractory systems have been studied using classical analysis, and classical creep mechanisms were determined to be operative in some systems. Snowden and Pask[13] found that, in the MgO-$CaMgSiO_4$ refractory system, both dislocation movement and viscous deformation of boundary regions were primary mechanisms. Burdick and Day[14] also used a classical approach to examine the creep behavior of direct-bonded, high-alumina refractories. Ainsworth and Kaniuk[15] and Beyer et al.[16] used a similar approach to examine creep of aluminosilicates, as did Bray et al.[17,18] on refractory concretes.

Creep of refractory concretes is particularly interesting in that many creep mechanisms are occurring simultaneously, so that the measured strain rate is a summation of separate strain rates.[19]

$$\dot{\epsilon}_{measured} = \Sigma \dot{\epsilon}_i \tag{2}$$

This analysis is generally written in the following form:

$$\dot{\epsilon}_{measured} = \Sigma_i A_i \sigma^n \exp\left(\frac{\Delta H_i}{RT}\right) \tag{3}$$

where the same form is assumed for each process. For refractory concretes, it has been shown that the activation energy for creep on initial heatup is equivalent to the activation energy for the formation of calcium aluminate (approximately 160 kJ/mol). It was also shown that stress-aided densification, sintering, and other mechanisms play a role in creep of these materials.[17,18]

Aluminosilicates held at high temperatures (>1200°C) can develop elongated mullite crystals which form an interlocking network of high strength. The presence of small amounts of Na_2O (approximately 0.5%) increases the rate of mullite formation, resulting in higher creep strengths. With increasing purity, mechanisms other than shear of amorphous phases can contribute to creep. Grain-boundary sliding has been suggested for high-alumina refractories, and a dislocation plastic flow mechanism for high-MgO refractories.[1]

Mechanical Analogs

It has been found that elastic-plastic and inelastic behavior of materials can be represented empirically by using mechanical analogs. Zener described anelasticity in metals by developing mechanical models consisting of arrays of elastic springs and viscous dashpots.[20] This type of analysis has been used to characterize diffusion of solutes in metals[21] and deformation behavior of stainless steels,[22,23] cadmium,[24] tungsten,[25] nickel,[26] zinc,[26] iron,[26] and copper-aluminum alloys.[27]

In these mechanical models, springs represent the elastic reponse of materials and behave according to Hooke's law (Fig. 2), where strain is linearly proportional to stress with a constant of proportionality (the spring constant, K, for springs, and elastic modulus, E, for materials). A viscous solid can be represented by a dashpot, where shear rate (γ) is proportional to shear stress (τ), with the constant of proportionality being viscosity (η) (Fig. 3). This element, then, represents the time-dependent deformation process, and in materials where a constant stress is applied, strain increases as a function of time proportional to applied stress over viscosity (σ/η).

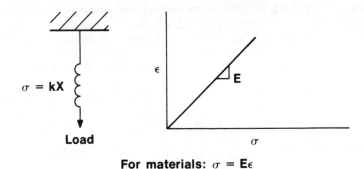

Fig. 2. Elastic behavior of materials and the mechanical analog.

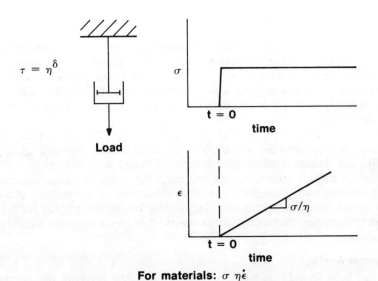

Fig. 3. Viscous behavior of materials and the mechanical analog.

Most materials, however, are neither purely elastic nor purely viscous in nature at high temperatures, particularly refractories, and therefore various combinations of springs and dashpots have been assembled to describe more complex creep behavior. Zener developed a model called the standard linear solid which describes the deformation behavior of many materials.[20] This model (Fig. 4) consists of a spring to describe the initial elastic strain of a material under an applied stress, in series with a Kelvin element (a spring and a dashpot in parallel), which describes the time-dependent portion of the creep curve where the strain rate is decreasing as a function of time, and a dashpot in series, which represents the time-independent strain rate (the spring and dashpot in series represent a Maxwell element). The strain response of this model then is:

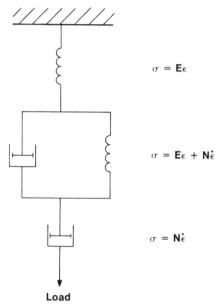

Fig. 4. Elastic-plastic material: standard linear solid.

$$\epsilon = \epsilon_S + \dot{\epsilon}_D t + \epsilon_K [(1 - \exp - t/\tau_K)] \quad (4)$$

where ϵ_S = displacement of the spring; $\dot{\epsilon}_D$ = displacement rate of the dashpot; t = time; ϵ_K = total possible displacement due to Kelvin element (Fig. 5); and τ_K = time constant of Kelvin element.

In deformation behavior of materials this corresponds to:

$$\epsilon = \epsilon_o + \dot{\epsilon}_{ss} t + \epsilon_p [1 - \exp(-t/\tau_K)] \quad (5)$$

where ϵ_o = elastic strain; $\dot{\epsilon}_{ss}$ = steady state strain rate; t = time; ϵ_p = total strain due to primary creep; and τ_K = empirical time constant.

For a given set of stress-temperature conditions, ϵ_o, $\dot{\epsilon}_{ss}$, ϵ_p, and τ_K are constants.

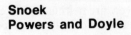

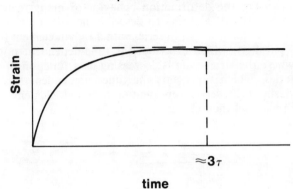

Fig. 5. Diffusion of solutes in metals behaves like the mechanical response of a Kelvin element. At $3\tau_K$, $\approx 99\%$ of the diffusion is complete.

Equation (5) represents the nonlinear, time-dependent, strain response of many materials subjected to an applied stress at high temperatures. The first term, ϵ_o, is not time-dependent and therefore is usually dropped from creep analysis. This relationship:

$$\epsilon = \dot{\epsilon}_{ss} t + \epsilon_p \left[1 - \exp\left(-t/\tau_K\right)\right] \tag{6}$$

cannot be linearized with respect to time, and therefore nonlinear least-squares techniques are required to fit time-strain data to this model. In this study, different nonlinear least-squares software computer routines were used with essentially identical results.

Experimental Procedure

Creep experiments in this study were performed in a standard constant load creep apparatus capable of temperatures to 1550°C and loads from approximately 0.1–10 kN. A linear, variable-differential transformer system was used to monitor specimen deformation. All experiments were performed in air using compressive loading. Sample size was a 5 by 5 by 11-cm prism.

Each specimen was brought to temperature with a subsequent 5–10-h soak to ensure thermal and compositional equilibrium with the environment. The standard creep experiment used in this study was 100 h at 1425°C under a stress of 1.72×10^5 (25 psi). (This test is an internal Alcoa standard for selecting materials.) Tests were performed at other temperatures in order to calculate the temperature dependence of the strain rate, $\dot{\epsilon}_{ss}$.

Once a curve was generated, the time-strain data were entered into a computer and fit to Eq. (5). A statistical fit of the line to the data according to Eq. (6) was then determined. This technique was also used to fit data from the literature.

Results and Discussion

Over 100 creep curves were developed principally on aluminosilicate, dry-pressed, refractory brick. In all cases, unless a discontinuity in the curve was encountered, the equation fit the data extremely well. In the majority of cases, the correlation coefficient raised to the second power (R^2) was greater than 0.99. Typical creep curves for these aluminosilicates are shown in Fig. 6. This graph points out a particularly interesting finding in this study. Both sample A and sample B are superduty-class bricks with almost identical chemistry and properties. If the total strain at the end of the test (100 h) was used to select materials by relative merit, these materials would be equivalent, with a strain of approximately 10.5%. If the test had ended at 50 h, however, sample A would have been the obvious choice since it exhibited a strain of approximately 6%, while sample B had deformed approximately 9%. It is visually clear, however, that if the intended use of this refractory is for long-time exposure to high temperatures, sample B is the obvious choice by examining the extrapolated curves.

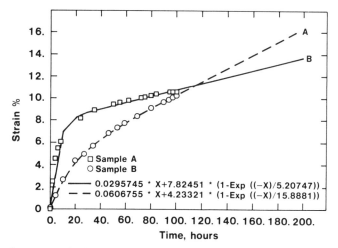

Fig. 6. Creep analysis at 1.72×10^5 Pa/1425°C·100 h for two superduty bricks, A and B.

The results of the nonlinear least-squares fit for both samples are represented in the equation at the bottom of Fig. 6. These results are summarized in Table I. The results in Table I quantify the difference between samples A

Table I. Results of Analysis on Superduty Bricks

Parameters	Brick A	Brick B
$\dot{\epsilon}_{ss}$, Steady-state strain rate (h)	0.029	0.060
ϵ_p, Total primary creep (%)	7.9	4.2
τ_K, Time constant (h)	5.3	15.9
ϵ_T, Total strain at 100 h (%)	10.6	10.2
R^2 Correlation coefficient^2	> 0.99	> 0.99

and B that can be seen in Fig. 6. The steady state strain rate, the most important parameter, for sample A is 0.029/h, whereas for sample B it is a factor of two higher at 0.060/h. It can also be seen that the time constant (after $3\tau_K$ the Kelvin element has exhausted $\approx 99\%$ of its contribution to deformation) is more than three times greater for sample B than A. This indicates that sample B does not reach steady state until roughly 50 h into the test.

The same analysis can be used on plastic refractories (Fig. 7), high-alumina castables (Fig. 8), and fireclay-based refractories (Figs. 9–11). In all but a few cases, the correlation coefficient raised to the second power (R^2)

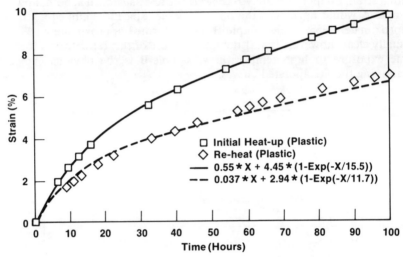

Fig. 7. Creep analysis for plastic A, a superduty-class refractory plastic, tested at 1425°C under 1.72×10^5 Pa stress for 100 h.

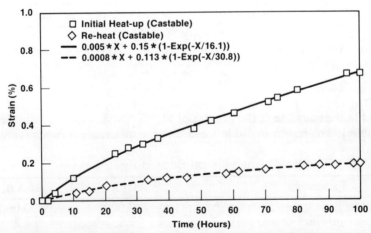

Fig. 8. Creep analysis for castable A, a high-alumina castable based on high-purity, CA cement and tabular Al_2O_3 tested at 1425°C under 1.72×10^5 Pa stress for 100 h.

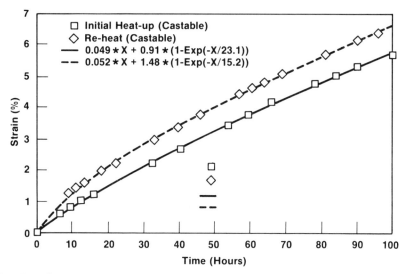

Fig. 9. Creep analysis for castable B, a fireclay castable with intermediate-purity, CA cement tested at 1425°C under 1.72×10^5 Pa for 100 h.

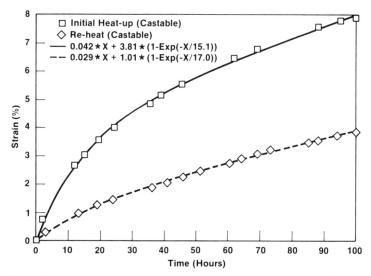

Fig. 10. Creep analysis for castable C, a lower quality castable based on intermediate-purity, CA cement tested at 1425°C under 1.72×10^5 Pa stress for 100 h.

was greater than 0.99. The results of the creep analyses are shown in Tables II and III. Creep of nonfired refractories is more complex than creep of fired refractories, and therefore creep curves are shown on initial heatup of the sample, and then on reheat.

The plastic A material exhibits a standard creep curve for both initial heatup and on reheat (Fig. 7). The steady state strain rate on initial heatup is

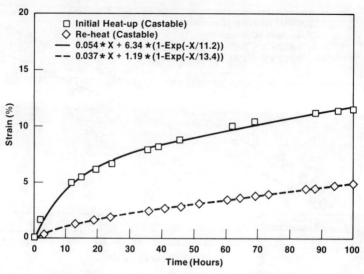

Fig. 11. Creep analysis for castable D, a still lower quality castable based on intermediate-purity CA cement tested at 1425°C under 1.72×10^5 Pa for 100 h.

Table II. Results of Analysis of Initial Heatup on Nonfired Refractories

Parameters	Plastic A	Castable A	Castable B	Castable C	Castable D
$\dot{\epsilon}_{ss}$, Steady-state strain rate (h)	0.055	0.005	0.049	0.042	0.053
ϵ_p, Total primary creep (%)	4.45	0.15	0.19	3.81	6.34
τ_K, Time constant (h)	15.47	16.14	23.13	15.06	11.16
ϵ_T, Total strain at 100 h (%)	9.9	0.64	6.6	8.0	11.3
R^2 Correlation coefficient²	> 0.99	> 0.99	> 0.99	> 0.99	> 0.99

approximately 0.055/h, whereas on reheat $\dot{\epsilon}_{ss}$ is approximately 0.037, which indicates that some densification and sintering is taking place. It can be concluded that the thermal history of the sample can be extremely important. This has been shown in the literature.[17]

Other parameters can also change from initial heatup to reheat, which indicates that the material's structure is now different. For castable A (Fig. 8), the change from initial heatup to reheat is quite significant for $\dot{\epsilon}_{ss}$, which is 0.0053/h on initial heatup and 0.0008 on reheat. This material also exhibits creep rates an order of magnitude lower than the other materials

Table III. Results of Analysis of Reheat on Nonfired Refractories

Parameters	Plastic A	Castable A	Castable B	Castable C	Castable D
$\dot{\epsilon}_{ss}$, Steady-state strain rate (h)	0.0370	0.0008	0.0510	0.0290	0.0370
ϵ_p, Total primary creep (%)	2.94	0.11	1.48	1.01	1.19
τ_K, Time constant (h)	11.65	30.83	15.17	16.98	13.37
ϵ_T, Total strain at 100 h (%)	6.4	0.18	5.8	3.8	5.0
R^2 Correlation coefficient[2]	0.98	> 0.99	> 0.99	> 0.99	> 0.99

tested due to its purity. The remaining parameters ($\dot{\epsilon}_p$, τ_K, ϵ_T) all change in the same manner as the other materials from initial heatup to reheat. Little change in $\dot{\epsilon}_{ss}$ is seen for castable B, which exhibits an $\dot{\epsilon}_{ss}$ of 0.049 on initial heatup and 0.051 on reheat. This indicates that no structural development occurred in the material during the test and that the creep rate remains high. Castable C behaves much like the superduty plastic, with the $\dot{\epsilon}_{ss}$ dropping from 0.042 on initial heatup to 0.029 when reheated, and ϵ_p changing from 3.81 to 1.01. Castable D again gives almost identical results to the superduty plastic. One other interesting point is that castable A does not reach steady state creep until > 90 h into the test. This clearly indicates that test duration is also very important when determining the creep resistance of refractories.

Conclusion

The fit of the data to the model is extremely good in all cases. Because of the range of materials evaluated, no general trends in creep changes from initial heatup to reheat can be seen. However, in evaluating a class of materials, e.g., superduty brick, this analysis can be extremely useful in understanding creep and determining relative rankings between materials.

This type of analysis determines the important parameters for creep of refractories, including the steady state creep rate ($\dot{\epsilon}_{ss}$), the total amount of primary creep (ϵ_p), and the time to reach steady state creep ($3\tau_K$). These factors, as a function of stress and temperature, can fully characterize the creep behavior of refractories.

Acknowledgments

The author wishes to thank Thomas Weinel and Karen Bowersox for their efforts in performing the experiments and analyzing the data.

References

[1] W.D. Kingery, H.K. Bowen, and D.R. Uhlmann, Introduction to Ceramics. John Wiley and Sons, Inc., New York, 1976.
[2] G.R. Terwilliger and K.C. Radford, "High Temperature Deformation of Ceramics: I," Am. Ceram. Soc. Bull., 53 [2] 172-79 (1974).

[3]K.C. Radford and G.R. Terwilliger, "High Temperature Deformation of Ceramics: II, Specific Behavior," *Am. Ceram. Soc. Bull.*, **53** [2] 465-72 (1974).
[4]R. Lagneborg, "Dislocation Mechanisms in Creep," *Int. Metall. Rev.*, Review No. 165; p. 130-46 (1972).
[5]A.G. Evans and T.G. Langdon, "Structural Ceramics," *Prog. Mater. Sci.*, **21** [3-4] 350-416 (1976).
[6]P.J. Pike. O. Buyukozturk, and J.J. Conner, "Thermomechanical Analysis of Refractory Lined Coal Gasification Vessels"; Department of Civil Engineering Research Rept. No. R80-2, Massachusetts Institute of Technology, January 1980.
[7]A. Crosby and P.E. Evens, "Creep in Non-Ductile Ceramics," *J. Mater. Sci.*, **80** 1759-64 (1973).
[8]J.H. Partridge, "Creep of Refractory Materials," *Trans. Br. Ceram. Soc.*, **53** [11] 731-41 (1954).
[9]C.O. Hulse and J.A. Pask, "Analysis of Deformation of a Fire Clay Refractory," *J. Am. Ceram. Soc.*, **49** [6] 312-18 (1966).
[10]B.A. Wiechula and A.L. Roberts, "The Elastic and Viscous Properties of Alumino-Silicate Refractories," *Trans. Br. Ceram. Soc.*, **51** [5] 173-97 (1952).
[11]E.V. Degtyareva, I.S. Kainarskii, and Y.Z. Shapiro, "Relation Between the Creep and Porosity of Refractories," *Ogneupory*, **11** 41-45 (1973).
[12]E.V. Degtyareva, I.S. Kainarskii, and I.I. Kabakova, "Structure and Creep of Corundum Refractories," *Ogneupory*, **4** 35-43 (1971).
[13]W.E. Snowden and J.A. Pask, "Creep Behavior of a Model Refractory System MgO-Ca Mg SiO$_4$," *J. Am. Ceram. Soc.*, **61** [5-6] 231-34 (1978).
[14]V.L. Burdick and D.E. Day, "Creep of High-Alumina Refractories," *Am. Ceram. Soc. Bull.*, **48** [12] 1109-13 (1967).
[15]J.H. Ainsworth and J.A. Kaniuk, "Creep of Refractories in High Temperature Blast Furnace Stoves," *Am. Ceram. Soc. Bull.*, **57** [7] 657-59 (1978).
[16]R.E. Beyer, R.D. Ek, J.L. Scott, G.A. Zeugner, and O.S. Whittemore, "Creep of Super-Duty Fire Clay Brick Under Compression," *Am. Ceram. Soc. Bull.*, **55** [12] 1049-51 (1976).
[17]D.J. Bray, J.R. Smyth, and T.D. McGee, "Creep of Monolithic Refractories." Quarterly Progress Reports submitted by Iowa State University to the DOE under Contract No. DE-AS05-78-OR13402.
[18]D.J. Bray, J.R. Smyth, and T.D. McGee, "Creep of a 90 + % Al$_2$O$_3$ Refractory Concrete," *Am. Ceram. Soc. Bull.*, **59** [6] 706-10 (1980).
[19]T.G. Langdon and F.A. Mohamed, "The Characteristics of Independent and Sequential Creep Processes," *J. Austral. Inst. Metals*, **22** [3-4] 189-99 (1977).
[20]C. Zener, Elasticity and Inelasticity of Metals. The University of Chicago Press, Chicago, 1948.
[21]R.W. Powers and M.V. Doyle, "Diffusion of Interstitial Solutes in the Group V Transition Metals," *J. Appl. Phys.*, **30** [4] 514-24 (1959).
[22]F. Garofalo, Fundamentals of Creep and Creep-Rupture in Metals, Macmillan, New York, 1965.
[23]F. Garofalo, O. Richmond, W.F. Domis, and F. Von Gremminger, Joint International Conference on Creep; p. 1. Inst. Mech. Eng., London, 1963.
[24]E.N. da C. Andvade and D.A. Aboav, *Proc. R. Soc. Sect. A;* pp. 203-352. London, 1964.
[25]J.B. Conway and M.J. Mullikin, *Trans. Met. Soc. AIME*, **236**, 1629 (1965).
[26]W.J. Evans and B. Wilshire, *Trans. Met. Soc. AIME*, **242**, 1303 (1968).
[27]W.J. Evans and B. Wilshire, *J. Metal Sci.*, **1** 2133 (1970).

Test Methods for Monolithic Materials

G.C. Padgett and F.T. Palin

British Ceramic Research Association Limited
Stoke-on-Trent, England

At the present time, there are several standard test methods designed specifically for monolithic refractories. Due to a number of circumstances, the test methods derived by the various standards organizations are different. This could be due to differences in philosophy regarding attitudes toward the materials and also to radical changes in materials themselves. These latter changes may be so significant that either new testing methods or new preparation techniques are required.

Recently, test methods have been developed within Europe by PRE (The European Federation for Refractories Producers). They have produced a comprehensive series of methods which have been submitted to ISO as a direct comparison to the complementary ASTM methods. The British Standards Institution (BSI) is currently revising its whole series of test methods, which will incorporate various aspects from ASTM and PRE. In addition, BSI intends to derive methods which are suitable for the new types of products which are now available in the market.

Testing Procedures

Increasing international importance attached to the commercial implications of quality assurance in most industries applies equally to refractory materials. For quality assessment of monolithic materials, the relevance of data obtained depends on the sampling procedure adopted. Although sampling is a very significant part of testing, it is outside the terms of reference of this paper. Attention is therefore concentrated on the more important aspects of testing as defined in a wide-ranging review carried out on behalf of sponsors from British industry. Consideration is therefore being given to test-piece preparation, strength, thermal conductivity, and dimensional stability.

Test-Piece Preparation

Test-piece preparation has a fundamental influence on properties determined in testing procedures and thus considerable effort was devoted to this topic.

Dense Castables: It was considered essential to replace existing manual techniques with vibrational procedures to simulate the service installation methods increasingly used with this type of product.

Practical trials of vibration techniques, covering a large number of commercially available products, indicated that manufacturers' recommended water content values were often unsuitable for compaction by vibration. Since it is important to use the minimum water content consistent with full consolidation of the material, the inclusion of a consistency test in a revised test-piece preparation procedure was considered a necessity.

Although the Pilny-Martienssen cone test, as utilized in PRE, gave an accurate value of water content, it was found too consuming in time and material to be employed in a standard test. A slightly modified ASTM "Ball in Hand" test which, after catching the ball, involves shaking the hand sideways about 2 cm., 20 times in five seconds was developed. The correct consistency is that which flows between, but not through, the fingers. This modification gives the drier consistency required for vibration and is very similar to the Pilny-Martienssen cone test. Good reproducibility was found in multioperator trials and in Round Robin tests on two different materials and the method was accordingly adopted.

Table I. Results of Round Robin Tests with Standard Deviations

Laboratory	Bulk Density (kg/m^3)		Cold Crushing Strength (MN/m^2)	
	Mean	Standard Deviation	Mean	Standard Deviation
A	2557	10	36.5	1.6
B	2512	8	36.5	1.4
C	2521	17	36.3	2.7
D	2517	10	38.7	1.9
F	2534	10	34.8	1.1
Grand mean	2528	20	36.5	2.2

The consolidation of the test piece with specialized vibrational equipment laid down in PRE/R26.1 was considered to be unnecessarily complicated. A simplified technique where the operator sets the vibrational frequency at the minimum value required to maintain surface flow for 1 min was evaluated in a Round Robin test. These results, determined at a fixed water content of 11.5% (from modified ASTM "Ball in Hand" test), are shown in Table I and show good reproducibility.

Additional experimental work on test-piece curing procedures, including a brief investigation of the effect of curing temperature, was also undertaken. This involved a comparison of three curing procedures: the present British Standard method, the ASTM method (C862-77), and a slightly modified version of the latter which used a curing temperature of 45°C. The comparison was made for dense castables containing both high-iron and low-iron cements, and the efficiency of curing was assessed on the basis of cold crushing strength at each of the three stages, "as cured," dried, and fired. The results of this investigation confirmed those of more detailed studies made elsewhere, and a recommended temperature range of 20°-25°C for curing all castable products has been adopted.

Current procedures are recommended only for hydraulically bonded castables; these procedures are often not suitable for a wide range of "special" castables currently available. Variations or modifications might be required to accommodate thixotropic properties, or special curing procedures for some chemically bonded castables. A note is included in the revised British Standard test to ensure that the procedure adopted is suitable to the material being tested.

Table II. Results Obtained Using the "Knocked Bowl" Technique for Water Content

Laboratory	Castable A		Castable B	
	Water Content (%)	Mean Bulk Density (kg/m³)	Water Content (%)	Mean Bulk Density (kg/m³)
A	120	587	60.0	879
B	122	550	61.5	818
C	123	546	58.5	852
D	104	627	59.0	964
E	112	614	54.6	930
Grand mean	116	584	58.7	889
Manufacturers' recommendations	100	630	69–71	809

Insulating Castables: The modified ASTM consistency test for dense castables is unsuitable for insulating castables primarily because of the large water additions, and the PRE cone test is intended for vibration compaction, which again should not be applied to insulating castables. A simple test was developed to determine the water content required for casting: Water is added to the refractory in increments and after each increment the bowl is knocked on a hard, rigid surface six times. The correct consistency is reached when the mix forms a shiny wet surface and flows slightly. Accordingly, this consistency test was included as part of a revised test-piece-preparation technique involving consolidation by rodding. In a subsequent Round Robin test involving five laboratories, two vermiculite-based, insulating castables were studied. The results are shown in Table II.

The variations between laboratories and the manufacturers' recommendations are considered acceptable in view of the inherent problems with this type of castable.

Moldables and Ramming Mixes: Moldables and ramming mixes are considered as equivalent for test-piece preparation. Manual methods, as presently incorporated in the British Standards, were considered to be subjective and too dependent on the operator. Another possibility is the use of hydraulic pressing as specified by ASTM; however, this is considered to be suitable for moldables only, and it was also felt that a forming method should reproduce characteristics shown in the installation. Installation by pneumatic ramming gives the material a component of lateral flow, which is not reproduced by uniaxial pressing. Some form of mechanical rammer, using a ramhead smaller in area than the mold itself, was therefore required. PRE could have adopted this technique with a purpose-built apparatus, but it was decided to adopt a simplified technique based on a modified AFA rammer. This modification is shown in Fig. 1. The validity of this technique was examined with three moldables and compared with existing British Standard (hand-ramming). A single specimen was prepared by a pneumatic procedure and the results are given in Table III. Clearly, lower standard deviations are recorded with the modified AFA rammer.

A test-piece-preparation procedure has been drawn up for consideration for a British Standard, which is extremely similar to that adopted by PRE. While the impact energy applied per unit area is only about one-third of that applied by the PRE test, the number of blows applied adjusts the situation so that the total energy applied is very similar. The pattern of blows applied in the ramming procedure has been adjusted to ensure similar bulk densities to those obtained by pneumatic techniques used on-site.

Strength Testing

Most British manufacturers base their description of the strength of their products on the results of cold crushing strength testing after drying or prefiring at specified temperatures. Testing carried out on dried castable specimens is important as an indication of the quantity and quality of the cement phases. However, results obtained at ambient temperature after prefiring to high temperatures are of no practical significance since any liquid phases formed at the higher temperatures will produce a glass when cooled and will impart a misleadingly high cold strength to the product. The situation is further confused by the more recent introduction of special types of material which have a genuinely higher than normal strength at intermediate tempera-

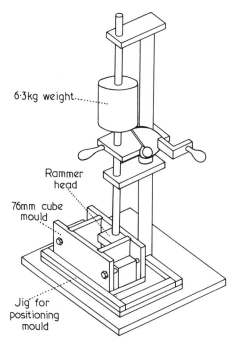

Fig. 1. Modified AFA rammer.

tures. Some uniform form of meaningful strength classification is therefore required. The most obvious tests available are crushing and modulus of rupture (MOR), which can be measured over a range of temperatures, and some form of refractoriness under load (RUL).

Testing at Ambient Temperature: A brief examination of the influence of specimen size showed that there are no advantages in changing from the

Table III. Comparison of Moldable Specimen Preparation Methods

Material	Ramming Method	Bulk Density as a Percentage of Manufacturer's Value	
		Mean	Standard Deviation
Moldable E	Hand	99.1	0.9
	AFA	100.3	0.6
	Pneumatic	96.1	
Moldable H	Hand	116.6	1.2
	AFA	116.3	0.4
	Pneumatic	109.1	
Ramming mix K	Hand	97.7	2.5
	AFA	94.2	0.7
	Pneumatic	98.3	

76 mm cube which is also used for fired shapes. The only changes recommended are faster loading rates to conform with those adopted for fired shapes and relaxation in tolerances for test pieces cut from shaped materials to allow the use of cut or rammed surfaces without further grinding.

Cold MOR testing has been discontinued for most types of monolithics, but should there be a requirement, the procedure and apparatus used for hot MOR should be followed.

Testing at High Temperature: No hot strength test is at present used as a British Standard, although a draft has been prepared for hot MOR of fired shapes. Although a hot crushing test has been examined, it is clearly too expensive to be considered on a regular basis. A comparison of results obtained from castable refractories in the refractoriness under load (RUL) and modulus of rupture (MOR) tests showed that the levels of strength in the MOR were similar for all types of product, but that the RUL test gave quite a reasonable differentiation. The variation between the two test methods is due to the sensitivity to small quantities of liquid phase. An example of such differentiation is shown in Table IV. More detailed RUL work which included the effects of prefiring time and a range of applied loads, showed that the test method is potentially rewarding. It was, however, not considered sufficiently simulative of service conditions and was abandoned in favor of the more flexible MOR test, despite the fact that the work carried out in Germany had led to a recommendation that this type of compressive test could be used well to distinguish between different types of monolithics.

Table IV. RUL Test Results

Material	Recommended Maximum Service Temperature (M.S.T.)	Temperature of Subsidence (°C)	
		Initial	10%
Castable J	1250	1200	1425
Castable B	1400	1100	1360
Castable K	1400	1150	1440
Castable A	1600	1300	1450
Castable T	1600	1420	1540
Castable W	1600	1540	1700
Castable N	1850	>1700	
Moldable G	1350	1070	1520
Moldable C	1700	1270	1700
Moldable D	1720	1380	1700

In fact, little research was required for development of a suitable MOR test, since this is one of the few areas where BSI, PRE, and ISO have found ready agreement for the testing of fired shapes and the method for unshaped materials had already been adopted by PRE. Accordingly, the test described in PRE/R18 (1978) incorporating 150 by 25 by 25 mm test bars was used for all subsequent work but, since the test does not specify prefiring time or minimum numbers of test pieces, comparative tests were first carried out to determine optimum requirements. They showed that little benefit was to be gained by extending the prefiring time beyond 2 h at any selected temperature and that a minimum of 6 test pieces was required to produce a reliable result. An example of the influence of time is given in Table V.

Table V. Comparison of Prefiring Times in MOR Test at 800 °C

Castable Type	2-h Prefire		5-h Prefire	
	MOR (MN/m²)	Standard Deviation	MOR (MN/m²)	Standard Deviation
Low-cement W	11.6	1.6	11.0	1.6
Dense, high-strength J2	15.7	2.3	16.3	1.5
Low-cement K2	17.5	1.4	18.5	2.1

A draft procedure has now been drawn up for presentation to BSI. An example of the pattern of results obtained using this procedure with a range of materials is shown in Fig. 2. The use of strength as a basis for classification was examined. From the results obtained, the selection of an intermediate temperature of 800 °C clearly differentiates between product types with traditional hydraulic bonds and those of genuinely high strength. However, more testing of moldables and ramming mixes confirmed the expected difficulties associated with test-piece preparation and very low strength levels. There does appear to be some merit in including strength in classification, but only for dense castables.

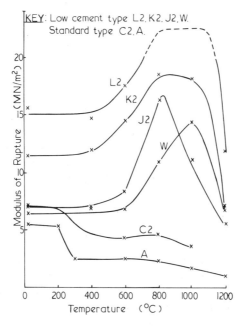

Fig. 2. Hot modulus of rupture of castables.

Thermal Conductivity Testing

The present British Standard includes two methods for moldable and castable material. Each employs the traditional method of heating on one face, but one uses specimens which have been prefired throughout and the other involves step-by-step heating of unfired specimens in the test apparatus itself. It would clearly be a major advantage if the method employed did not operate under a specific temperature gradient since this would allow much greater freedom for applying results to actual service situations. Consequently, attention was diverted to the development of an alternative test method known as the "hot-wire" test, variations of which are the welded-wire and parallel-wire techniques. In these tests, conductivity is determined by measuring the temperature rise caused by a linear heat source embedded in the material. It is a dynamic method and virtually isothermal, so determination is at specific temperatures. The test method was already being developed for fired shapes and appeared ideal for monolithic materials where small differences in temperature might produce considerable differences in thermal conductivity. Since the test procedure would be identical for shaped or unshaped refractories, development had to be restricted only to preparation and adoption of appropriate test pieces.

Initial work was carried out using the welded-wire technique. Full-sized bricks (230 by 114 by 76 mm) were prepared by casting or ramming and, by taking suitable precautions, flat faces were obtained. Tests were carried out on two dense castables, four insulating castables, and one ramming mix, and good agreement was obtained between the welded-wire method and the British Standard method involving prefiring throughout. A typical example is shown in Fig. 3. As a result of further development studies, the parallel-wire method became the preferred version, and further tests were then carried out. At this stage, it became necessary to determine the feasibility and necessity of embedding the heater wires into the test blocks, rather than assembling them dry as in the earlier work. Although this could be done satisfactorily, comparisons showed that "wet" and "dry" assembly of the heaters gave similar results for both the insulating castable and the moldable investigated, both sets of results also agreeing with the British Standard method. There was thus no need to embed the heaters.

The parallel-wire version is the recommended technique with the welded-wire offered as an alternative which, however, tends to suffer from reduced accuracy for materials with conductivity <2 $W/m\cdot K$. For the present, the existing test method, whereby a prefired block is heated from one face, will be retained as an alternative in the standard for companies or organizations which have not yet acquired the hot-wire apparatus.

Dimensional Stability

The determination of permanent linear change on firing (PLC) is probably the single most important test of quality of an unshaped refractory. It is clearly undesirable to use a material which will shrink or expand excessively in service since severe cracking or disruption could result. Inevitably, dimensional stability will be involved to some degree in classification of a material. A comparison of results obtained from isothermal testing with those obtained under temperature gradient conditions (described later under Simulative Testing) is required. On the bases of such comparison, consideration of other classification systems, and general background experience, the

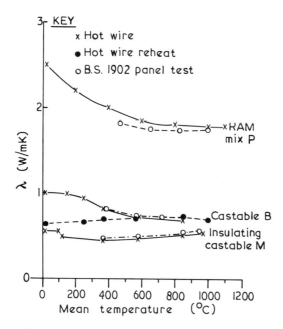

Fig. 3. Comparison of hot-wire and panel test.

decisions on classification were made, as described later.

In terms of possible revision of the test method itself, work has been restricted to examination of the effects of test-piece size and duration of heating at the test temperature. In these respects the PRE and ASTM procedures specify larger test pieces and longer times at the test temperature than those used in the British Standard; the intention was to determine if there was any justification for this procedure.

Effect of Test-Piece Size: The ASTM specifies brick specimens (230 by 114 by 64 mm), the PRE requires either a brick or half-brick size, and the British Standard uses a 76 mm cube. Comparative test work using these 3 sizes and an additional bar-shaped specimen 230 by 50 by 50 mm (as used in petrochemical specifications) was carried out on 8 castable products, 6 moldables, and 1 ramming mix, covering a wide variety of qualities. Typical results obtained on castable material are shown in Table VI and indicate that in one instance only (bars-castable T) was there any significant difference between cubes and other sizes. Results obtained in comparison of cubes and bricks after a longer soak at temperature (5 h) gave similar levels of significance to those shown for the bars, but the differences between cubes and bricks were variable in direction. On the basis of such results, the additional material required for the larger test pieces did not appear justified, and the 76 mm cube should be retained.

Effect of Duration of Heating at the Test Temperature under Isothermal Conditions: Having decided on the 76 mm cube as the test-piece size, it was used in a similar program conducted to determine the effect of time at the test temperature (soaking period). A wide range of castable, moldable, and ram-

Table VI. Results of Tests on Castable Material

Castable	Temperature (3-h Soak)	PLC (%) ± Standard Deviation (Significance Level vs Cubes in Parentheses)		
		BSI Cube (mean of 3 dimensions)	PRE Half-Bricks (One Dimension)	230 by 50 by 50 mm Bar (One Dimension)
C	1000	−0.58 ± 0.07		−0.44 ± 0.04 (5%)
P	1100	−0.64 ± 0.16	−0.54 ± 0.11 (not sig.)	
R	1300	−0.38 ± 0.11	−0.42 ± 0.10 (not sig.)	
S	1100	−0.35 ± 0.11	−0.27 ± 0.07 (not sig.)	−0.32 ± 0.04 (not sig.)
T	1550	+2.11 ± 0.28	+1.84 ± 0.35 (not sig.)	+1.48 ± 0.20 (1%)
Y	1200	−0.58 ± 0.10		−0.40 ± 0.09 (not sig.)

ming mix products were used in investigations covering soaking periods up to 100 h at temperatures at or close to the service temperature limit of the materials. Except for one or two products which gave no distinct pattern, the results obtained after various soak periods were almost identical. Examples of three materials are given in Table VII. Extension of the British Standard soaking period from 3 h to the 5 h used in ASTM and PRE would in many instances lead to an extension of firing schedules beyond the length of a normal working day. It was agreed that the results obtained from the investigation did not justify such an extension. Accordingly, it is recommended that the soaking period be retained at 3 h.

Table VII. PLC—Effect of Soak Time

Material	Test Temperature	PLC (%) 3 h Mean	Standard Deviation	100 h Mean	Standard Deviation
Dense castable	1200	0.52	0.04	0.61	0.02
Medium-weight castable	1100	0.31	0.13	0.36	0.01
Insulating castable	1000	2.09	0.16	1.80	0.22

Simulative Testing Involving a Temperature Gradient: Simulative testing is not designed for use in standard test procedures but to give an indication of the relationship between laboratory results and service behavior, which then becomes an aid to the derivation of limits in classification of materials. Three areas were considered: (1) an investigation of the relationship between the change in dimensions of specimens fired isothermally and under a thermal gradient; (2) comparison of results for dimensional change of specimens varying in their free oxygen and carbon monoxide contents fired in different atmospheres; and (3) an investigation of thermomechanical behavior.

Effect of Temperature Gradient: The permanent dimensional change on firing was measured on the hot face of test pieces fired under a temperature gradient similar to the gradient that might be found in service applications. These results were compared with those from the isothermal test described earlier. Four castables and four moldables covering a range of recommended maximum service temperatures were examined, the testing being carried out at 50°–100°C below this maximum. For each material examined, the isothermal test showed greater dimensional change than the gradient test, but no consistent relationship was found, as shown in Table VIII.

Although the results showed the expected reduction in dimensional change under the gradient conditions, it was not possible to derive an isothermal test limit for classification purposes because of the inconsistent relationship found. Background experience of shrinkage under service conditions (i.e., temperature gradient) suggested that this was likely to become serious when greater than 2%. On the basis of the figures shown in Table VIII, it was thought that a test result of ±2% dimensional change obtained under isothermal conditions would give an adequate safety margin as the service shrinkage would invariably be less. In the case of moldables, which are sup-

Table VIII. Relationship between Isothermal and Gradient PLC

Material	Ratio of Isothermal to Gradient PLC
Castable A	2.9
Castable Y	1.3
Castable E2	1.7
Castable F2	2.0
Moldable B	2.5
Moldable D	1.3
Moldable E	1.9
Moldable N	1.8

plied and installed wet, this would be increased to ±3% to allow for drying shrinkage, which was not taken into account in the figures shown in the table. These derived limits are similar to those used by ASTM and PRE; the resulting classification is dealt with more thoroughly later in this paper.

Effect of Atmosphere: Information on the effects of firing atmosphere was required so that potentially any classification could include an indication of the suitability of materials for use in reducing atmospheres. Such a classification would augment the PLC-based temperature classification, which relates only to performance in oxidizing atmospheres.

Published literature and background experience indicate that only materials high in iron oxide are grossly affected by reducing atmospheres. Therefore, experimental work was concentrated on iron-containing materials which would be expected to exhibit increased shrinkage under reducing conditions, but two low-iron materials were also examined to verify that they were unaffected. PLC values were determined after firing in various atmospheres ranging from highly oxidizing (21% O_2) to highly reducing (6% CO). The expected increase in PLC under reducing conditions was found for the high-iron materials, while no significant effect was found for the low-iron types. The shrinkage of monolithic refractories containing up to about 3% iron oxide (Fe_2O_3) at temperatures close to their recommended maximum service temperatures were relatively unaffected by reducing atmospheres. When more than 3% iron oxide was present, shrinkage of the materials was increased to such an extent that an appropriate item should be included in the classification to bring it to notice. However, the type of classification item should be extended to cover, specifically, atmospheres of unusually high carbon monoxide content. These aspects are dealt with in greater detail later.

Investigation of Thermomechanical Behavior: The results obtained from hot-strength testing described earlier relate to loading rates and stress distributions which may be totally different from those in a practical service application. Castables rarely support much load since load is usually transmitted via the anchoring system to the shell of the vessel. Any stress generated in the material is more likely to be due to strain caused by differences in thermal expansion. These thermally produced strains can be simulated by suppressing the thermal expansion of a specimen while its temperature is slowly raised. This part of the work program was included to derive a relationship between the results of this more sophisticated type of thermomechanical testing and the simpler MOR, and hence to relate the latter test to behavior in service. This information could then be used to formulate a meaningful strength classification.

A thermomechanical investigation of two dense castables indicated that at high temperatures the thermal expansion could be totally suppressed without the generation of large stresses in the material. Therefore, no simple strength test would provide useful information about behavior in typical service conditions, as such tests involve high stress levels. In the light of these results the study of thermomechanical behavior was discontinued.

Classification

The formulation of a classification scheme for unshaped refractories was included in the work program, primarily to rectify the confusing situation created by manufacturers' adopted classifications. In addition, consideration of the classifications already adopted or being discussed in ISO led to dissatisfaction for several reasons, including the arbitrary nature of the shrinkage limits set by ASTM, PRE, and the dangerously broad ISO chemistry classification. An alternative British scheme was therefore an essential requirement.

The first stage in the development of a suitable scheme was a survey of manufacturers' data sheets, conversion of all data to standard units, and storage of the standardized data in a word processor to enable rapid evaluation of possible classifications. Two important facts were immediately apparent from the survey:

(1) If the limiting figure for shrinkage allowed by PRE were also applied to expansion, about 20% of the products surveyed would have a classification temperature lower than the manufacturer's recommended maximum service temperature.

(2) There was no overlap in bulk density for dense and insulating castables so that the latter could be defined by bulk density rather than porosity, as used in PRE.

Regarding fact (1), the work on PLC under gradient conditions indicated that shrinkage in service could be much less than in an isothermal test at the same temperature, although a precise relationship was not found. On the basis of this work, it was agreed that a PLC limit of $\pm 2\%$ (as measured in the isothermal test) would be set for dense and insulating castables. The PRE shrinkage limits for moldables and ramming mixes (3 and 2%, respectively) were considered acceptable, and it was agreed that equivalent limits should be applied to expansion. The PLC limits for all 4 types of material coincide with the unofficial classification used by most manufacturers, since the classification temperature now equals the quoted maximum service temperature for 244 of the 262 products surveyed.

Classification by bulk density proved relatively easy by virtue of the obvious demarcation lines noted in fact (2), and also because it was seen that such demarcation fitted quite well with those used by the petrochemical industry, which had tended to be adopted elsewhere in UK industry. Accordingly, castable refractories were separated into dense and insulating varieties with further subdivision of the latter into medium-weight, lightweight-, and extra-lightweight-grades.

Initial attempts to formulate a chemistry classification proved difficult

Table IX. Summary of Coding Used in Classification

Type of Product	General Classification			Temperature Classification	Chemistry Classification		Strength Classification	
	Bulk Density Class	Product Definition	Suitability for Gunning		Alumina	Iron Oxide	Cold	Hot
Dense castables	No classification	D	Use of the Letter G where appropriate	Classification Temperature (°C) given in full, but rounded down to the nearest 50°C	Group number based on increments of 10% Al_2O_3 content, starting at 5%; e.g., Group 50 = 45 to 55% Al_2O_3	Very low iron oxide content (<1.5% Fe_2O_3) = A. Low iron-oxide content (<3% Fe_2O_3) = B <3% Fe_2O_3 = —	High cold strength = C	High hot strength = H

*

Insulating castables	Medium weight (<1600 kg/m³) = 160 Lightweight (<1250 kg/m³) = 125 Extra lightweight (<900 kg/m³) = 090	L	High cold strength relative to bulk density = S	No classification
Moldables	No classification	M		
Ramming mixes		R		

			G			
Example 1	125	L	1100	40	—	S
Example 2		D	1750	90	A	C	H

*Apply to all types.

as a result of trying to relate chemistry (alumina content) to classification or maximum service temperature, but the problem was eventually resolved by totally divorcing the three elements. It was agreed that the chemistry classification should consist basically of increments of 10% alumina content, but that classification temperature, in increments of 50°C, should be based solely on results of the isothermal PLC test as noted above. It was also decided that this classification temperature should be isolated from maximum service temperature, which is dependent on service conditions and should be left entirely to the manufacturer's discretion. A further chemistry classification, based on iron oxide content, was agreed to give some indication of resistance to various types of service atmosphere. Under this system, products would be split into three groups to denote resistance to carbon monoxide in particular ($<1.5\%$ Fe_2O_3), reducing conditions in general ($<3\%$), or use in oxidizing conditions only ($>3\%$). The 3% limit was chosen on the basis of work carried out under Dimensional Stability, while the 1.5% limit was chosen purely on the basis of arbitration. All the chemistry and temperature classifications were to apply to all unshaped refractories.

The remaining requirement for the classification scheme was some form of strength classification. A cold-strength classification was considered commercially desirable for both dense and insulating castables, while a hot-strength classification was required only for dense castables. It was decided that no strength classification was necessary for moldables and ramming mixes.

For cold-strength classification, based on crushing strength, a lower limit of 40 MN/m^2 was adopted for dense castables to be classified as having "high cold strength." This value was derived from the survey of manufacturers' data. Two limits were agreed for insulating castables, 12 MN/m^2 for medium-weight products and 4 MN/m^2 for lightweight products. These values were taken from a petrochemical specification since the petrochemical industry was said to be the main customer for high-strength insulating castables. No strength classification was required for extra-lightweight products.

The hot-strength classification was based on work carried out on MOR at 800°C. After examination of results for a large number of castables, covering normal grades, conventional "strong" types, and the newer "low-cement" and chemically bonded types, a lower limit of 8 MN/m^2 was adopted for "high hot-strength" materials.

A classification scheme was drafted that incorporated all the described systems, and a relatively simple coding scheme was introduced to summarize the classification of any particular product. In addition, several standard descriptive terms were defined in terms of the classification (e.g., "high alumina," "medium weight"), which would provide better comparability between manufacturers' data sheets. These uniform descriptive terms should be used in manufacturers' trade literature, and the coding scheme applied in such applications as computer systems and labeling of product containers. The coding scheme is shown in Table IX. The first example shown in the table indicates an insulating castable, classified lightweight (<1250 kg/m^3), suitable for gunning, classification temperature 1100°C, with around 40% alumina, and $>3\%$ iron oxide, and a high cold crushing strength for its density (>4 MN/m^2). This castable would appear as 125LG/1100/40-/S.

The final version of the classification and coding scheme for all monolithic refractories is to be presented to BSI for their consideration.

Compressive Stress/Strain Measurement of Monolithic Refractories At Elevated Temperatures

W. R. Alder and J. S. Masaryk

Kaiser Refractories Company
Pleasanton, CA 94566

Modeling thermomechanical behavior of refractories is becoming more important in industry, which has created a need for more complete physical property information on refractory materials. Application of currently available techniques for nonlinear modeling and finite element analysis to refractory lining design problems has been limited by the scarcity of adequate data for these materials.[1,2]

Good temperature-dependent engineering data, of which the most important are thermal conductivity, specific heat, thermal expansion, and stress/strain behavior, enable a model to be used with confidence. The authors' laboratory has already developed equipment and procedures to generate data for the first three categories: (1) Thermal conductivity by a hot-wire method to 1500°C, which yields "at temperature" data[3]; (2) specific heat by drop calorimeter to 1450°C; and (3) thermal expansion to 1650°C, which uses capacitive displacement transducers for better resolution and more stability.[4]

With respect to stress/strain measurement at high temperatures, others have used a modified modulus-of-rupture apparatus to obtain modulus of elasticity (but not Poisson's ratio).[5,6] Elastic modulus at high temperatures has also been determined by many investigators using resonant frequency (dynamic) methods. Rigby's work typifies this approach.[7]

More recently, the axial and diametral strain of monolithic refractory concrete specimens was measured while subjected to both uniaxial and biaxial compressive loading at high temperatures.[8] Poisson's ratio and elastic modulus were derived from these measurements. The method of determining stress/strain behavior here described uses a similar approach.

Equipment Development

The general design parameters chosen for the equipment used for stress/strain measurement included: (1) the ability to test ceramic-refractory materials to 1650°C; (2) the capability of handling both plastic and brittle materials with precision; (3) simultaneous measurement of axial and diametral strain while under compressive loading; and (4) the flexibility to do stress relaxation studies.

A summary of the resultant high-temperature material test system* used to generate the data is presented in the appendix. Figure 1 shows the basic equipment layout.

---
*MTS Systems Corp., Minneapolis, MN

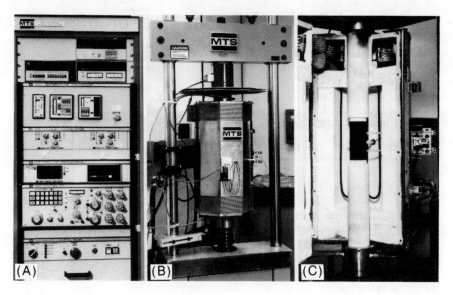

Fig. 1. (A) Control console; (B) furnace closed around test zone with hydraulic load frame; (C) furnace open, showing specimen between alumina rams with axial and diametral extensometers in place.

Generation of Data

Data were generated using this system by applying compressive stress while under strain control (constant strain rate). Data obtained from the continuous measurement of axial stress (σ), axial strain (ϵ_a), and diametral strain (ϵ_d) were used to plot stress/strain curves and to compute modulus of elasticity and Poisson's ratio. They were also used to derive values for yield strength (σ_y) and ultimate strength (σ_{max}). Figure 2 presents a typical stress/strain

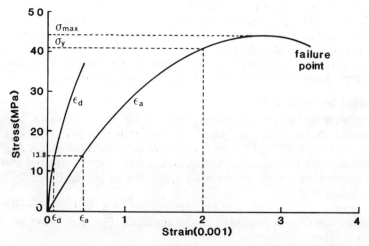

Fig. 2. Typical stress/strain behavior of a monolithic refractory.

curve together with graphic representations of the various derived values.

Elastic modulus, Poisson's ratio, and yield strength were obtained from cylindrical (core-drilled) specimens 5.1 cm (2 in.) in diameter and 11.4 cm (4.5 in.) long. The ultimate strength of monolithic refractory concrete specimens exceeded the capacity of our test system at lower temperatures. For this reason, we used specimens measuring 3.3 cm (1.3 in.) in diameter and 8.9 cm (3.5 in.) long to obtain σ_{max}.

All specimens were soaked a minimum of two hours at temperature before testing. For the monolithic refractory concrete only, strain rate ($\dot{\epsilon}_a$) was set at 0.15%/min for the 5.1-cm diameter specimens. Except when testing for σ_{max}, all other compositions were run at $\dot{\epsilon}_a = 0.02$–0.04%/min. The 3.3-cm diameter specimens for σ_{max} were run at $\dot{\epsilon}_a = 0.008$%/min (0.08%/min for 1316°C runs) to minimize catastrophic failure of specimens.

The modulus of elasticity was calculated on the basis defined for measurement of elastic modulus in concrete.[9] Known as the secant modulus, it is calculated by dividing a selected stress value (σ) by the axial strain (ϵ_a) observed at this stress. A value of 13.8 MPa (2000 psi) was used for most of this work.

Poisson's ratio was calculated according to the conventional relationship: ϵ_d/ϵ_a, where both strain values are measured at 13.8 MPa (2000 psi), unless σ_{max} dictates the use of a lower value. Yield strength in this work is defined as the stress at which 0.2% axial strain occurs. It is a modified version of a yield-strength determination described elsewhere,[10] and is an extrapolated value in the accompanying tables if higher than 41.4 MPa (6000 psi).

Behavior of Monolithics under Load at Elevated Temperatures

Effect of Measurement Technique

All of the values obtained by the method described in this paper were obtained at the temperature indicated. Two other methods of obtaining elastic modulus data yield somewhat different results, as shown in Table I. In addition to testing gunned, 105°C-dried specimens of monolithic refractory concrete at temperature, we reheated more specimens to the same test temperatures, then ran stress/strain tests on these specimens at room temperature. Dynamic (resonant frequency) data were also obtained at temperature for other, similar specimens.

The difference between data obtained from the two static (compressive loading) test procedures is apparent: The elastic modulus of reheated specimens tested at room temperature reaches a minimum for those reheated between 538°C (1000°F) and 816°C (1500°F). It then moves back up to its highest value for the specimen reheated to 1316°C (2400°F). For specimens tested at reheat temperatures, we see a continuous decline in the modulus as a function of increased temperature. This phenomenon was noted by others.[11,12] Refractories do not follow Hooke's law at high temperatures.[13]

The dynamic modulus data obtained at temperature show only a slight downward trend with respect to increased temperature because modulus measured in this way reflects almost purely elastic behavior. Since negligible stress is applied during dynamic testing, modulus is unaffected by creep, particularly at high temperatures. For the same reason, dynamic modulus will be appreciably higher than the secant (static) modulus.[14]

Table I. Modulus of Elasticity Comparison (Monolithic, Gunned, 105°C-Dried)

Reheat Temp.		Static MOE* of Reheated Specimens Tested at Room Temp.		Static MOE* of 105°C-Dried Specimens Tested at Reheat Temp.		Dynamic MOE† of 105°C-Dried Specimens Tested at Reheat Temp.	
(°C)	(°F)	($\times 10^4$ MPa)	($\times 10^6$ psi)	($\times 10^4$ MPa)	($\times 10^6$ psi)	($\times 10^4$ MPa)	($\times 10^6$ psi)
260	(500)	2.1	3.0	2.1	3.0	3.8	5.5
538	(1000)	1.4	2.1	1.4	2.0	3.7	5.3
816	(1500)	1.4	2.1	1.2	1.7	3.7	5.3
1093	(2000)	1.8	2.6	0.57	0.83	3.5	5.1
1316	(2400)	2.2	3.2	0.09‡	0.13‡		

*Secant modulus at $\sigma = 13.8$ MPa (2000 psi).
†Resonance method.
‡Determined at $\sigma = 2.4$ MPa (348 psi) to stay within σ_{max}.

Table II. Compressive Stress/Strain Data

Temp. of Test (°C)	(°F)	Modulus of Elasticity* (×10⁴ MPa)	(×10⁶ psi)	Poisson's Ratio (μ)	Yield Strength MPa	psi	Ultimate Strength MPa	psi
				Vibration-Cast, 105°C-Dried				
22	(72)	4.3	6.3	0.16	86.9	12 600	103.4	15 000
260	(500)	3.7	5.3	0.10	72.4	10 500	>107.6	>15 600
538	(1000)	3.1	4.5	0.19	61.9	8 970	104.1	15 100
816	(1500)	3.0	4.4	0.18	58.2	8 440	104.8	15 200
1093	(2000)	2.2	3.2	0.24	†	†	25.9	3 750
1316	(2400)	0.53	0.77	0.19	4.14	600	4.31	625
				Gunned, 105°C-Dried				
22	(72)	2.7	3.9	0.18	41.2	5 980	41.7	6 040
260	(500)	2.1	3.0	0.19	34.3	4 980	51.6	7 480
538	(1000)	1.4	2.0	0.17	25.0	3 630	42.3	6 140
816	(1500)	1.2	1.7	0.17	21.4	3 110	45.7	6 620
1093	(2000)	0.57	0.83	0.19	12.4	1 800	16.2	2 350
1316	(2400)	0.09	0.13	0.22	2.34	340	2.76	400

*Secant modulus at $\sigma = 13.8$ MPa (2000 psi); at service limit of 1316°C (2400°F), modulus determined at $\sigma = 2.4$ MPa (348 psi) to stay within σ_{max}.
†0.00180 strain at failure.

At the time of his analysis of refractory linings, Coatney used available elastic modulus values obtained by the resonant frequency method at room temperature.[15] In later unpublished work, predicted brick joint stress in the lining of a rotary coke calciner at high temperature was much higher using these values than those from a repeat analysis using data generated by the current method (compressive loading at temperature).

Effect of Emplacement Method

Vibration-cast, monolithic refractory concrete stress/strain behavior differs significantly from that of a gunned installation at every temperature with respect to magnitude of elastic modulus, yield strength, and ultimate strength. For example (referring to Table II), the vibration-cast material has an elastic modulus which ranges from $1^1/_2$ to 4 times higher than its gunned counterpart. The largest differences are observed at temperatures $\geq 538\,°C$ (1000°F). Figure 3 displays a graphic comparison.

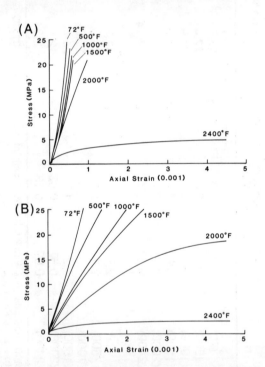

Fig. 3. (A) Stress/strain of vibration-cast monolithic; (B) stress/strain of gunned monolithic.

Effect of Stainless Steel Fiber Addition

The addition of 3 wt% stainless steel fibers significantly increases the room-temperature elastic modulus of gunned monolithic refractory concrete. This effect is shown in Table III. Little or no improvement to elastic modulus or yield strength is observed at other temperatures however. Fiber addition does increase the ultimate strength for all gunned samples up to 1093°C

Table III. Compressive Stress/Strain Data

Temp. of Test (°C) (°F)	Modulus of Elasticity* (×10⁴ MPa)	(×10⁶ psi)	Poisson's Ratio (μ)	Yield Strength MPa	psi	Ultimate Strength MPa	psi
			Gunned, 105°C-Dried				
22 (72)	2.7	3.9	0.18	41.2	5 980	41.7	6 040
260 (500)	2.1	3.0	0.19	34.3	4 980	51.6	7 480
538 (1000)	1.4	2.0	0.17	25.0	3 630	42.3	6 140
816 (1500)	1.2	1.7	0.17	21.4	3 110	45.7	6 620
1093 (2000)	0.57	0.83	0.19	12.4	1 800	16.2	2 350
1316 (2400)	0.09	0.13	0.22	2.34	340	2.76	400
			Gunned with Fibers, 105°C-Dried				
22 (72)	4.3	6.3	0.21	†	†	52.6	7 630
260 (500)	2.4	3.5	0.21	42.3	6 140	56.5	8 190
538 (1000)	1.0	1.5	0.16	21.7	3 150	49.7	7 210
816 (1500)	0.90	1.3	0.15	16.9	2 450	58.3	8 460
1093 (2000)	0.57	0.83	0.21	11.9	1 720	17.2	2 490
1316 (2400)	0.09	0.13	0.26	2.34	340	2.59	375

*Secant modulus at σ = 13.8 MPa (2000 psi); at service limit of 1316°C (2400°F), modulus determined at σ = 2.4 MPa (348 psi) to stay within σ_{max}.
†0.00148 strain at failure.

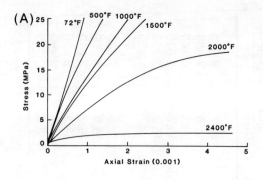

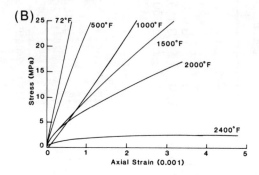

Fig. 4. (A) Stress/strain of gunned monolithic; (B) stress/strain of gunned-fibers monolithic.

Table IV. Compressive Stress/Strain Data

Temperature of Test		Modulus of Elasticity*		Poisson's Ratio*
(°C)	(°F)	(× 10⁴ MPa)	(× 10⁶ psi)	
		Low-Cement Castable		
22	72	5.9	8.6	0.15
538	1000	4.8	7.0	0.17
816	1500	3.1	4.5	0.17
1093	2000	3.8	5.5	0.23
1260	2300	0.82†	1.2†	0.22†
1399	2550	1.1§	1.6§	0.17§
		Monolithic: Vibration-Cast		
22	72	4.3	6.3	0.16
260	500	3.7	5.3	0.10
538	1000	3.1	4.5	0.19
816	1500	3.0	4.4	0.18
1093	2000	2.2	3.2	0.24
1316	2400	0.53‡	0.77‡	0.19‡

*Secant modulus and Poisson's ratio at 13.8 MPa (2000 psi); lower stress calculation points were used as indicated to stay within the ultimate stress boundary: †6.9 MPa (1000 psi), ‡2.4 MPa (348 psi), and §1.38 MPa (200 psi).

(2000°F). Figure 4 illustrates the effect of fiber addition on the stress/strain behavior of this material at each test temperature.

Low-Cement Castable vs Regular Castable

Table IV compares elastic moduli of a vibration-cast, low-cement castable with those of a regular ("high" cement) monolithic refractory concrete. These data and the stress/strain curves of Fig. 5 show that the low-cement castable has a 27–42% higher elastic modulus at test temperatures ranging up to 1093°C (2000°F). Also note that the 816°C (1500°F) modulus of the low-cement castable is lower than that determined at 1093°C (2000°F), suggesting the development of a more refractory bond at 1093°C, which is less subject to viscoelastic behavior.

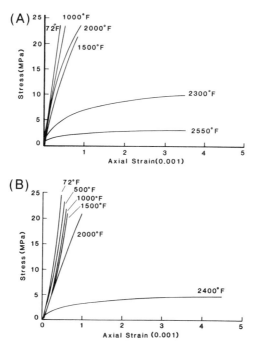

Fig. 5. (A) Stress/strain of 3200°F, low-cement castable; (B) stress/strain of vibration-cast monolithic.

Comparison of Basic and Clay-Alumina Brick vs Monolithics

Comparison of the data for 70% and 90% Al_2O_3 clay-alumina brick in Table V with data for the low-cement castable and vibration-cast monolithic presented in Table IV is highly revealing. The elastic modulus of the low-cement castable is seen to be equal to or higher than even the 90% Al_2O_3 brick up to 1093°C (2000°F). The modulus of the monolithic also exceeds that of the 70% Al_2O_3 brick up to 1093°C (2000°F). Beyond this temperature, the increased refractoriness of the brick compositions influences results, since the low-cement castable and monolithic are only ~60% Al_2O_3. Figure 6 presents stress/strain curves for the 70% and 90% Al_2O_3 brick, showing how stress/strain response is influenced by temperature for these two compositions. With respect to basic brick, the low-cement castable has a higher elastic modulus at room temperature; however, the reverse is the case at 1093°C (2000°F) and higher. Stress/strain curves for a 60% MgO, direct-bonded brick are presented in Fig. 7.

Table V. Compressive Stress/Strain Data

Temperature of Test		Modulus of Elasticity*		Poisson's Ratio*
(°C)	(°F)	(× 10⁴ MPa)	(× 10⁶ psi)	
70% Alumina Brick				
22	72	2.1	3.1	0.08
816	1500	2.6	3.8	0.13
1093	2000	1.5	2.2	0.17
1204	2200	0.19	0.28	0.59
90% Alumina Brick				
22	72	3.1	4.5	0.09
1093	2000	3.8	5.5	0.10
1371	2500	3.1	4.5	0.30
1482	2700	0.78	1.1	0.37

*Secant modulus and Poisson's ratio at 13.8 MPa (2000 psi).

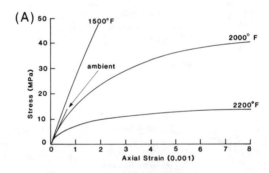

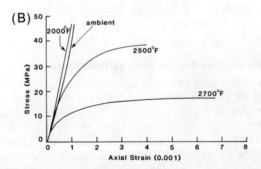

Fig. 6. (A) Stress/strain of 70% alumina brick; (B) stress/strain of 90% alumina brick.

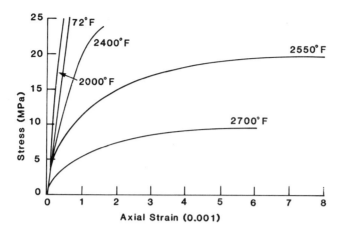

Fig. 7. Stress/strain of basic brick (60% MgO direct-bonded).

Even the phosphate-bonded patching mix (Table VII) has a higher elastic modulus than both clay-alumina and basic brick at room temperature. Its stress/strain behavior with respect to temperature is shown in Fig. 8.

Conclusion

We now have the ability to generate critical engineering data for use in finite element modeling of refractory systems. The stress/strain response vs temperature that we have observed for various products can be used to make better judgments for specific applications, which ultimately should lead to improved refractory life. In addition, the capability of running stress relaxation measurements with this equipment will give us the opportunity to better predict expansion allowance.[16] We will also be able to obtain data on viscoelastic behavior of refractory materials.[17]

Table VI. Compressive Stress/Strain Data (60% MgO, Direct-Bonded Basic Brick)

Temperature of Test		Modulus of Elasticity*		Poisson's Ratio*
(°C)	(°F)	($\times 10^4$ MPa)	($\times 10^6$ psi)	
22	72	3.8	5.5	0.07
1093	2000	6.2	9.0	0.28
1316	2400	2.4	3.5	0.38
1399	2550	0.94	1.4	0.83
1482	2700	0.40†	0.58†	0.40†

*Secant modulus and Poisson's ratio at 13.8 MPa (2000 psi); lower stress calculation point was used as indicated to stay within the ultimate stress boundary.
†6.9 MPa (1000 psi).

Table VII. Compressive Stress/Strain Data (Phosphate-Bonded Clay-Alumina Patching Mix)

Temperature of Test		Modulus of Elasticity*		Poisson's Ratio*
(°C)	(°F)	(× 10⁴ MPa)	(× 10⁶ psi)	
22	72	4.9	7.1	0.10
816	1500	2.4	3.5	0.09
960	1760	1.2	1.7	0.27
1093	2000	0.37†	0.53†	0.43†

*Secant modulus and Poisson's ratio at 13.8 MPa (2000 psi); lower stress calculation point was used as indicated to stay within the ultimate stress boundary.
†6.9 MPa (1000 psi).

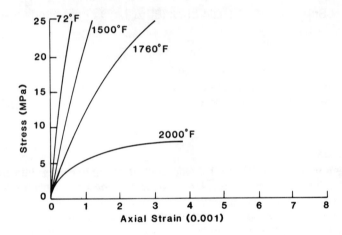

Fig. 8. Stress/strain of patching mix.

APPENDIX

MTS High-Temperature Material Test System

Test Capabilities
    (1) Measures axial and diametral stress/strain behavior of refractory materials from ambient to maximum temperature. Stress/strain curves are generated. Elastic modulus, Poisson's ratio, yield strength, and ultimate strength are calculated.
    (2) Provides data for calculating primary and secondary creep rates.
    (3) Measures stress and diametral strain during stress relaxation (under axial strain control).
Equipment Capabilities
    (1) Furnace
        (a) Maximum temperature: 1650°C (3000°F).
        (b) Maximum heating rate: 260°C/h.
        (c) Maximum cyclic variation at soak: ±2.5°C.

(2) Hydraulic Testing Machine Specifications
  (a) Maximum compressive load: 100 kN (22 000 lbs).
  (b) Maximum stroke: 15 cm (6 in.).
  (c) Load train actuator drift: ±0.1% of range (7-day period).
  (d) Load cell variation: 0.5% of full scale.
(3) Extensometers
  (a) Axial: maximum travel: ±0.5 cm (±0.2 in.). Equivalent to 10% strain. Maximum sensitivity (range 4): ±2.5 × 10^{-5} cm.
  (b) Diametral: maximum travel: ±0.25 cm (±0.1 in.). Equivalent to 5% strain. Maximum sensitivity (range 4): ±1.25 × 10^{-5} cm.

References

[1] Oral Buyukozturk and Tsi-Ming Tseng, "Thermomechanical Behavior of Refractory Concrete Linings," *J. Am. Ceram. Soc.*, **65** [6] 301 (1982).

[2] A. A. Patuzzi, "Influence of Refractory Lining on Converter Vessel Construction," *Iron Steel Eng.*, **59** [6] 53-58 (1982).

[3] G. R. Angell, "Thermal Conductivity of Refractories by an Automated Hot Wire Technique"; for abstract see *Am. Ceram. Soc. Bull.*, **59** [8] 859 (1980).

[4] S. V. Gilbert, "Modeling of Subsidence in Glass Furnace Checkers Using Creep Equations"; for abstract see *Am. Ceram. Soc. Bull.*, **62** [3] 429 (1983).

[5] B. Jackson and T. J. Partridge, "Properties Required for Arc-Furnace Roof Refractories," *Trans. Br. Ceram. Soc.*, **72** 143 (1973).

[6] E. I. Greaves, "Stress Strain Properties of High Alumina and Basic Refractories and Their Use in Arc Furnace Roofs," *Refract. J.*, [No. 3] 13-22 (1978).

[7] G. R. Rigby, "Mechanical Properties of Basic Bricks," *Trans. Br. Ceram. Soc.*, **69**, 189-98 (1970).

[8] Tsi-Ming Tseng, "Thermomechanical Behavior of Refractory Concrete-Lined Vessels"; Res. Rept. No. R82-44, MIT Dept. of Civil Eng., August 1982.

[9] A. M. Neville, pp. 310-11 in Properties of Concrete, 2d ed. Hallstead Press, New York, 1975.

[10] Van Nostrand's Scientific Encyclopedia, 3rd ed. D. Van Nostrand Co., Princeton, NJ, 1958; pp. 1667-68.

[11] C. A. Schacht, "Fundamental Considerations in the Structural Evaluation of Refractory-Lined Cylindrical Shells," *Iron Steel Eng.*, **59** [6] 45 (1982).

[12] O. Buyukozturk, and Tsi-Ming Tseng, *J. Am. Ceram. Soc.*, **65** [6] 302-303 (1982).

[13] J. F. Wygant and M. S. Crowley, "Designing Monolithic Refractory Vessel Linings," *Am. Ceram. Soc. Bull.*, **43** [3] 180 (1964).

[14] A. M. Neville, p. 318 in Properties of Concrete, 2d ed. Hallstead Press, New York, 1975.

[15] R. L. Coatney, "Thermal and Stress Analysis of Refractory Linings"; for abstract see *Am. Ceram. Soc. Bull.*, **58** [8] 800 (1979).

[16] G. C. Padgett and D. J. Bettany, "The Relaxation Behavior of Refractories," *Trans. Br. Ceram. Soc.*, **73** 153-65 (1974).

[17] J. Sweeney and M. Cross, "Analyzing the Stress Response of Commercial Refractory Structures in Service at High Temperatures: I. A Simple Model of Viscoelastic Stress Response," *Trans. Br. Ceram. Soc.*, **81**, 25 (1982).

Aggregate Distribution Effects on the Mechanical Properties and Thermal Shock Behavior of Model Monolithic Refractory Systems

J. Homeny

University of Illinois at Urbana-Champaign
Department of Ceramic Engineering
Urbana, IL 61801

R. C. Bradt

University of Washington
Department of Materials Science and Engineering
Seattle, WA 98105

The mechanical properties and resistance to thermal shock damage of seven specially designed calcium aluminate-bonded, high-alumina aggregate refractory concretes were investigated. It was found that variation of the cement contents and aggregate distributions was an effective means of producing substantial modifications of certain physical and mechanical properties. Theoretical thermal shock damage resistance parameters derived from the mechanical properties were correlated with the experimentally observed trends in the thermal shock behavior.

Refractory concretes utilized in high-temperature applications typically consist of two constituents, an aggregate and a cement. The most common refractory concretes consist of an aggregate of a heat-resistant nature, such as calcined clays, bauxites, or high-purity aluminas, bonded with calcium aluminate cement. In these systems, chemical reactions with water at room temperature form hydration products which are ultimately related to the bonding and subsequent property development. Since refractory concretes are utilized at high temperatures, the nature of the room-temperature hydration products changes in a complex manner with the application of heat. At low temperatures (100°–400°C), a dehydration process occurs that usually results in a decrease in the effective bonding of the cement constituent. At moderately high temperatures (400°–900°C), dehydration is virtually complete, and at high temperatures (>900°C), solid-state chemical reactions occur between the cement matrix and the aggregate that further modify the nature of the bonds in the refractory.

Presently and in the recent past, both the volume and varieties of refractory concretes have been increasing. These increases are a combination of new applications and, in some instances, the replacement of conventional burned refractory bricks by concretes. Although refractory concretes are receiving this increased attention and use, relatively little has been published concerning their mechanical properties and thermal shock damage characteristics. Unlike conventional refractory bricks, whose final properties are to

Table I. T-61 Aggregate Size Distributions

U.S. Sieve Number	Generic (%)	Continuous (%)	Gap-Sized (%)
4 × 6	1.0	7.4	61.5
6 × 8	14.0	7.2	
8 × 12	15.0	6.6	
12 × 16	10.0	7.3	
16 × 20	6.0	6.5	24.5
20 × 30	4.0	6.7	
30 × 50	15.0	13.1	
50 × 70	12.0	6.4	
70 × 100	1.0	6.3	9.8
100 × 200	9.5	12.3	
200 × 325	2.5	8.8	
− 325	10.0	11.4	4.2

a large extent fixed before use, refractory concretes have properties that are continually changing during their use. For example, the thermal gradient between the hot and cold faces of a refractory lining produces chemical, phase, and property gradients that may be attributed to the structural transition from hydraulic to chemical bond formation. Superimposed on these developing property gradients are the effects of the aggregate-cement matrix interaction on the mechanical properties.

This paper investigates the fracture behavior of a model refractory concrete system, specifically the CA-25/T-61 concrete, where CA-25* is a high-purity, calcium aluminate cement and T-61* is a high-purity, tabular alumina aggregate. Further, this study makes appropriate property measurements at room and high temperatures in an attempt to understand the specific roles of the cement and aggregate constituents as they affect the fracture behavior when the volume fractions and the aggregate size distributions are varied in accordance with fundamental particle packing theory. It is an additional objective to experimentally examine the thermal shock damage resistance of these concretes to attempt to ascertain whether their thermal shock performance can be described in terms of their physical properties as for other refractory materials. Basic to these objectives is understanding how the microstructures of refractory concretes can be better designed to improve their physical properties and to better withstand severe thermal shock conditions.

Experimental Procedures

Materials and Sample Preparation

For this research, the concretes consisted of a high-purity, calcium aluminate cement (CA-25) and a high-purity, tabular alumina aggregate (T-61). Three distinct types of aggregate size distributions were utilized to constitute the model concrete systems studied. The aggregate distributions are presented in Table I and are labeled as generic, continuous, and gap-

---

*Alcoa; Aluminum Company of America, Pittsburgh, PA.

sized. The generic distribution simulates that of a typical standard industrial product. The gap-sized and the continuous formulations are based on the particle packing theories of Furnas[1] and are designed to achieve a maximum density of the aggregates.

To produce a series of concrete systems, the aggregate distributions were combined with various levels of cement content. The cement contents were 20, 25, and 30% by weight for both the gap-sized and continuous systems, whereas the generic system was produced only at the 25% cement level. In all, seven different concrete systems were produced. The generic concrete, since it simulated a standard commercial product, was essentially used as a baseline for comparative purposes. In this manner, the effects of microstructural variation on the properties of the gap-sized and the continuously graded systems could be evaluated by comparison with those of the generic concrete.

To prepare specimens for testing, a series of steps was necessary, including dry mixing, wet mixing, casting, curing, drying, and firing. The preparation steps are outlined below:

(1) Dry mixing—For each concrete system, the correct proportions of the sized T-61 aggregate and the CA-25 cement were dry-mixed in a V blender for five minutes.

(2) Wet mixing—After dry mixing, the individual batches were transferred to a paddle mixer, distilled water was added, and the batch was blended for an additional two minutes. Typical water levels were about 10%.

(3) Casting—Individual specimens were cast into 2.54 by 2.54 by 17.78 cm aluminum molds placed on glass surfaces. An excess of wet concrete was cast into the molds and then vibrated on a standard laboratory vibratory table for one minute prior to leveling the top surface with a straightedge.

(4) Curing—Immediately after casting, the specimens and molds were sealed in plastic containers in order to maintain a high relative humidity. Under these conditions, the specimens were cured at room temperature for the first 24 hours. After this initial curing period, the specimens were removed from the molds and then cured for an additional 24 hours exposed to air at room temperature.

(5) Drying—Samples were then dried at 100°C for 24 hours.

(6) Firing—After drying, samples were fired in an electric furnace at 1200°C. The heating rate was 60°C/h with a 5-h hold at the maximum temperature.

Property Measurements

The physical properties of interest were measured as follows:

(1) Bulk density—Bulk densities were determined on all specimens immediately after firing at 1200°C. They were calculated by dividing the sample weights by the measured volumes of the specimens.

(2) Thermal expansion—Thermal expansion was measured using an automatic recording dilatometer.[†] The specimens, 1.27 by 1.27 by 5.08 cm bars cut from the 2.54 by 2.54 by 17.78 cm bars with a diamond saw, were heated at a rate of 3°C/min to 1200°C. A plot of percent expansion vs temperature was obtained from which the linear coefficient of thermal expansion was calculated.

---

†Orton Co., Columbus, OH.

For the concrete systems, all mechanical property measurements were performed either at room temperature or at 1200°C. All experimental values are reported as the average of 5 to 10 test points. The test procedures are described below:

(1) Elastic moduli—The elastic moduli, Young's elastic modulus, shear modulus, and Poisson's ratio were measured by the dynamic mechanical resonance method.[2-4] Specimen dimensions were 0.64 by 2.54 by 17.78 cm and were cut from 2.54 by 2.54 by 17.78 cm cast bars with a diamond saw. Using a commercial acoustic spectrometer,‡ samples were excited to their primary flexural and torsional resonance frequencies by means of a driver-pickup transducer system. Utilizing the resonant frequencies, sample dimensions, and weights, the room-temperature elastic moduli were calculated from the following equations:

$$E = \frac{f_f^2 CM(0.94645)}{B(10)^7} \quad (1)$$

$$G = \frac{f_t^2 LRM(4)}{S(10)^7} \quad (2)$$

$$\nu = \frac{E}{2G} - 1 \quad (3)$$

where E is Young's elastic modulus (MN/m²), f_f the resonant frequency of the first mode of flexural vibration, C a shape factor dependent on sample dimensions and ν, M the mass (grams), B the specimen width (cm), G the shear modulus (MN/m²), f_t the resonant frequency of the first mode of torsional vibration, L the length of the sample (cm), R a shape factor dependent on sample dimensions, S the cross-sectional area of the specimen (cm²), and ν Poisson's ratio.

To determine the high-temperature elastic moduli, the 0.64 by 2.54 by 17.78 cm specimens were suspended in an electrically heated furnace by wires which were attached to the driver-pickup transducer system.[5,6] The high-temperature E and G were then calculated from

$$E' = E \frac{(f_f')^2}{(f_f)^2} \left(\frac{1}{1+\alpha T}\right) \quad (4)$$

$$G' = G \frac{(f_t')^2}{(f_t)^2} \left(\frac{1}{1+\alpha T}\right) \quad (5)$$

where E' is Young's elastic modulus at high temperature, f_f' the resonant frequency of the first mode of flexural vibration at high temperature, G' the shear modulus at high temperature, f_t' the resonant frequency of the first mode of torsional vibration at high temperature, α the linear coefficient of thermal expansion, and T the temperature difference between the room and test temperature.

(2) Fracture stress—All fracture stress measurements were performed on

---

‡Nametre Co., Edison, NJ.

simple three-point bend bars on a commercial mechanical testing machine.§ The 2.54 by 2.54 by 17.78 cm bars were broken on a 15.24-cm lower span at a crosshead speed of 1.27 mm/min. High-alumina testing rams with machined knife edges were used. For high-temperature testing, the sample and rams were surrounded by an electrically heated furnace. The flexural strengths were calculated using the standard engineering mechanics formula:

$$\sigma_f = \frac{3PL}{2BH^2}\left(\frac{\frac{L}{2}-X}{\frac{L}{2}}\right) \tag{6}$$

where P is the load at failure, L the lower span between knife edges, B the sample width, H the sample height, and X a correction factor for off-center fractures, simply the distance between the center loading point and the fracture origin on the tensile surface.

(3) *Notched-beam fracture surface energy*—The notched-beam fracture surface energy was measured by the single-edge, notched-beam method using three-point bending. The 2.54 by 2.54 by 17.78 cm specimens were center-notched to one-half their thickness with a 0.76-mm-thick diamond blade. Specimens were tested on the same apparatus as the fracture stress measurements at a crosshead speed of 1.27 mm/min on a 15.24-cm lower knife-edge span. The notched-beam fracture surface energy was then calculated from the following equation:[7]

$$\gamma_{NBT} = \frac{9P^2L^2C(1-\nu^2)}{8W^2D^4E}\left[A_0 + A_1\left(\frac{C}{D}\right) + A_2\left(\frac{C}{D}\right)^2 + A_3\left(\frac{C}{D}\right)^3 + A_4\left(\frac{C}{D}\right)^4\right]^2 \tag{7}$$

where P is the applied load for crack initiation, L the span, C the notch depth, W the sample width, D the sample thickness, $A_0 = 1.90 + 0.0075(X)$, $A_1 = -3.39 + 0.08(X)$, $A_2 = 15.40 - 0.2175(X)$, $A_3 = -26.24 + 0.2815(X)$, $A_4 = 26.28 - 0.145(X)$, and $X = (L/D)$.

(4) *Work of fracture*—For the work-of-fracture measurements, the 2.54 by 2.54 by 17.78 cm specimens were center-notched with a 0.76-mm-thick diamond saw so that a triangular or chevron-shaped cross section remained. Specimens were broken at a crosshead speed of 0.05 mm/min on a 15.25-cm span on the same apparatus used for the fracture stress measurements. An automatic integrator was used to determine the area under the load displacement curve to calculate the total energy required to create the new crack surfaces. The following equation was then applied to calculate the work of fracture:[8-10]

$$\gamma_{WOF} = \frac{\int F dx}{2A} \tag{8}$$

where $\int F dx$ is the work required for formation of new surfaces and A is the cross-sectional area of the fracture surface.

---

§Instron Corp., Canton, MA.

Thermal Shock Behavior

Two types of prism-quench, thermal-shock tests were performed to ascertain the performance of the various concretes during severe temperature changes. The two tests used were based on single and multiple cooling cycles (quenches). The procedures for the single-quench, thermal-shock tests were as follows:

(1) Samples, previously fired at 1200°C, were placed in an electrically heated furnace at temperatures of 225°, 425°, 625°, 825°, 1025°, or 1225°C and equilibrated for 15 min.

(2) The samples were removed from the furnace and rapidly air-cooled or quenched to room temperature, approximately 25°C. (This provided ΔTs of 200°, 400°, 600°, 800°, 1000°, and 1200°C.)

(3) The damage resulting from the various temperature changes was monitored by three-point, flexural-strength tests. A set of unshocked specimens was used to establish the control baselines.

The multiple thermal shock tests were performed as described below:

(1) Samples, previously fired at 1200°C, were placed in an electrically heated furnace at a temperature of 1225°C and equilibrated for 15 min.

(2) The samples were removed from the furnace and air-cooled at approximately 25°C for 15 min.

(3) Cooling cycles of 15 min at 1225°C and 15 min at 25°C were performed 1, 3, 5, 7, or 10 times, each on different specimens.

(4) As for the single thermal shock tests, the damage sustained during this multiple thermal shocking was monitored by three-point, flexural-strength tests.

Results and Discussion

Property Trends

The room- and high-temperature test results are presented in Tables II and III, respectively. The property trends as a function of microstructure follow:

(1) Bulk density—The density results clearly indicate that the gap-sized concrete systems are more efficient in terms of particle packing than the continuously sized systems. At identical cement contents the gap-sized systems exhibited higher densities than the continuously graded and the generic systems. Additionally, for both the continuous and the gap-sized systems, the densities decreased with increasing cement content.

(2) Thermal expansion coefficient—The coefficients of thermal expansion were fairly constant for all seven systems, indicating that the microstructural variations had very little effect on this property. However, the coefficients increased slightly at the 1200°C test temperatures as compared with the room-temperature results.

(3) Elastic moduli—Both the Young's elastic modulus and the shear modulus exhibited significant increases for the gap-sized concretes as compared with the generic and continuous systems. These increases were approximately a factor of 2. The combination of the coarser aggregate-size distribution and the increased bulk densities were responsible for these observed increases in E and G. For both the continuous and the gap-sized systems, the values of E and G decreased with increasing cement content. The previously described bulk density trends are clearly related to these variations in E and G

Table II. Room-Temperature Data (Samples Fired at 1200°C and Tested at Room Temperature)

Aggregate Dist.—Cement Content	ρ (g/cm³)	α (°C⁻¹)	E (MN/m²)	G (MN/m²)	ν	σ_f (MN/m²)	γ_{NBT} (J/m²)	γ_{WOF} (J/m²)
Generic-25	2.59	7.0×10⁻⁶	2.09×10⁴	1.01×10⁴	0.04	8.40 (±1.34)*	7.14 (±2.76)	44.22 (±5.61)
Continuous-20	2.69	6.3×10⁻⁶	3.44×10⁴	1.58×10⁴	0.09	6.66 (±0.81)	2.80 (±0.25)	27.91 (±2.37)
Continuous-25	2.54	6.4×10⁻⁶	2.79×10⁴	1.28×10⁴	0.09	9.14 (±0.89)	10.10 (±1.31)	61.85 (±4.78)
Continuous-30	2.51	6.5×10⁻⁶	2.51×10⁴	1.18×10⁴	0.06	9.52 (±0.71)	10.52 (±1.26)	51.65 (±3.88)
Gap-Sized-20	2.83	6.5×10⁻⁶	6.51×10⁴	3.08×10⁴	0.06	6.18 (±0.80)	4.85 (±0.79)	88.23 (±9.62)
Gap-Sized-25	2.72	6.5×10⁻⁶	6.41×10⁴	3.02×10⁴	0.06	5.10 (±0.64)	2.76 (±0.27)	70.29 (±7.51)
Gap-Sized-30	2.62	6.8×10⁻⁶	4.70×10⁴	2.12×10⁴	0.11	4.71 (±0.51)	2.58 (±0.31)	79.18 (±11.11)

*95% Confidence Intervals

Table III. Elevated Temperature Data (Samples Fired and Tested at 1200°C)

Aggregate Dist.—Cement Content	α (°C^{-1})	E (MN/m^2)	G (MN/m^2)	ν	σ_f (MN/m^2)	γ_{NBT} (J/m^2)	γ_{WOF} (J/m^2)
Generic–25	8.3×10^{-6}	1.50×10^4	0.78×10^4	0.05	6.86 (±0.90)*	6.52 (±2.08)	118.72 (±14.05)
Continuous–20	7.9×10^{-6}	2.47×10^4	1.13×10^4	0.09	7.65 (±0.89)	5.31 (±1.38)	196.51 (±43.31)
Continuous–25	7.6×10^{-6}	2.00×10^4	0.91×10^4	0.09	8.88 (±1.11)	6.86 (±1.86)	205.67 (±44.41)
Continuous–30	7.7×10^{-6}	1.83×10^4	0.86×10^4	0.06	10.72 (±0.97)	9.78 (±2.41)	233.50 (±24.55)
Gap-Sized–20	8.3×10^{-6}	4.71×10^4	2.23×10^4	0.05	3.96 (±0.84)	0.98 (±0.26)	110.02 (±28.64)
Gap-Sized–25	9.1×10^{-6}	4.60×10^4	2.16×10^4	0.07	6.49 (±0.82)	2.00 (±0.29)	251.49 (±56.75)
Gap-Sized–30	8.2×10^{-6}	3.37×10^4	1.50×10^4	0.11	5.34 (±0.80)	2.77 (±0.59)	234.60 (±41.27)

*95% Confidence Intervals

with cement content. For all systems, the high-temperature values of E and G decreased from the room-temperature values by approximately 25%. Additionally, the values of Poisson's ratio were unaffected by variations in microstructures or by test temperature.

(4) Fracture stress—The fracture stress values for the generic and continuously graded systems were generally greater than the values obtained for the gap-sized systems. Apparently, the coarser aggregate-size distribution of the gap-sized systems limited the strength by creating a larger flaw size. The aggregate particles can generate microcracks on the same order of magnitude as their size, due to aggregate-matrix interactions. Thus the greater quantity of the coarser aggregate fractions in the gap-sized systems produced a greater density of larger flaws. Increasing the cement content in the continuously graded systems produced only minor increases in strength, whereas increasing the cement content in the gap-sized system revealed no clear trend. Also, testing at 1200°C, as compared with room temperature, had little effect on the fracture stress.

(5) Notched-beam fracture surface energy—The notched-beam fracture surface energy is a measurement of the energy necessary to initiate movement of a preexisting flaw. The γ_{NBT} values for the gap-sized systems were generally much lower than those of either the generic or the continuously graded systems. This indicates that fracture initiation is a much easier process in the concrete systems containing coarser aggregates. As far as the effects of cement content and test temperature are concerned, no clear trends could be seen.

(6) Work of fracture—The work of fracture is the energy necessary to propagate a slowly moving crack in a stable manner. It is a measure of the total energy dissipated during fracture. At room-temperature testing conditions, the γ_{WOF} values for the gap-sized systems were larger than those for either the generic or continuously graded systems. Evidently, the increase in γ_{WOF} for the coarser aggregate microstructures is a consequence of an increased degree of microcracking and the formation of more tortuous fracture paths. At the elevated test temperatures, all of the concrete systems exhibited large increases in γ_{WOF} values. At 1200°C, plastic behavior due to softening of the structure is responsible for this increased ability to dissipate energy. As for the effects of cement content, no obvious trends were observed; the only exception was the gap-sized system at the 20% cement content where the total fines content was much too low to provide for effective bonding in the structure, resulting in a low γ_{WOF} value.

Thermal Shock Behavior

The single-quench, thermal-shock tests for all seven of the refractory concretes are illustrated in Figs. 1–7. Flexural strengths are plotted vs the temperature differences of the shock quenching. All of the concretes exhibited gradual strength losses with increasing temperature differences, signifying quasi-static crack growth. No rapid strength decreases, indicative of kinetic crack growth, were noted. A straight line was fit to the data to indicate the gradual strength losses.

The multiple thermal shock test results are illustrated in Figs. 8–14. Flexural strengths are plotted vs the number of quenching or cooling cycles at the temperature difference of 1200°C. Strength losses generally leveled off

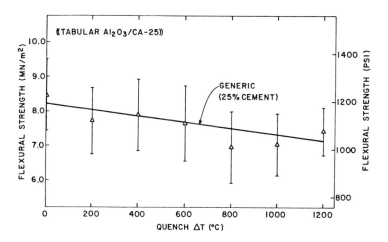

Fig. 1. Single quench thermal shock test for the generic concrete.

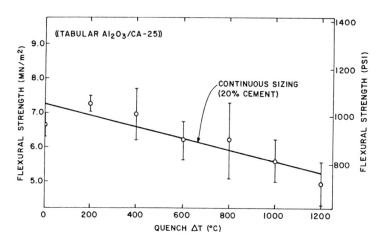

Fig. 2. Single quench thermal shock test for the continuous concrete at 20% cement.

after three to five cycles. Because of the gradual or continual nature of the thermal shock damage, this type of plot provides a more reliable measure of the resistance of the concrete to thermal shock damage. Since the damage levels become constant after about five cycles, a relatively accurate percentage strength loss can be calculated to provide a basis for comparison. Comparing the seven figures, the percentage strength losses are the greatest for the generic and continuously graded concretes, approximately 30 to 40%, and lowest for the gap-sized concretes, only about 15 to 20%.

For refractory materials, two common theoretical parameters often applied to describe thermal shock damage resistance are R'''' and R_{ST}. Both parameters are based on energy balance criteria.[11-13] R'''' applies specifically to

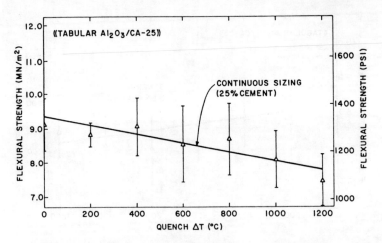

Fig. 3. Single quench thermal shock test for the continuous concrete at 25% cement.

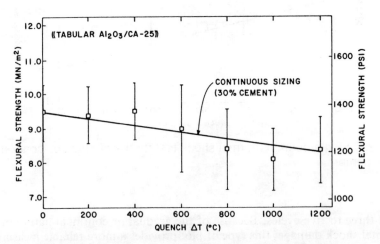

Fig. 4. Single quench thermal shock test for the continuous concrete at 30% cement.

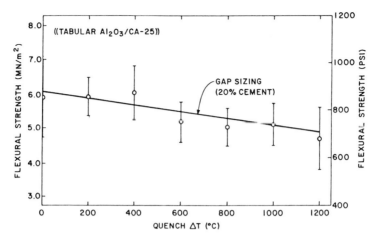

Fig. 5. Single quench thermal shock test for the gap-sized concrete at 20% cement.

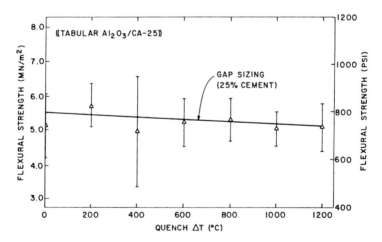

Fig. 6. Single quench thermal shock test for the gap-sized concrete at 25% cement.

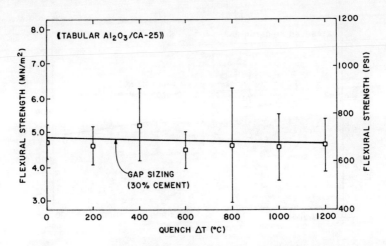

Fig. 7. Single quench thermal shock test for the gap-sized concrete at 30% cement.

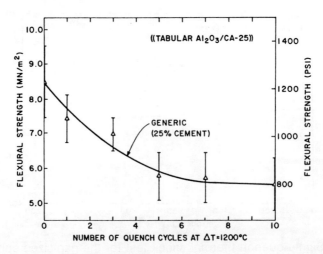

Fig. 8. Multiple quench thermal shock test for the generic concrete.

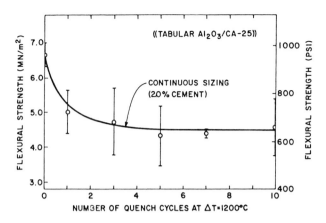

Fig. 9. Multiple quench thermal shock test for the continuous concrete at 20% cement.

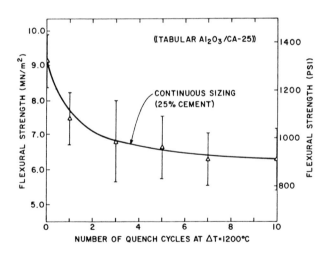

Fig. 10. Multiple quench thermal shock test for the continuous concrete at 25% cement.

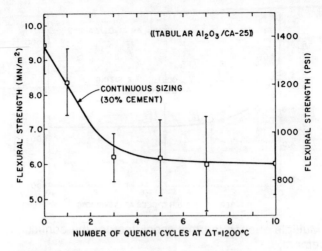

Fig. 11. Multiple quench thermal shock test for the continuous concrete at 30% cement.

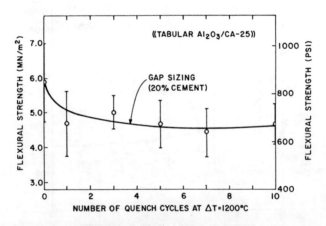

Fig. 12. Multiple quench thermal shock test for the gap-sized concrete at 20% cement.

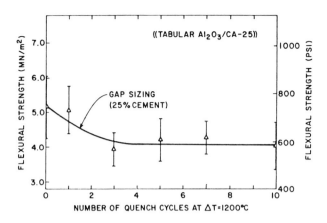

Fig. 13. Multiple quench thermal shock test for the gap-sized concrete at 25% cement.

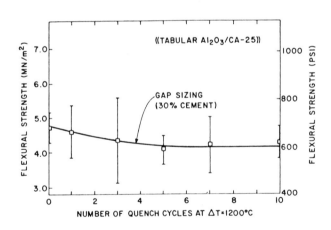

Fig. 14. Multiple quench thermal shock test for the gap-sized concrete at 30% cement.

high-strength materials where the initial crack length is relatively short and crack propagation occurs kinetically. R'''' is defined as follows:

$$R'''' = \frac{E\gamma_{WOF}}{\sigma_f^2(1-\nu)} \tag{9}$$

R_{ST} applies to lower strength materials where the initial crack length is relatively large and crack propagation occurs in a quasi-static manner. R_{ST} is given by:

$$R_{ST} = \left(\frac{\gamma_{WOF}}{\alpha^2 E}\right)^{1/2} \tag{10}$$

Maximizing either R'''' or R_{ST} should result in an increased resistance to thermal shock damage for the appropriate conditions.

The thermal shock damage resistance parameter R_{ST} is normally correlated with the single-quench, thermal-shock tests.[14] R_{ST} can frequently be correlated with the slope (designated as ψ) derived from the strength vs the temperature difference (ΔT) diagram. Table IV contains the calculated ψ values from the data illustrated in Figs. 1–7. R_{ST} parameters are calculated from both room-temperature and 1200°C data and are shown graphically in Fig. 15. The behavior exhibited by these refractory concretes is exactly the opposite of other refractory materials. Normally, the higher the R_{ST} value, the better the resistance of the material to thermal shock damage but, in this instance, the higher R_{ST} values are associated with the higher slopes, and the lower R_{ST} values with the lower slope values. The continuous and generic concretes have the larger calculated R_{ST} values but have the higher slopes, indicating a greater strength loss during the initial thermal shocks. The gap-sized systems have the lower calculated R_{ST} values but exhibit the smaller slope values or lower strength losses. It should also be noted that the R_{ST} parameters, calculated from either the room-temperature or 1200°C data, exhibit exactly the same trends and actually separate the concretes into two distinct groups according to their thermal shock damage resistance.

It is a normal practice to correlate the R'''' thermal shock damage resistance parameter with the results of the multiple quench tests.[15-16] For R'''', the results are presented in Table V and are plotted in Fig. 16. The R'''' values calculated from both the room-temperature and 1200°C measurements are contrasted with the strength losses from the multiple quench results in Figs. 8–14.

Table IV. Relationship between Strength Loss Slope (ψ) and R_{ST}

Concrete Systems	R_{ST} at RT (°C·cm$^{1/2}$)	R_{ST} at 1200°C (°C·cm$^{1/2}$)	ψ (× 10^{-3} MN/m^2 °C)
Continuous–30% cement	61.34	146.39	−1.24
Generic–25% cement	55.38	107.24	−1.15
Continuous–25% cement	60.67	133.24	−1.23
Continuous–20% cement	47.71	112.78	−1.65
Gap-Sized–20% cement	42.82	85.18	−0.99
Gap-Sized–25% cement	37.33	81.14	−0.20
Gap-Sized–30% cement	41.86	101.67	−0.18

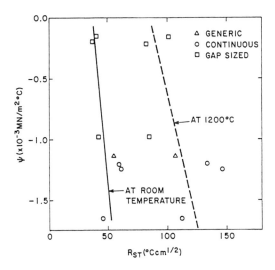

Fig. 15. Comparison between strength loss slope ψ and R_{ST}.

As shown, the concretes are again separated into two distinct groups according to their thermal shock resistance behavior. The generic and continuously graded concretes have the lowest R'''' values and, as expected, have the higher strength losses after thermal shocking. The gap-sized systems have the highest R'''' values and the lower strength losses. As shown in Table V, the R'''' values for both thermal shock test conditions correctly separate the concretes in terms of their thermal shock damage resistance. The R'''' values calculated from the 1200°C measurements exhibit higher values than when calculated from the room-temperature data.

From the above discussion, it is apparent that the trends of the R_{ST} values do not accurately describe the thermal shock damage resistance of these concretes, while the R'''' values correlate fairly well with the thermal shock behavior. Figure 17 illustrates the R_{ST} values plotted vs the R'''' values for room-temperature and 1200°C measurements. Positive slopes are normally expected, but negative slopes are shown. The negative slopes indicate that the two parameters show completely opposite behavior. This behavior can be understood by considering the stored elastic strain energy term. From

Table V. Relationship Between Strength Loss and R''''

Concrete Systems	R'''' at RT (cm)	R'''' at 1200°C (cm)	% Strength Loss
Continuous-30% cement	1.54	3.98	37.0
Generic-25% cement	1.89	3.96	35.2
Continuous-25% cement	2.29	5.73	32.2
Continuous-20% cement	2.40	9.10	31.2
Gap-Sized-20% cement	16.02	34.84	19.7
Gap-Sized-25% cement	18.69	29.38	18.9
Gap-Sized-30% cement	19.08	31.23	14.3

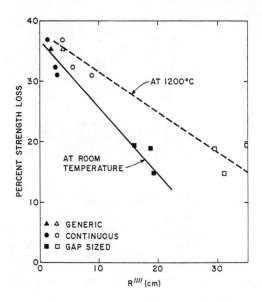

Fig. 16. Comparison between strength loss and R''''.

Eqs. (9) and (10), it can be seen that R'''' is proportional to E/σ_f^2 and R_{ST} to $(1/\alpha^2 E)^{1/2}$. Both parameters are inversely related to the stored elastic strain energy which is directly proportional to σ_f^2/E. The R'''' parameter includes both E and σ_f. R_{ST} does not include the σ_f term, which perhaps makes it invalid for these concretes. Typically for most refractory materials, a relationship exists between E and σ_f: As one parameter increases, usually the other increases as well. However, that usual trend does not hold for these refractory concretes, where E and σ_f are not directly related. As already shown, a low σ_f in the gap-sized systems is related to a higher E value. This occurs because E is primarily dependent on the bulk density, while σ_f depends primarily on the cement content and the aggregate-size distribution. Thus, E and σ_f in these concretes are related in a complex manner to the bulk density, cement content, and aggregate-size distribution. Since R_{ST} does not include the σ_f term, it should predict trends opposite to the actual observed thermal shock behavior of these concretes.

Conclusions

Microstructural variation due to cement content and aggregate-size distribution modifications can produce significant changes in certain mechanical properties of refractory concretes: Bulk densities, Young's elastic moduli, shear moduli, notched-beam fracture surface energies, and the work of fracture varied significantly with these microstructural modifications. Fracture strengths, Poisson's ratios, and the thermal expansion coefficients were affected to a much lesser degree.

These microstructural variations also modified the behavior of the different concrete systems during thermal shock testing. In laboratory thermal shock tests, the gap-sized systems showed increased resistance to thermal shock damage compared with the generic or continuously graded concretes.

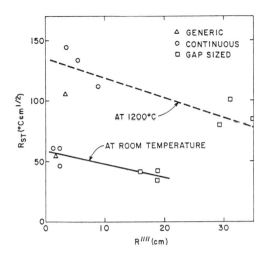

Fig. 17. Comparison between R_{ST} and R''''.

The thermal shock damage resistance parameter R_{ST} was found to be inversely related to single-cycle thermal shock trends, while R'''' correlated directly with trends observed in the multiple cycle thermal shock tests.

Acknowledgment

The authors wish to thank the U.S. Department of Energy, whose financial support made this research possible.

References

[1] C.C. Furnas, "Grading Aggregates, Mathematical Relations for Beds of Broken Solids of Maximum Density," *Ind. Eng. Chem.*, **23** [9] 1052-58 (1931).

[2] S. Spinner and W.E. Tefft, "A Method for Determining Mechanical Resonance Frequencies and for Calculating Elastic Moduli From These Frequencies," *Proc. ASTM*, **61**, 1221-36 (1961).

[3] W.R. Davis, "Measurement of the Elastic Constants of Ceramics by Resonant Frequency Methods," *Trans. Brit. Ceram. Soc.*, **67**, 515-41 (1968).

[4] D.P.H. Hasselman, Tables for the Computation of the Shear Modulus and Young's Modulus of Elasticity from the Resonant Frequencies of Rectangular Prisms; pp. 1-95. The Carborundum Company, New York, 1961.

[5] J.B. Wachtman and D.G. Lam, "Young's Modulus of Various Refractory Materials as a Function of Temperature," *J. Am. Ceram. Soc.*, **42** [5] 254-60 (1959).

[6] S. Spinner, "Elastic Moduli of Glasses at Elevated Temperatures by a Dynamic Method," *J. Am. Ceram. Soc.*, **39** [3] 113-18 (1956).

[7] W.F. Brown and J.E. Srawley, pp.13-15 in Plane Strain Crack Toughness Testing of High-Strength Metallic Materials, ASTM STP 410. American Society for Testing and Materials, 1966.

[8] J. Nayakama, "Direct Measurement of Fracture Energies of Brittle Heterogeneous Materials," *J. Am. Ceram. Soc.*, **48** [11] 583-87 (1965).

[9] H.G. Tattersall and G. Tappin, "Work of Fracture and Its Measurement in Metals, Ceramics and Other Materials," *J. Mater. Sci.*, **1** [3] 296-301 (1966).

[10] J.A. Coppola, D.P.H. Hasselman, and R.C. Bradt, "On the Measurement of the Work-of-Fracture of Refractories," *Am. Ceram. Soc. Bull.*, **52** [7] 578 (1973).

[11] D.P.H. Hasselman, "Thermal Stress Resistance Parameters for Brittle Refractory Ceramics: A Compendium," *Am. Ceram. Soc. Bull.*, **49** [12] 1033-37 (1970).

[12] D.P.H. Hasselman, "Elastic Energy at Fracture and Surface Energy as Design Criteria for Thermal Shock," *J. Am. Ceram. Soc.*, **46** [11] 535-40 (1963).

[13] D.P.H. Hasselman, "Unified Theory of Thermal Shock Fracture Initiation and Crack Propagation in Brittle Ceramics," *J. Am. Ceram. Soc.*, **52** [11] 600-604 (1969).

[14] S.B.S. Persson, "Thermal Shock Resistance of Refractories: Correlation Between Relative Loss of Strength and the Thermal Stress Resistance Parameter, R_{ST}," Proceedings of the 3rd CIMTEC 3rd International Meeting on Modern Ceramic Technologies; pp. 325-28, Rimini, 1976.

[15] J.H. Ainsworth and R.H. Herron, "Thermal Shock Damage Resistance of Refractories," *Am. Ceram. Soc. Bull.*, **53** [7] 533-38 (1974).

[16] T.A. Beals, "Thermal Shock Damage of Alumina Refractories"; B.S. Thesis. The Pennsylvania State University, University Park, May 1978.

The Heat Evolution Test for Setting Time of Cements and Castables

C.H. Fentiman

Lafarge Aluminous Cement Co., Ltd.
Essex, UK

C.M. George*

Lafarge Fondu International
Neuilly S/Seine, France

R.G.J. Montgomery[†]

Lafarge Aluminous Cement Co., Ltd.
Essex, UK

An initial set of hydraulically bonded castables coincides with the first heat evolution from bulk hydrate formation. This suggests a simple, automated detection method using thermal sensors, a technique that would enable false setting to be disregarded. Setting time results by Vicat penetration and heat evolution techniques are compared.

Such penetration tests as the Gilmore or Vicat type were developed for measuring the hardening or setting time of portland cements. By standardizing the diameter of the penetration needle and the force applied, fairly reproducible estimates are possible of the time taken for hydrating cement to reach some arbitrarily chosen level of mechanical rigidity. These same procedures are used for aluminous cements and give information of comparable quality.

Penetrometer tests are discontinuous. Also, a new surface must be found for each individual measurement of mechanical resistance. Consequently, frequent operator attendance is required, or specially constructed automatic equipment is necessary.

A penetration test works best with neat cement pastes, somewhat less well with mortars, and not at all with concrete. This is a matter of some practical inconvenience, since cement is employed most widely in concrete, least alone, and the actual setting time differs significantly depending on whether the cement is used neat or mixed with aggregate.

In the case of aluminous cements, a further difficulty may be encountered. Premature stiffening of the cement paste, mortar, or concrete, sufficient to resist entry of a needle or probe, can sometimes develop without the attendant cohesiveness which accompanies true setting. Under these circumstances, the Gilmore or Vicat test may indicate, for example, that a casting is ready for demolding when this is not in fact the case. There is thus some incentive for seeking an alternative test for setting. Recording heat evolution during the setting process offers one alternative with several advantages.

---

*Currently with Lone Star Lafarge, Inc., Norfolk, VA
[†]Currently with Lafarge Coppee Research, Viviers S/Rhone, France.

With hydraulic cements, the development of mechanical properties results from the formation of hydrates that create bonds between the anhydrous grains in the system. This is an exothermic process and the heat evolved may therefore be detected. Under adiabatic conditions, an increase in temperature can thus be registered using a simple thermal device.

With calcium silicate cements, the hydration process is relatively slow. In calcium aluminate cements, which differ from calcium silicates in the mechanism of hydration, hydrate formation, once initiated, accelerates rapidly.[1] Consequently, although the total heat of hydration in both cases is substantially the same, the adiabatic temperature rise displayed by aluminous cements is much greater. The use of heat evolution to measure setting time is thus far more readily accomplished with aluminous cements.

Adiabatic calorimetric studies of aluminous cement in contact with water show two exotherms. The first occurs immediately on contact with water, due to heat of wetting and rapid dissolution of cement to form a solution saturated in lime and alumina. A dormant period then follows during which hydrate nuclei form and develop. Once critical nuclei have been formed, bulk precipitation of hydrate occurs, giving rise to a second major exotherm and initiating the hardening process.[2] Under the conditions used in the present work, only the second major exotherm is detected.

Experimental Procedure

In developing the heat evolution (H.E.) technique, thermocouples linked to a conventional millivolt chart recorder were chosen for simplicity and convenience. The iron-constantan couple was found to give adequate sensitivity over the temperature range involved. Reproducible response was obtained from junctions formed by twisting clean wires together. The hot junction was embedded in the fresh paste or mortar (concretes could have been used, but were not systematically tested in the present work).

Initially, the Vicat test and the H.E. test were compared using separate samples for each, but a much more reliable procedure of using the same sample for both measurements simultaneously was rapidly adopted. The cold junction of the thermocouple device was placed in the water at 20°C, surrounding the setting-time mold.

Both normal production samples and specially selected nontypical samples of two industrial aluminous cements were used in the study. Cement A (Cement Fondu) was a 40% Al_2O_3, gray, aluminous cement, and cement B (Secar 71) was a pure, white, aluminous cement containing 70% Al_2O_3.

Neat cement pastes prepared according to NF P15-403, and mortars made according to BS 4550 were generally used, as well as some nonstandard mixes, to demonstrate the effect of cement content on time-temperature profiles.

Although some authors[3] have preferred larger samples for quantitative studies of heat evolution, the above procedure proved quite adequate in achieving sufficiently adiabatic conditions of measurement. The work was compared only if initial set was indicated by the two methods.

Results and Discussion

Table I compares the rate of temperature rise with the total temperature rise for specimens containing various percentages of cement A. Except at the

Table I. Exotherms Related to Cement Content

Cement in Dry Mix (%) I	Exotherm Height (arbitrary units) II	Exotherm Slope III	I/II	I/III
12.5	15.5	0.62	0.8	20.1
27	24.5	0.84	1.2	32.0
33	27.5	0.90	1.1	36.7
50	46.0	1.43	1.1	35.0
100	100.0	2.60	1.0	38.5

lowest cement content (12.5%), both rate of temperature rise and maximum recorded temperature are linearly related to cement content.

Figure 1 shows typical results obtained with 27% cement B used in a siliceous sand mortar. There is good correspondence between initial set as measured by the Vicat and H.E. methods.

The choice of the point on the temperature-time profile recorded by the H.E. method which best represents the initial set as indicated by the Vicat needle was made by calculating correlation coefficients at successive intervals along the profile. From a calibration of the thermocouples, it was apparent

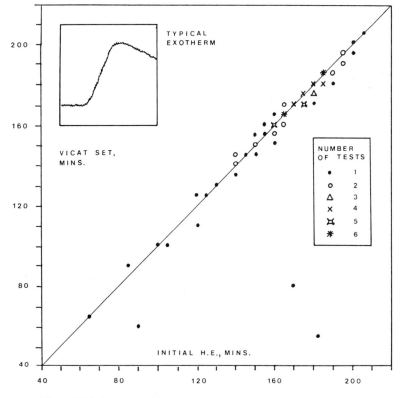

Fig. 1. Vicat initial set and initial H.E. for cement B mortars.

that the best fit is obtained when a 0.5 °C temperature rise has occurred. This is about the minimum detectable displacement of the chart pen above the base line. It follows that initial set, as recorded by the penetration technique, corresponds to the beginning of heat evolution, and thus hydrate formation, as was suggested earlier.[3]

The ability of the H.E. test to predict initial set as shown by the Vicat method is demonstrated in Table II. An excellent agreement is evident between the two methods. The results for cement B shown in Table II are those plotted in Fig. 1, with the exception of the three points which fall well outside the normal distribution. These three results relate to specially selected samples obtained under recognizably abnormal production conditions. These samples exhibited premature stiffening, which was recorded as an apparent very fast set by the penetration test. These same samples showed H.E. curves, indicating that true setting occurred much later.

Table II. Correlation between Vicat and H.E. Methods

System Tested	Ratio of Vicat Setting Time to H.E. Setting Time			Setting Time Range (min)
	Number	Mean	Coefficient of Variation (%)	
Cement A Neat paste	26	0.992	7.84	180 -300
Cement A Sand mortar	90	0.998	1.69	130 -230
Cement B Sand mortar	80	1.088	5.97	70 -230

Intrinsic premature stiffening or false set is rarely encountered in practice, since such off-specification material is normally eliminated by production quality control. However, where it does arise, and this may also occur due to such extrinsic factors as chemical additions to castable formulations, the H.E. method offers a means of ensuring that adequate curing has been carried out before molds are struck.

Some attempts were made in the present study to relate setting time, as shown by heat evolution, to the mineralogy of the cement. This was confined to cement B, where the mineralogical composition is relatively simple. There is some evidence that the ratio of combined lime to combined alumina in the cement affects the time-temperature profiles recorded by the H.E. method (for example, as $CA_2/C_{12}A_7$ increases, initial setting time becomes longer), but more work is needed to establish reliable relationships.

The nature of the H.E. test is readily adaptable to automatic procedures, and an extension of the present study may also enable final setting and binder capacity to be predicted.

Conclusions

The H.E. method is a reliable alternative to penetration tests for estimating initial set of aluminous cement binders and castables. The H.E. test enables false setting to be disregarded; the test method is simple and readily automated; and it has been adopted as a routine quality control procedure for aluminous cement manufacture.

References

[1] T.D. Robson, High Alumina Cements and Concretes; p. 57 in Contractors Record Ltd., London, 1962.

[2] L.S. Wells and E.T. Carlson, J. Res. Natl. Bur. Stand., RP 2723, **57**, 335 (1956).

[3] W.S. Treffner and R.M. Williams, "Heat Evolution Tests with Calcium Aluminate Binders and Castables," J. Am. Ceram. Soc., **46** [8] 399–406 (1963).

Section III
State-of-the-Art Activities in Installation and Bakeout of Monolithics

Introduction of Automatic Gunning Machines for Tundish Linings 139
    T. Morimoto, K. Ogasahara, A. Matsuo, and S. Miyagawa

Properties and Service Experience of Organic Fiber-Containing Monoliths 149
    T. R. Kleeb and J. A. Caprio

The Development of Dry Refractory Technology in the United States 161
    J. L. Turner, Jr. and D. M. Myers

A Review of International Experiences in Plastic Gunning ... 165
    L. P. Krietz, G. Wilson, D. Hofmann, and M. Tsukino

Viscosity and Gunning of Basic Specialties 175
    W. Siegl

Dryouts and Heatups of Refractory Monoliths 192
    N. W. Severin

Introduction of Automatic Gunning Machines for Tundish Linings

T. Morimoto, K. Ogasahara, A. Matsuo, and S. Miyagawa

Kawasaki Steel Corp.
Kurashiki, Japan

An automatic gunning machine for the monolithic lining of a tundish is introduced. Development of the technology, especially with regard to the development of monolithic material and a spray nozzle for this gunning machine, is described. Characteristics of this process are compared with the board lining.

In the continuous casting process, either monolithic refractories consisting of high alumina or magnesia or insulating boards are being used extensively for the layer covering the tundish lining. Insulating board, with high thermal insulating properties, has been used only when preheating cannot be provided for an adequate period of time. The representative characteristics and properties of such materials are shown in Table I.

At the Mizushima Works of Kawasaki Steel Corporation, revamping or modifications are carried out at the tundish maintenance yard. The choice of the preferred material for coating the tundish lining determines the selection

Table I. Characteristics and Properties of the Coating Materials for the Tundish

Properties	Monolithic Refractory A	Monolithic Refractory B	Insulating Board
Chemical composition (%)			
MgO	89	91	45
SiO_2	5	6	34
LOI			16
Water added (%)	14	12	
Thermal conductivity at room temp. ($kcal \cdot m \cdot h \cdot °C$)	2.5	2.5	0.4
Bending strength (kgf/cm^2)			
110°C × 10 h	19	31	20
900°C × 3 h	12	15	
1300°C × 3 h	18	30	
1500°C × 3 h	50	35	

of equipment and facilities to be employed and installed at the tundish maintenance yard. We ran comparative tests and experiments on the abovementioned materials and concluded that employing a monolithic refractory as the coating material was the preferred choice. We then developed an automatic gunning machine for mechanically applying the refractory coating onto the inside of the tundish lining.

This report covers the results of the experiments conducted to select the material for coating the tundish lining, the development of materials carried out during development of the automatic gunning machine, and the effects attained by the use of the gunning machine.

Material Tests, Experiments, and Selection

As part of the process of selecting equipment and facilities required for tundish maintenance purposes, tundishes provided with either monolithic refractories or insulating board were put to various comparative tests to determine their properties. The results of the tests are given in Table II.

Molten steel temperatures in the tundishes using insulating board and monolithic refractories are shown in Fig. 1. The decrease in temperature is expressed by the difference between the temperature of molten steel at the completion of degassing or bubbling and that registered when the tundish is filled after the start of teeming from the ladle in the continuous casting process. The temperature decrease of the tundish covered with the insulating board is about 3.5 °C greater than that of the tundish covered with the monolithic refractory. This corresponds very well to the calculation results.[1]

The total production cost of the tundish with the insulating board is about 1.4 times higher than that of the tundish with the monolithic refractory. (The total cost consists of the refractory cost plus the cost of the coke oven gas required for preheating.) The cost of the insulating board is even higher when one takes into account the lower tapping temperature based on the thermal insulating property.

The tundish with the insulating board is more advantageous because its preheating time can be shortened. At the Mizushima Works, we have recently begun to perform continuous-continuous casting of different steels by exchanging the tundishes, which obviates the need for resetting the dummy bar. In this process, a tundish to be exchanged after casting 1–3 heats will no longer need preheating time and, therefore, we are providing a tundish with insulating board, which does not require 3 h or more of preheating.

The tundish coated with monolithic refractory is more workable. The manpower required for the tundish with the insulating board is about 1.5 times greater than that for the tundish with the monolithic refractory. Further, changes in the profile of the refractory bricks of the tundish that result from erosion make mechanized application of insulating board difficult.

In the comparative test for slag resistance, the erosion rate of the slag line of the tundish with insulating board (at about 10 mm/heat) was far greater than that of the tundish with monolithic refractory (at 1.4 mm/heat). The addition of molten steel over a number of heats results in erosion of the slag line of a tundish with insulating board which can lead to penetration of molten steel behind the board. Such penetration obviates the advantage of the insulating board in terms of easy removal of drain steel.

With regard to the quality of the steel obtained, no differences were seen in the supersonic test or the level of H content of the steel. To revamp and modify the tundish maintenance yard, we needed to develop a monolithic refractory capable of withstanding quick temperature changes as a coating for the lining of the tundish and to develop an automated spraying process to reduce labor power and refractory cost.

Table II. Comparison of Tundishes Provided with Monolithic Refractories and Insulating Board

	Monolithic Refractory	Insulating Board	Remarks
Thermal insulation	Fair	Excellent	The temperature of molten steel in the tundish is higher by 3.5°C when the insulating board is used.
Cost of refractory	Good	Bad	The cost of the insulating board is 1.5 times greater than that of the monolithic refractory.
Preheating	Bad	Excellent	No preheating is possible when the insulating board is used. About 3 h are necessary when the monolithic board is used.
Workability	Excellent	Bad	The man-hours needed for the board are 1.5 times greater than those needed for the monolithic refractory. Mechanized application of the insulating board is difficult.
Slag resistance	Excellent	Fair	Erosion rate of slag-line: Insulating board, 10.0 mm/heat; monolithic refractory, 1.4 mm/heat.
Quality of steel obtained	Good	Good	No difference

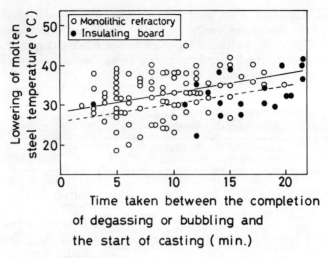

Fig. 1. Lowering of the temperature of molten steel in the tundishes provided with monolithic refractory and insulating board.

Development of Monolithic Material Resistant to Quick Temperature Changes

To improve the explosion resistance of the material, we conducted an explosion resistance test using numerous test samples with materials of different particle size and permeability, binders of different types, and varying quantities of water.[2] Test pieces measuring 40 by 40 by 80 mm were cast and, after 45 min, they were heated to 1000°C within 3 min.

The results indicated that test pieces that contained a different binder and that offered improvements in permeability and bending strength proved to be free from the danger of explosion even when heated quickly. Table III shows the properties of the newly developed monolithic refractories (D and E) and the conventional monolithic refractory (A). During heating, the monolithic refractory coating emits water vapor and generates vapor pressure inside the layer; whenever a part of the layer becomes unable to resist such pressure, an explosion occurs.

Because the generated vapor is diffused during the drying process, it is difficult to correlate the inner vapor pressure with the saturated vapor pressure. Toei studied the relationship between the vapor pressure and the temperature by dividing the applied layer into two models of a dry and a wet area.[3] According to his theoretical equation, the vapor pressures generated inside may be expressed by:

$$P_s = [(T_a - T_s)(keRw/rpD_e) + P_a/T_a]T_s \qquad (1)$$

$$D_e = D_v^1/\mu_{ed} \qquad (2)$$

$$\mu_{ed} = 1.5 - \epsilon/2 \qquad (3)$$

Table III. Properties of the Newly Developed and Conventional Monolithic Refractories

	Refractory (A)	Refractory (D)	Refractory (E)
Chemical composition (%)			
MgO	89	92	91
SiO_2	5	5	6
Al_2O_3			
Thermal conductivity at room temp. (kcal/m·h·°C)	2.5	2.0	2.5
Bulk density			
110°C for 10 h	2.45 ≈ 2.46	2.40 ≈ 2.42	2.43 ≈ 2.44
1500°C for 3 h	2.46 ≈ 2.47	2.44 ≈ 2.45	2.44 ≈ 2.45
Bending strength (kgf/cm²)			
110°C × 10 h	18 ≈ 20	16 ≈ 18	40 ≈ 43
900°C × 3 h	11 ≈ 12	10 ≈ 12	14 ≈ 26
1500°C × 3 h	30 ≈ 37	34 ≈ 36	51 ≈ 57
Slag erosion ratio* (1600°C for 2 h)	100	112	105
Permeability (cm²/s H_2O·cm)	1.79×10^{-3}	6.55×10^{-3}	3.80×10^{-3}
Remarks	Conventional material	With fiber	With a different binder

*Compounded slag (c/s 0.44%; Fe_2O_3 4.0%; MnO 12.4%; Al_2O_3 15.4%).

where

T_a = Absolute temperature of the heated air flow over the surface (K)
T_s = Absolute temperature on the vapor surface (K)
ke = Thermal conductivity (kcal/m·h·°C)
Rw = Coefficient of the gas (kgf/m·kg·K)
r_p = Evaporation latent heat of the water (kcal/kg)
D_v^1 = Coefficient of diffusion of effective water vapor in the porous body (m²/h)
P_s = Inner vapor pressure (kgf/m²)
P_a = Vapor pressure over the surface (kgf/m²)
ϵ = Porosity
μ_{ed} = Tortuosity factor

Further, according to the studies by Yamamoto et al.,[4] the strength of the applied layer after its initial drying period depends on the tensile strength of the material.

To study the relative explosion resistance of the test material, P_s was calculated by Eq. (1), and E, obtained by dividing P_s by the tensile strength at

110°C, was taken to represent the explosion resistance (refer to Eq. (4)). No explosion would occur when E is smaller than 1.

$$E = P_s/\sigma t \tag{4}$$

where σ_t is tensile strength (kgf/m²).

To introduce a factor of permeability under D_e, we correlated Eq. (4) to obtain

$$D_e = (\epsilon_0 D_v / \mu_{ed})(\mu/\mu_0) \tag{5}$$

where

μ = Permeability (cm²/s H₂O cm)
μ_0 = Permeability of the standard material (cm²/s H₂O cm)
ϵ_0 = Porosity of the standard material

The relationship between E and the explosion frequency in the explosion resistance test is shown in Fig. 2.

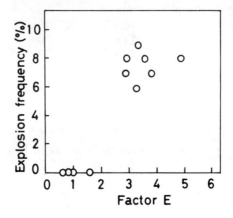

Fig. 2. Factor E for the evaluation of explosion resistance and explosion frequency.

It is clear that a material with a factor of 1 or less exhibits good explosion resistance. Placing more emphasis on the explosion resistance and slag resistance, we have put the new technology to work in a tundish using monolithic refractory (E), the strength of which has been improved by a new binder. We applied the monolithic refractory to part of the tundish covered with insulating board, heated it to the standard for tundishes with insulating board (heating up to 1000°C for 45 min), and examined the tundish for explosion resistance. The results showed that this tundish was explosion-resistant and could be heated in the same way as a tundish with insulating board. Further, there was no difference in the reduction rate of molten steel temperature compared with a tundish coated with traditional monolithic refractory, as long as it is heated quickly and gas is fed at a proper rate.

Development of this material has reduced the number of tundishes requiring insulating board to 2-3%. By preheating, the balance of the tundishes formerly using insulating board may be provided with monolithic refractory. For this reason, a fully automatic gunning machine for applying the coat over the lining of the tundish with monolithic refractory was developed.

Development of the Fully Automatic Gunning Machine

Specifications of the Gunning Machine

There are two possible methods for applying the coat: dry gunning and wet gunning. The wet gunning method calls for premixing the material, as well as cleaning the nozzle and the hose after use. In addition, space is needed for storing necessary equipment and materials, and difficulties related to dust generated during mixing and problems connected to waste water need to be settled. In dry gunning, on the other hand, although dust problems may be inevitable, many problems inherent to the wet gunning process can be avoided and substantial advantages can be expected with respect to labor savings.

Due to these problems, we introduced a pretesting machine to develop a proper coating material and to find solutions to problems that may be encountered in mechanizing the gunning process. The gunning system consists of an automatic gunning machine and a nozzle supporting device, which are mounted on a car traveling over the tundish to mix the gunning material with water at the end of the nozzle and to apply the material onto the interior surface of the tundish.

Development of the Coating Material

When applying the coating material onto the interior surface of the tundish, proper means must be available to provide satisfactory coating despite the short distance between the tip of the nozzle and the interior surface. Developmental efforts were made to develop a coating material that would minimize rebound loss even when applied under the preceding conditions.

To reduce rebound loss as much as possible, gunning tests were conducted using coating materials of different particle sizes. Figure 3 illustrates

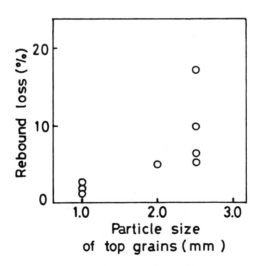

Fig. 3. Particle size of top grains vs rebound loss.

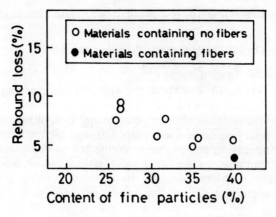

Fig. 4. Content of fine particle (under 70 μm) and rebound loss.

the percent rebound loss in the case of a material of top-grain particle size, and Fig. 4 plots the rebound loss when materials with different contents of fine particles are used. It is evident from such tests that smaller particle sizes of top grains and higher contents of fine particles reduce the rebound loss. It was also found that rebound loss could be reduced by employing a material containing fibers.

It was possible through proper blending of particles to obtain a material with a depositing ratio of 95% or higher, and it was also possible to come up with a coating material with resistance to quick temperature changes by mixing a binder with a monolithic refractory that is resistant to quick temperature changes.

The developmental work showed that, although it was possible to improve the depositing ratio by using a blend consisting of top grains of smaller particle size with a high content of fine particles, blocking of the nozzle would develop when attempting to apply this material at a high rate. This problem, however, has been solved by intensified stirring of the material with water and by impeding growth of deposits of the material inside the nozzle.

Introduction of a Fully Automatic Gunning Machine

As a result of the tests conducted on an experimental machine, a fully automatic gunning machine was introduced.

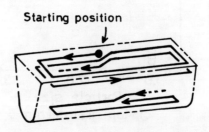

Fig. 5. Movement of the gunning nozzle (schematic diagram).

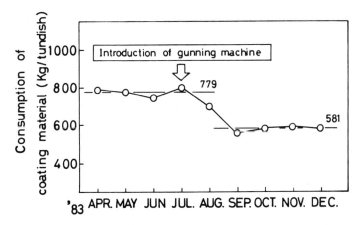

Fig. 6. Trends of the consumption of the coating material after the fully automatic gunning machine was introduced.

The movement of the gunning nozzle is shown in Fig. 5. The angle of the nozzle is freely adjustable to keep it positioned at a right angle to the internal wall of the tundish at all times. This fully automatic machine can be run by just one person, and the whole process of coating the internal wall of the tundish can be completed automatically simply by setting the starting position and pressing the START push button.

Plotted in Fig. 6 are the trends for consumption of the coating material after the fully automated gunning machine was introduced. Use of the

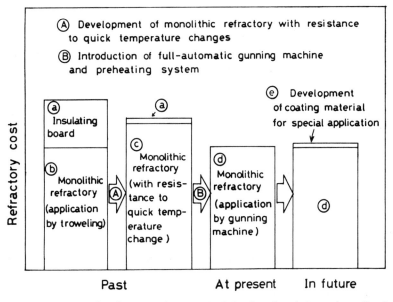

Fig. 7. Changes in the coating materials for the internal wall of the tundishes.

machine has resulted in a reduction of the material by about 25% vs that recorded in the trowel application. Further, this new process has made it possible to reduce manpower requirements by one person × 3 shifts.

Conclusion

Figure 7 is a schematic of the changes that have taken place in the application of the materials for coating the inner walls of the tundishes. Thanks to the introduction of a monolithic refractory with resistance to quick temperature changes, the tundishes (with the exception of just a small fraction of the total) are now coated with the monolithic refractory.

Through optimization of the temperature rise pattern, the rate of temperature decrease of molten steel inside a tundish with an inner coating of monolithic refractory has been lowered to a negligible level, in contrast to those tundishes with the insulating board. What is more, the monolithic refractory is more cost-effective than the insulating board. Using the monolithic refractory for coating the inner wall of the tundish, it has become possible to apply the material by mechanized means.

At the Mizushima Works, as part of the program of revamping and modifying the tundish maintenance yard, a gunning machine capable of automatically applying the coating material over the inner wall of the tundish has been installed, which has resulted in both reduced manpower requirements and refractory cost.

In addition, it is now possible to apply the material while the tundish is still warm, which has led to a substantial reduction in the tundish recycling needs due to the synergistic efforts afforded through the use of other mechanical systems.

Since the monolithic refractory has proved advantageous in terms of cost as the preferred coating material for the inner wall of the tundish for making plain carbon steel, the gunning application method of this material is expected to be used increasingly. Further, it has been found that blending fibers with the coating material produces a monolithic refractory with particularly high thermal insulation. Based on this finding, further development will work out proper coating materials with such features as high slag resistance and high thermal insulation.

In the meantime, the application method for insulating boards is expected to find use in steel casting, where low superheat casting is required because the air gap formed behind the insulating board provides good heat insulating effects. Since a coating material for making high-grade steel with very low levels of H and inclusions is required, future developmental efforts will be made to make the best use of the advantages of both methods of applying the coating materials.

References

[1]Ohishi, Ogasahara, and Nanbu, *Taikabutsu Overseas*, **3** [3] 33 (19??).
[2]Ohishi, Ogasahara, Nanbu, Ninomiya, and Yoshimura, *Taikabutsu*, **33** [2] 105 (1981).
[3]Toei, Saikin no kagaku kohgaku; p. 203. Maruzen Co., Ltd., 1968.
[4]Yamamoto, Yamane, and Mitsui, *Taikabutsu*, **33** [8] 445 (1981).

Properties and Service Experience of Organic, Fiber-Containing Monoliths

T.R. Kleeb and J.A. Caprio

Harbison-Walker Refractories Co.
Pittsburgh, PA 15222

Laboratory testing of Harbison-Walker's proprietary castables and gunning mixes containing organic, copolymer fibers showed that they had increased permeabilities that resulted in lower internal pressures during heatup and reduced tendencies to steam spall compared with similar fiber-free compositions. The fiber-containing monoliths had slightly lower strengths after drying and reheating, but were found to have higher residual strengths when heated rapidly. The proprietary mixes containing copolymer fibers have also been found to be suitable for use in metal contact refractories. Experience has shown improved service life with the use of these fibers in monolithic refractories.

It has long been recognized that refractory castables will steam spall during heatup if they are improperly cured or heated rapidly. Extensive published work has shown that a major contributing factor to steam spalling is the calcium aluminate cement phases formed when castables are cured at less than 21°C (70°F).[1-4] It has been shown that CAH_{10} (C = CaO, A = Al_2O_3, H = H_2O) and alumina gel are the primary bonding phases formed at these curing temperatures. As the castable is heated, CAH_{10} dehydrates to form C_2AH_8 (C = CaO, A = Al_2O_3, H = H_2O) and additional gel at about 21°C (70°F). This gel is generally blamed for the low permeability and difficulty in drying refractory castables cured at low temperatures.

One method of increasing the permeability of cement-bonded monoliths is by adding natural or synthetic fibrous or straw-shaped materials. Success in increasing permeability and reducing steam spalling has been achieved by adding seeds, fish scales, wood fibers, cotton fibers, shredded paper, roots, pine needles, and natural and synthetic straws. Because these materials have relatively large diameters, recent interest has been in such synthetic fibers as polyethylene and polypropylene, where smaller fiber diameters are available. The proprietary synthetic fiber composition evaluated in the current study was a copolymer of two thermoplastic resins. This polymer was chosen because of its unique property of shrinking 50% linearly when heated to 60°–66°C (140°–150°F). Figure 1 is a scanning electron micrograph of a shrunken fiber in the channel that it created after the refractory castable had been heated to 77°C (170°F). This shrinkage opens channels to aid in drying at temperatures well below the vaporization temperature of water. The copolymer absorbs less than 0.1 wt% water, so its use does not increase the water requirements of most refractory monoliths. Fibers of this material can be manufactured to diameters of 15 μm and cut to 6-mm lengths, dimensions that have been found suitable for use in refractory monoliths.

Fig. 1. SEM showing shrunken organic copolymer fiber.

General Testing of Properties

For most applications, copolymer fiber additions of 0.2–1.6 vol% have effectively improved a monolith's resistance to steam spalling. The adverse effect of fiber additions on density and strengths is minimal, particularly for high-alumina monoliths (Table I). Moreover, monoliths that were manufactured with fiber additions have withstood thermal cycling better than similar fiber-free products. The ability to use copolymer, fiber-containing refractories in metal contact applications is illustrated in Fig. 2. These samples held 7075 aluminum alloy at 816°C (1500°F) for 5 days. There was no difference in aluminum penetration between the fiber-free castable on the left and the fiber-containing castable on the right. The residual aluminum was more easily removed from the fiber-containing cup than the standard cup. This unexpected non-wetting phenomenon has been seen in most high-alumina, fireclay, and chrome ore castables tested to date and is a subject of continuing laboratory investigation.

Table I. Physical Properties: 90% Al_2O_3 Castable (−13 mm Aggregate)

Copolymer addition	No	Yes
Bulk density (kg/m^3 × 10^3)	2.82	2.82
Modulus of rupture (MPa)	14.1	13.9
Compressive strength (MPa)	93.1	84.7
Prism spalling test (1200°C/water cycles to failure)	7.0	11.0

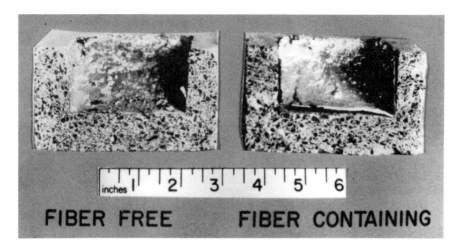

Fig. 2. Aluminum cup tests.

Copolymer fiber additions to gunning mixes have been found to have no effect on installation properties. In most cases, density and strength losses have been negligible (Table II).

Table II. Physical Properties: Fireclay Gunning Mix

Copolymer Addition	No	Yes
Bulk density (kg/m^3 × 10^3)	2.24	2.22
Modulus of rupture (MPa)	15.0	14.1
Compressive strength (MPa)	61.5	60.2

Steam-Spalling Testing

In a test to show the effect of copolymer fiber additions on resistance to steam spalling, undried 46 by 30 by 23-cm test blocks (cured for 24 h at 21 °C) were heated on one side to 1400 °C (2550 °F) in 2 h. For a commercial, 90% alumina castable based on − 13 mm sintered alumina, the fiber-free castable spalled on the hot face to a depth of about 8 cm at 1040 °C (1900 °F) (Fig. 3). The same product containing the copolymer fibers did not steam spall and showed only minor surface crazing (Fig. 4).

This same spalling test was run on another 90% alumina castable based on − 4 mm sintered alumina with similar results. The fiber-free castable lost about 60% of its hot face due to steam spalling at 1060 °C (1940 °F) (Fig. 5). When 0.3 vol% copolymer fibers were added, steam spalling did not occur, but numerous 1-mm cracks occurred on the hot face (Fig. 6). At a fiber level of 0.5 vol%, the only hot-face imperfection was slight crazing (Fig. 7). After steam-spalling testing, 5-cm cubes for compressive testing were cut from each block at depths of 9–14 cm and 18–23 cm. Cubes were also cut from the hot faces of the two blocks which did not steam spall. The residual compressive strengths of the fiber-containing castables were found to increase with in-

Fig. 3. Steam-spalling test: fiber-free castable.

Fig. 4. Steam-spalling test: fiber-containing castable.

Fig. 5. Hot face, 0.0% fibers: 60% spalling.

Fig. 6. Hot face, 0.3 vol% fibers: 1-mm cracks.

Fig. 7. Hot face, 0.5 vol% fibers: crazing.

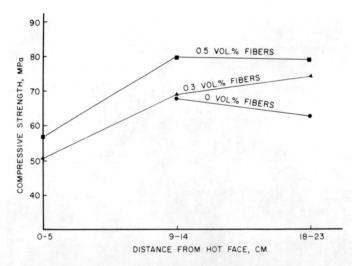

Fig. 8. Residual compressive strength vs distance from the hot face.

creasing fiber content over the fiber-free product (Fig. 8), which suggests that even when steam spalling does not occur in castables subjected to excessively high heating rates, bond disruption may reduce strengths and result in shorter service life.

In 1982, the development of an internal pressure probe capable of measuring relative internal pressures of cast refractory specimens during heatup was reported.[5] This probe was used to study internal pressures of 90% alumina castables with and without copolymer fiber additions. Test blocks measuring 23 by 18 by 15 cm were heated at 275°C/h to 1370°C (2500°F) from one side. At distances of 2.5 cm and 7.6 cm from the hot face, the maximum internal pressures of the fiber-free block were measured with the probe and found to be the same—723 kPa. At these same depths, the maximum internal pressures of the sample containing copolymer fibers were 15–20% lower: 572 kPa at a distance of 2.5 cm and 614 kPa at 7.6 cm.

Gitzen and Hart showed that castable curing temperature greatly influences the refractory's tendency to steam spall during heatup.[6] A curing temperature of 32°C (90°F) was found to be optimum for providing high permeability, good resistance to steam spalling, and the best strengths. A sintered alumina-based castable containing 30% high-purity calcium aluminate cement was tested with and without copolymer fiber additions for resistance to steam spalling.

In this test, 63.5-cm cubes of the castables were cast and cured at 4.5°C (40°F), 32°C (90°F), and 43°C (110°F). The undried cubes were placed in a muffle furnace at various temperatures until they steam spalled or remained intact for 20 min. The calculated steam-spalling temperature was the average of the highest temperature at which the sample did not spall and the lowest temperature of failure. The test-temperature intervals were usually 55°–111°C (100°–200°F). The steam-spalling temperature of the fiber-free castable was below 427°C (800°F) when cured at 4.5°C (40°F) (Fig. 9). This temperature increased to 1243°C (2270°F) when cured at 32°C (90°F), and to over 1630°C (2966°F) when cured at 43°C (110°F). The addition of 0.5 vol% fibers increased the steam-spalling temperature of this castable after curing at 4.5°C (40°F) and 32°C (90°F). At 0.8 vol% fibers, the effect of the fiber addition on steam-spalling temperature was dramatic. The steam-spalling temperature of this castable after curing at 4.5°C (40°F) was 1454°C

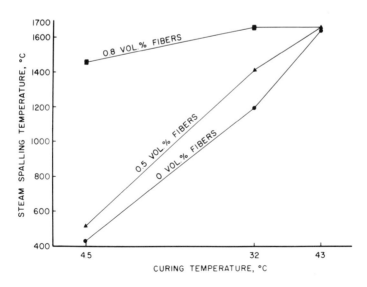

Fig. 9. Steam-spalling temperature vs curing temperature.

(2650°F) and it increased to over 1630°C (2966°F) when cured at 32°C (90°F) and above. Even though the use of copolymer fibers increased the steam-spalling temperature after curing at 4.5°C (40°F) and 32°C (90°F), the benefits of proper curing are also evident.

Permeability Testing

In another series of tests, a sintered alumina-based castable containing 15% high-purity cement was studied with and without the addition of copolymer fibers. In this work, the dry castable and tempering water were conditioned to 20°C (68°F), mixed with 8.4% tempering water, and cast into standard straights and 6.35-cm diameter cylinders with heights of 1.25-cm using ASTM C-862 procedures. Samples of each castable were cured for 24 h from 10°–43°C (50°–110°F) in 5.5°C (10°F) intervals. Samples were tested after curing and heating for 5 h at 816°C (1500°F). The cylinders were used for permeability testing, and 6.35-cm cubes were cut from the brick for compressive strength testing.

The permeabilities of the fiber-free castable and the fiber-containing castable were similar after all curing temperatures (Fig. 10) as was expected, since the fibers totally fill the channels they create before heat is applied. Permeability was very low at curing temperatures below 26.7°C (80°F). At 32°C (90°F), the permeability increased considerably as the hydrated bond became more crystalline, decreased at 37.8°C (100°F), and increased again at 43°C (110°F)—similar to Gitzen and Hart's results. The permeability decrease at 37.7°C (100°F) is thought to be a result of crystal growth of the hydrated cement phases.

After heating to 816°C (1500°F), the permeability increases due to the fiber addition are obvious (Fig. 11). Again, the permeabilities of the fiber-free and the fiber-containing castables increase with increasing curing tem-

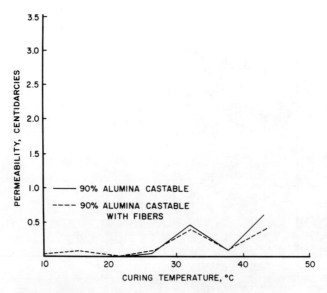

Fig. 10. Permeability vs curing temperature after curing.

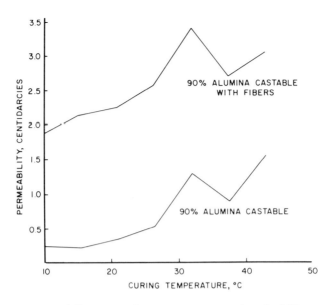

Fig. 11. Permeability vs curing temperature after 816°C.

perature up to 32°C (90°F), decrease at 37.8°C (100°F), and increase at 43°C (110°F). In these curves, however, the benefits of curing at increasingly higher temperatures above 10°C are more apparent.

After curing, the compressive strengths of both castables were nearly identical (Fig. 12). Each showed the strongest compressive strengths after curing at 26.5°C (80°F), but no strong trends were evident from these data.

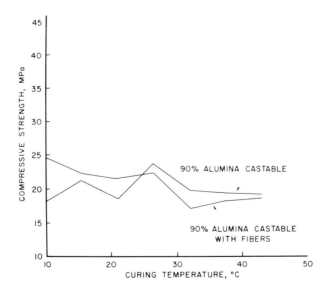

Fig. 12. Compressive strength after curing vs curing temperature.

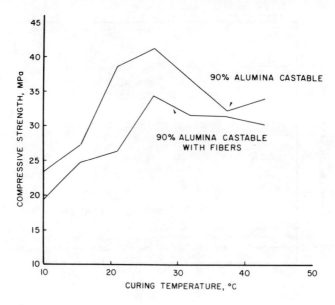

Fig. 13. Compressive strength after 816°C vs curing temperature.

Plots of the compressive strengths after heating to 816°C (1500°F) of the fiber-free castable and the copolymer fiber-containing castable against curing temperature showed the fiber-free castable to have a slight to moderate strength advantage at all curing temperatures (Fig. 13). The maximum strengths for both castables were again obtained after curing at 26.7°C (80°F).

Service Experiences

The copolymer fiber-based technology has been commercially used for over two years. The following service experiences reflect the properties studied in various laboratories and discussed above.

Fiber Properties

A receiving spout of a channel induction furnace was subjected to molten iron from a cupola. Approximately 10 900 kilos of a dense, high-purity 90% Al_2O_3 castable were required. In one campaign, a polypropylene additive with a diameter of 15–20 μm was included, which resulted in a spout life of four months. Under similar conditions, a spout containing the organic copolymer additive experienced an eight-month campaign with little or no cracking. Both linings were conservatively heated using a baffled gas torch, with ambient conditions of 5°–10°C (40°–50°F) during installation.

Heatup and Steam Spalling

A 5450-kilo induction furnace holder required 400 kilos of a dense, high-purity 90% Al_2O_3 castable in the dome. Molten brass was poured from melters to the holder through pour tubes in the dome. During the last year, refractories containing the copolymer fiber were used in the dome. Table III summarizes the dome refractory details. In the above unit, preheat and fur-

Table III. Heatup/Steam-Spalling Analysis in Commercial Induction Furnace

	High-Purity, 90% Al$_2$O$_3$ Castable	High-Purity, 90% Al$_2$O$_3$ Castable with Fiber Addition
Dome preheat	260 °C (500 °F) for 3 weeks	260 °C (500 °F) for 4-5 days
Heatup on furnace	55 °C (100 °F)/h	167 °C (300 °F)/h
Activity (482 °C–649 °C)	Steam spall or crack	No steam spall or crack
Typical life	4-5 weeks if no steam spalling	Other furnace problems dictate life

nace heatup times were minimized, and the integrity and service life were maximized.

Metal Handling

The previously mentioned laboratory aluminum tests with metal contact refractories containing copolymer fibers indicate excellent compatibility with metal. Moreover, a commercial aluminum induction furnace has successfully operated with a high-purity 90% Al$_2$O$_3$ castable lining with these fiber additions.

Originally, the induction furnace was lined with a 90% Al$_2$O$_3$ ramming mix especially designed for aluminum applications. On initial heatup, the lining cracked badly and was replaced with the fiber-containing, high-purity 90% Al$_2$O$_3$ castable. Sidewalls and the inductor were cast with the fiber-containing castable and subjected to a similar heatup. The original lining with the fiber addition has now run successfully for over one year.

Crack Minimization

Figures 5-7 showed the reduction of steam spalling and cracking in a copolymer fiber-containing castable after a catastrophic heatup. In addition, field evidence indicates a reduction of standard craze and dryout cracks after conservative air drying and heating procedures with the use of the fiber additions.

In January 1984, an industrial incinerator was gunited with an extra-strength, low-iron, fireclay gunning mix. Approximately 50% of the primary combustion chamber was gunned with a fiber-free material, and the remainder was gunited with the copolymer fiber additive. After an ambient air cure and a low-temperature heatup, the material with the additive appeared to be more monolithic, exhibiting less than the typical amount of craze cracking found in the fiber-free material. The material was installed in a temperature range of 0 °C (32 °F) to 4 °C (40 °F).

In 1982, a large 15.2-m diameter by 24.4-m high petrochemical vessel was gunned with a high-purity, high-strength, fireclay gun mix with the addition of the copolymer fibers and stainless steel fibers. The unit was inspected before and after a controlled cure and dryout procedure. After one year of service, the refractory contractor's inspection showed an unusual lack of dryout and shrinkage cracks.

Finally, the qualitative aspect of crack minimization was quantified by a Midwest contractor. After using a high-purity, 90% Al_2O_3 castable with the copolymer fiber addition in a channel induction furnace throat lining, the lack of craze or dryout cracks indicated a reduction in crack propagation and spalling. The contractor estimated that the throat lining life is 100% longer with the additive.

Conclusion

The state of the art is rich with approaches and patents on the use of nonmetallic additions to promote moisture release in refractory castables and gunning mixes. The proprietary organic copolymer under evaluation combines unique shrinkage characteristics with appropriate dimensional and combustion properties to control steam spalling and to allow heatup acceleration. Laboratory tests involving explosion block analysis, internal pressure measurement, and permeability studies indicate that the efficiency of moisture removal is increased. In addition, such advantages as cold weather insurance and crack minimization, without significantly affecting physical properties and metal handling, are indicated by steam-spalling tests after various ambient conditions and metal cup tests.

In summary, the above work depicts the formation of a network of fine channels within a refractory system. Although the channels increase permeability, the increase has a minimal effect on measured physical properties. More importantly, the permeability is sufficient to increase moisture removal, thus minimizing the bond breakage and crack formation that routinely occur in standard castable linings during ambient cures and high-temperature dryouts. The qualitative field data indicate that the above benefits are typically realized in actual operation, with the added benefit of longer service life.

Acknowledgments

The authors would like to thank George MacZura and Frank Rohr of Alcoa Technical Center and Dr. Jesse Brown of Virginia Polytechnic Institute for their contributions in conducting some of the steam-spalling testing reported in this work.

References

[1]T. D. Robson; pp. 51-64 in High Alumina Cements and Concretes. Wiley & Sons, New York, 1962.

[2]S. J. Schneider, "Effect of Heat Treatment on the Constitution and Mechanical Properties of Some Hydrated Aluminous Cements," *J. Am. Ceram. Soc.*, **42** [4] 184-93 (1959).

[3]P. K. Mehta, "Retrogression in the Hydraulic Strength of Calcium Aluminate Cement Structures," *Miner. Process.*, **2** [11] 16-19 (1964).

[4]G. V. Givan, D. Hart, P. Heilich, and G. MacZura, "Curing and Firing High Purity Calcium Aluminate-Bonded Tabular Alumina Castables," *Am. Ceram. Soc. Bull.*, **54** [8] 710-13 (1975).

[5]D. L. Hipps and J. J. Brown, "Measurement of Internal Pressure in High Alumina Castables During Curing"; for abstract see *Am. Ceram. Soc. Bull.*, **61** [3] 395 (1982).

[6]W. H. Gitzen and L. D. Hart, "Explosive Spalling of Refractory Castables Bonded With Calcium Aluminate Cement," *Am. Ceram. Soc. Bull.*, **40** [8] 503-07, 510 (1961).

The Development of Dry Refractory Technology in the United States

JOHN L. TURNER, JR. AND DAVID M. MYERS

Allied Mineral Products, Inc.
Columbus, OH 43220

Early growth in the use of electric, coreless induction furnaces in the U.S. foundry industry during the 1960s placed a new set of demands on refractory technology and application procedures. The unusual combination of an electrically powered, water-cooled coil, and a bath of molten metal, separated by only a few inches of refractory, required careful design and quality control of the refractories. Also, conventional refractory designs of mortared brick, wet rams, or castables presented a number of physical and logistical problems that reduced the effectiveness of the coreless induction furnace as a production tool. The combination of these factors gave rise to a new physical form of refractories known as dry rams, now commonly known as dry-vibration refractories. The discussion that follows will describe what dry refractories are, their advantages, and their history of development and application in the United States.

Definition and Description of Dry Refractories

Dry refractories are monolithic refractory materials that contain absolutely no water or liquid chemical binder. As are other monolithic refractories, they are formed on-site. Moreover, the dry refractories do not depend on the addition of water or other chemicals to produce a wet, green-strength bond, hydraulic bond, or other types of room-temperature bonds or strengths. Dry refractories rely on heat setting or ceramic sintering mechanisms. Depending on the application, dry refractories can be designed to begin developing strengths at temperatures as low as 200°C or as high as 1100°C. Final sintering can be designed to occur at temperatures from 425°–1800°C. The temperature at which an initial sinter or initial strength takes place is dependent on the type and amount of bond added, which, in turn, is dependent on the temperature of the operation and whether a consumable form, a removable form, or an application with no form is used. The dry refractory should never have more bond than is required for an application. It is desirable to maintain as much unsintered material as possible in the cooler areas of the refractory structure to provide better relief from mechanical and thermal stresses.

One of the critical aspects of dry refractory design is grain-size distribution. The maintenance of a tightly controlled grain-size distribution allows for the installation of dry refractories with a minimum of mechanical effort; thus, installation by the use of external vibration methods, as well as by direct ramming, results in a uniform, high-density mass. Maximum densification of dry refractories is dependent both on the quality control of the manufacturer and on proper handling techniques of the user.

History of Dry Refractory Development

Dry refractories have been developed and improved in response to the increased size of induction furnaces, particularly the coreless induction furnace. The first coreless furnaces in the United States were small and utilized wet ramming mixes and prefired crucibles. Since the furnaces were small, the wet ramming mix could be properly dried in a reasonable time, and the crucibles were readily available and easy to manufacture. As coreless furnaces became larger and started to replace cupolas for ironmelting, the limitations of the wet ramming mixes and crucibles became more apparent. As the furnaces became larger, it also became more difficult and more expensive to obtain crucibles, and the time necessary to properly dry wet ramming mixes severely interfered with the furnace's ability to be used as a production melting tool. Four or more days of downtime, coupled with problems due to rushing the dryout, gave birth to the use of dry ram refractories.

Early dry rams were actually very low moisture wet rams. It was soon discovered that completely dry refractories rammed around a melt-out form would develop adequate sintered strength on the hot face before the form softened and melted in. The initial refractory used for this purpose was a high-purity quartzite from Sweden, mixed with a boric acid bond that acted as a flux to aid in hardening of the hot face.

In the early 1960s, the dry-ram silica was installed by hand-tamping around the melt-out form. The installation time of 16-24 hours for a medium-size furnace was not any better than that for wet ram refractories, but the drying time was eliminated. In the mid-1960s, the Bosch electric tamper was introduced. This ramming/vibrating tool reduced installation time to eight hours or less. Further improvements were made during the 1970s with the development of internal, form vibration techniques that utilized pneumatic and electric vibrators. Direct ramming of the refractory was virtually eliminated, and installation time was further reduced to four hours or less for most medium-size furnaces. It should be noted that the above developments occurred with high-purity silica, which was being manufactured domestically by the mid-1960s.

The success of the Bosch tamper for installing silica caused U.S. refractory manufacturers to pursue the use of this tool on other chemical forms of dry ram refractories. By the mid-1970s, alumina, magnesia, zircon, and mullite dry refractories were being installed in coreless furnaces using both the Bosch and the internal form vibration techniques. It has become possible to completely turn around the largest coreless furnace in less than two days.

The channel type of induction furnace has been in use in the United States since the 1930s, primarily in the nonferrous metal industry. Its use in iron foundries started developing during the 1960s, with significant advancements occurring in the late 1960s and 1970s. Conventional refractories were originally used in the construction of all areas of channel furnaces. Brick and mortar were used in the upper case sidewalls, and wet ramming mixes for the throat and floor. Inductors were lined with either wet ram alumina mixes or magnesia castables. In nonferrous applications, alumina wet rams and castables were predominantly used. Plastics and gunning mixes were used extensively for refractory maintenance, as they are today. Starting in the late 1960s, the advantages observed with dry-vibrated refractories in coreless furnaces led to their application in channel furnaces.

Typically, channel furnaces are designed with detachable inductors that can be prepared separately and then kept on standby for attachment to the furnace when required. Therefore, the limiting factor in the rapid turnaround of a channel furnace has been the removal of moisture from refractories in the floor, throat, and upper case. Since the throat refractory is changed more often than the sidewalls, it became the area where dry vibratables were first applied.

Successes in the throat with high-alumina dry mixes led to their use in the floor and sidewalls of vertical-style channel furnaces. Dry refractories in vertical channel furnaces have been most successful when replacing castable refractories. Since castables contain 7–10% moisture, they require 6–12 days for complete curing and sintering. Dry refractories have reduced that time to three days, resulting in the inductor becoming the limiting factor in turnaround time.

Most inductors lined with magnesia castables or wet ram aluminas require a minimum of three days for curing and sintering after they are attached to the furnace. To reduce this time, dry alumina inductors were tried several years ago with good success in nonferrous applications, but only moderate success in ironmelting. Approximately five years ago, U.S. refractory manufacturers began working on magnesia-based dry refractories due to their success in Europe. The dry magnesias have been applied successfully in all types of inductors melting iron, and the new application techniques have led to the successful application of dry aluminas in inductors as well.

Most dry refractories in vessels other than coreless furnaces have been installed using a conventional Bosch electric tamper. However, over the last three years, substantial use has been made of external pneumatic vibrators for densification of the refractory in inductors. The principle is the same as with coreless induction furnaces except that the vibrator is attached externally rather than mounted on the form. Experimentation continues to apply the technique to throat, floor, and upper case sections of the vertical channel furnace. Dry refractories installed with these new vibrating procedures make it possible to bring in a new inductor refractory lining within 24 hours or complete a hot inductor change in only a few hours.

The ability to successfully apply dry refractories to channel furnaces is directly related to the development of low- and medium-temperature ceramic binders which permit the use of removable forms. Since the various parts of channel furnaces do not have the cylindrical shape of coreless induction furnaces, the forms used for installing wet rams and castables are expensive to construct, which makes the use of melt-out forms cost-prohibitive. By utilizing small amounts of low- and medium-temperature binders, the dry refractory may be hardened by low-temperature heat (usually less than 550°C) which allows forms to be heated without damage. The forms may then be removed, since the refractory is strong enough to support itself.

Applications beyond Induction Furnaces

The ability to obtain structural strength in dry refractories at low temperatures has enabled their use in applications other than induction furnaces. The first area of expansion of dry refractory use came about during the mid-1970s, when they were first applied to ladles. As long as a suitable steel form is used, dry refractories may be applied to ladles using conventional ramming or vibrating techniques. The complete absence of moisture allows turnaround in just a few hours. All heat applied to a ladle is used primarily to raise the temperature in preparation for use and not for removal of moisture. A castable refractory with 8% water replaced with a dry refractory results in a 30% energy saving.

Application of dry refractories has been successful in the aluminum industry, where they have replaced castables, ramming mixes, and brick linings in reverberatory and other gas-fired and resistance-heated furnaces. There are indications that the low-temperature, ceramic binder produces a lower porosity hot face, which has led to less dross and metal buildup during the furnace campaign.

In 1981, dry refractories were introduced in U.S. blast furnace casthouses in troughs as well as in iron and slag runners. The dry refractories used generally consist of fused alumina with silicon carbide and carbon additions, as well as low-temperature, ceramic binders.

The first successful application of dry refractories was at the iron and slag runner joints with removable troughs. The dry refractories were then applied to the furnace joint of the main trough. The joints had been a difficult problem since they are constructed at the same time a new trough is brought on-line. Prior to dry refractories, wet graphitic plastics were used which required several hours of drying, and often complete drying did not occur. Insufficient drying led to joint deterioration due to steam evolution and runouts would occur. The dry joints eliminated this problem, and their term of service was doubled, allowing them to run the life of the trough.

The next successful application of dry ram refractories in the casthouse was in the iron and slag runners. By using external vibration of the bottom as well as internal vibration of the removable runner forms, installation time was greatly reduced compared with conventional wet rams. Furthermore, the dry ram refractories have the ability to be shave-patched, that is, only the hot face needs to be removed and new dry material once again installed and vibrated around the removable forms.

The advantage of rapid turnaround time has made dry ram refractories of particular interest in single-taphole blast furnaces. They have been successfully applied in the bottoms of single-taphole main troughs and have been shave-patched successfully. Dry ram refractories have shown economical performance in most casthouse applications where they have been tried.

Further investigation of the steel industry has found dry ram refractories applied in many soaking pit floors and reheat furnace hearths. Bulk handling of the dry refractories, along with improved vibration techniques, enable rapid installation of large hearth areas. Initial reports indicate that the dry ram refractories equal the performance of the wet rams, plastics, and brick they have replaced and reduce the downtime of these furnaces.

Summary

Conventional wet rams, castables, and brick and mortar refractories require tedious installation, long drying and curing times, or both. Improvements in the technology gained from the use of dry ram refractories in coreless induction furnaces has led to their use in many other applications, and the advantages of fast turnaround, energy savings, and improved performance have been realized.

A Review of International Experiences in Plastic Gunning

Leonard P. Krietz, Glenn Wilson, Dieter Hofmann, and Mitsuaki Tsukino

Plibrico Co.
Chicago, IL 60614

In 1980, a technique was developed whereby gunning of plastic refractories is possible without the addition of water at the nozzle. Since then, widespread field experiences have been gained around the world. These experiences are discussed, along with performance histories on the completed jobs.

The pneumatic gunning of plastic refractories is a new installation method that combines the ease of installation and high production rates associated with conventional gunning, and the physical property advantages and bond system choices available with plastic refractories. Plastic gunning allows plastic refractories to be placed pneumatically without added water at the nozzle. Material consistency and moisture content are controlled at manufacture. This method allows for uniform, high placement rates of consistent quality material.

The gunning of plastic refractory materials began in Germany in early 1979 when a granular, carbon SiC-based ramming mix was pneumatically gunned to repair a blast furnace runner. The success of this method of installation led to further development. In 1980, trials were carried out in Germany and Austria gunning soaking pits and reheat furnace walls with superduty and 70% air-bonded, plastic materials. The continued successful application of plastic materials in this manner in Europe led to field trials in Japan in 1981, and in Canada and the United States in early 1982.

Initial trials in these countries met with much success and, to date, in excess of 15 000 tons of refractory plastic has been successfully gunned worldwide. Gunned plastic refractory installations now exist that have been in service in conjunction with conventional rammed installations for four years, and the service performance of the gunned material has been comparable to that of the rammed plastic. Applications have included soaking pit walls and covers, reheat furnace roofs, walls, flues and hearths, aluminum reverbatory furnace sidewalls and hearths, rotary kilns in cement and mineral processing industries, boilers, and municipal incinerators, to name a few.

Advantages

The advantages of plastic refractories installed by gunning are (1) high material placement rates, (2) reduced man-hours for installation, (3) simplified installation logistics, (4) uniform installation with no lamination, and (5) good physical properties. Typically, rammed plastic is installed at an average of 0.5–0.7 m^3 per man per day (assuming an 8-h shift). This rate includes both bricklayers ramming the plastic and support laborers. Gunning installation rates for a plastic gunning crew (typically six men) are between 1.0 and 1.4 m³ per man per day. This is a 100% increase in installation rate. Installation logistics are simplified, since the gunning equipment and material can be placed in a convenient location up to 60 m horizontally and 12 m vertically from the job site, an obvious advantage for difficult to reach locations. Figure 1 shows what happens when a plastic is not adequately rammed. Note the visible laminations present after firing, caused by an improper installation technique that did not allow the individual slabs of plastic to knit together. This problem will not occur with gunned plastic installations. Laminations are virtually eliminated, since the plastic, during gunning, is essentially applied very uniformly. The physical properties of gunned plastic are also equivalent to the properties that can be obtained by conventional ramming techniques, which will be covered more fully later.

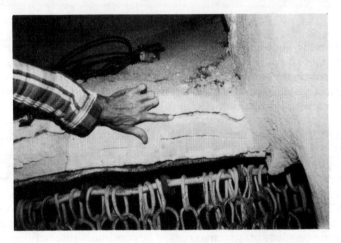

Fig. 1. Laminations present in rammed plastic installations due to improper installation technique.

Equipment and Processes

Figure 2 is a schematic of a typical plastic gunning system, which consists of a shredder-granulator, a feed conveyor, and an extensively modified rotary gun. The shredder-granulator reduces the charged refractory to pieces suitable for introduction to the gun. These pieces are 1.25–2.00 cm in diameter. The conveyor has a variable speed control and regulates the amount of feed (and hence the installation rate) to the gun. Several types of rotary guns have been successfully used, but they do require modifications to the drive and feed mechanisms and to the hopper arrangement to make them suitable for gunning plastics. A system, therefore, can be as simple as a shredder, conveyor, and gun. However, the gunning equipment can be very sophisticated, such as the equipment shown in Fig. 3. This equipment, currently being used

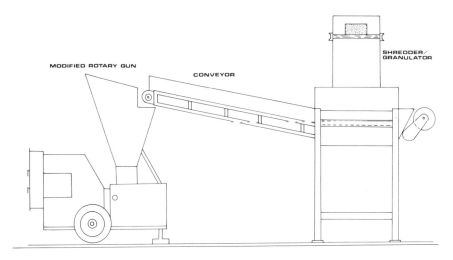

Fig. 2. Plastic gunning equipment arrangement.

in Japan, is automated to allow the nozzleman to start, stop, and control the installation rate remotely. If a granular plastic material is being used, it is possible to forego the shredder and charge the material directly into the gun. In this case, further modifications to the gun are also required.

Naturally, in plastic gunning it is necessary to have sufficient air pressure and volume. The momentum of the refractory exiting the nozzle section of hose is very important for good consolidation and placed density. An adequate compressor is therefore essential. Compressor requirements can vary from job to job, depending on the logistics of the installation and the type of gun being used. Typically, compressor volume ranges between 20 and 30 m^3 air/min at 7-8 bar air pressure.

Fig. 3. Automated plastic gunning equipment.

The hose size used from the gun is typically 5 cm in diameter with a short, 3.8 cm-diameter nozzle section attached. Since no water is added during gunning, no separate nozzle is required. This gunning system is capable of delivering material at high rates. Actual job rates are lower due to installation, rebound, and maintenance considerations. It is practical for a 6-man crew to install 18–22 tons of plastic refractory in an 8-h shift.

Plastic Refractory

Plastic refractories with both clay-air bond and phosphate-bond systems have been successfully gunned. These plastic materials can be manufactured in both granular and extruded forms. Currently, granular materials are being gunned in Europe and Japan, whereas extruded materials are used in Canada and the United States, mostly due to the type of equipment being used in the respective locations.

In all cases, the plastics used for gunning applications are based on conventional plastics that are modified for gunning use. While it is possible to ram a plastic designed for gunning, experience has shown it is not possible to install an off-the-shelf conventional plastic by gunning. A conventional material will cause buildup in the gunning system and clog the hose in a very short time.

Rebound loss is to be expected when gunning plastics, and formulas are designed to minimize rebound. The percentage of rebound expected on a job varies according to installation parameters and lining design considerations. Generally, it will be 15–20%; however, plastic gunning rebound, unlike rebound associated with conventional castable gunning, can be recycled. With recycling, it is possible to reduce the amount of throw-away rebound to 5–15%.

As was briefly mentioned previously, the physical properties of gunned plastics are very good, evidenced by results obtained on specimens cut from

Table I. Superduty Plastic

	Gunned	Field-Rammed
Weight in service (kg/m^3)	2050	2019
Linear change (%)		
815°C	−0.5	−0.7
1090°C	−0.6	−0.7
1340°C	+0.5	+0.2
Cold MOR (MPa)		
110°C	3.2	2.3
815°C	2.9	1.9
1090°C	4.5	3.5
1340°C	6.3	5.2
Hot MOR (MPa)		
815°C	4.3	3.2
1340°C	1.1	0.8

large gunned panels. For conventional plastics, physical properties are obtained from specimens pressed to 6.89 MPa in the laboratory. For comparison, large panels of plastic were rammed to simulate a field installation. Samples were then cut from these panels in a manner similar to the gunned samples. Table I compares the physical properties obtained from gunned, superduty plastic vs a field-rammed, superduty plastic. As can be seen, the properties are very similar, with the hot strengths of the gunned samples being slightly better. A similar comparison is made with a 70% alumina phosphate-bonded material (Table II).

Table II. 70% Alumina Phosphate-Bonded Plastic Refractory

	Gunned	Field-Rammed
Weight in service (kg/m^3)	2211	2242
Linear change (%)		
815°C	− 0.2	− 0.4
1090°C	− 0.3	− 0.4
1340°C	+ 0.2	+ 0.1
Cold MOR (MPa)		
110°C	4.0	4.4
815°C	5.1	5.5
1090°C	6.5	7.8
1340°C	12.8	13.5
Hot MOR (MPa)		
815°C	6.3	6.2
1340°C	4.1	4.3

Creep tests have also indicated that gunned plastic exhibits less creep than conventional rammed materials. Figure 4 shows the results obtained on a gunned, superduty plastic vs a rammed one. In this case, the gunned material exhibited approximately 50% less creep than its rammed counterpart.

Recent developments have led to a medium-weight, insulating, superduty gunnable plastic. The physical properties are shown in Table III. This material exhibits very good physical properties and offers a weight advantage; it is approximately 25% less dense than its gunned, superduty counterpart. Thermal conductivity testing has shown that the K factors of this material are 30% lower than its denser gunned counterpart.

Installation Review

The actual gunning installation of plastic is very similar to that of conventional gunned castable. The nozzleman's job is much simplified; since he need not be concerned with controlling water additions, he can concentrate

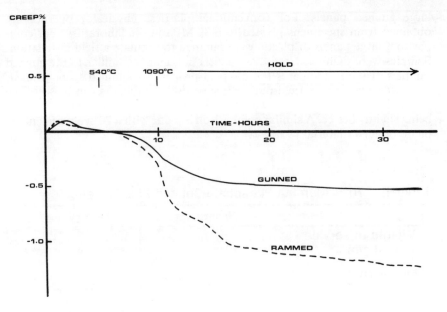

Fig. 4. Superduty plastic (0.17 MPa load).

Table III. Medium-Weight Superduty Plastic

	Gunned
Weight in service (kg/m³)	1618
Linear change (%)	
815°C	−0.6
1090°C	−0.8
1340°C	−1.0
Cold MOR (MPa)	
110°C	2.1
815°C	2.2
1090°C	5.8
1340°C	5.7
Hot MOR (MPa)	
815°C	2.2
1090°C	3.3

more fully on installation techniques. Lining designs and anchoring systems are also very similar to those of rammed installations. With proper precautions, it is possible to gun against board insulation and ceramic fiber, as well as against insulating firebrick and castables. Figure 5 shows how well gunned plastic consolidates around an anchor tile. The impression in the plastic panel of the anchor tile ribs is very evident after removal of the anchor. Figure 6

Fig. 5. Anchor tile removed from gunned plastic panel.

Fig. 6. Simulated plastic roof construction supported solely by anchor tiles.

shows a roof construction simulation. The gunned plastic is supported solely by the anchor tiles on 30 cm centers. Conventional plastics are installed overhead by gunning perpendicular to the hot face, which allows all anchor tiles to be set before the installation. Figure 7 shows the preparatory work for a

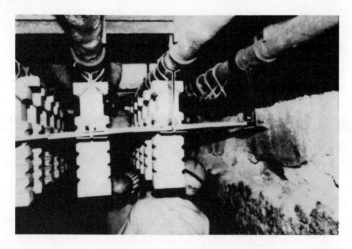

Fig. 7. Preparatory work prior to gunning plastic in a reheat furnace roof.

roof installation in a reheat furnace. Note that the anchors are in place, with a board backup against which to gun. After gunning, the backup boards will be removed, and insulation cast into place. It is also possible to install the insulation prior to gunning and to gun against it rather than boards. Figure 8 shows the plastic gunning of the roof. This particular roof section required approximately 55 tons of superduty plastic which was installed in approximately 18 h.

Another example is an aluminum reverberatory furnace upper sidewall. Figure 9 shows the anchor and board insulation installation. Figure 10 shows the plastic gunning installation underway. Note the absence of dusting. The finished installation prior to bakeout is shown in Fig. 11, and after bakeout in Fig. 12.

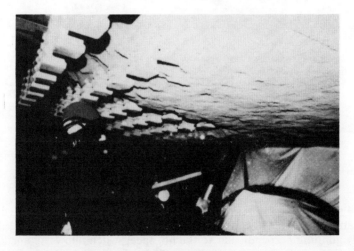

Fig. 8. Plastic gunning of a reheat furnace roof section.

Fig. 9. Aluminum reverberatory furnace upper sidewall prior to plastic gunning with block insulation and anchors in place.

Fig. 10. Plastic gunning of aluminum reverberatory furnace upper sidewall (no dusting).

Fig. 11. Plastic gunned aluminum reverberatory furnace upper sidewall prior to bakeout.

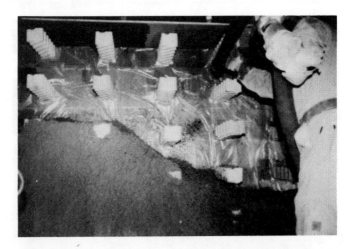

Fig. 12. Plastic gunned aluminum reverberatory furnace upper sidewall after bakeout.

In summary, the pneumatic gunning of plastic refractory material is a very viable alternative to conventional refractory installation techniques. The main advantage is that a quality installation is possible in less time and at less cost than with conventional ramming practice. To date, gunned plastic installations are performing as well as conventional rammed ones in a wide variety of applications around the world.

Viscosity and Gunning of Basic Specialties

WALTER SIEGL

Österreichisch Amerikanische Magnesit AG
Radenthein, Austria

Basic gunning mixes have been used for a long time for repair and maintenance of metallurgical furnaces and vessels. The largest consumers are EAF, basic oxygen furnaces, and tundishes; a smaller amount of gunning is done in iron mixers, steel ladles, open hearth furnaces, and in other units in the nonferrous industry.

The present quality of the basic gunning mixes is mainly a result of a learning process between field experience and laboratory development. Very often, theoretical understanding of the gunning process and gunning mixes remained behind practical experience. Some points of gunning technology are now understood; however, there still remain many dark areas which have to be explained. Many of the factors affecting the gunning process are independent of the gunning mix quality and can be called "external factors"; to this group belong the influences of gunning equipment, handling, temperature, and surface quality of the wall. There is plenty of literature describing these phenomena, and a key to the specific reference that deals with each of the external factors is given in Table I. The second group of factors, called "internal factors," includes the properties of a gunning mix that are necessary for its successful application and, again, a key to the reference literature is given in Table II.

From this summary, it is obvious that most of the authors concentrate, in addition to the mix compositions, on the physical properties of a gunned layer, including its adherence to the lining. In respect to the actual gunning,

Table I. External Factors Affecting the Gunning Process with Key to Reference Literature

External Factors	References
Gunning machine	1–3
Gunning rate	4
Air pressure	2–5
Water addition	2, 4–7
Water temperature	1, 6, 8, 9
Length of nozzle	4
Handling of the nozzle	3, 4
Distance from the wall	4, 5, 7
Gunning angle	4, 7
Wall temperature	4–7
Adherence of mix to the lining	7–15
Furnace atmosphere	7, 10
Metallurgical conditions	7

Table II. Internal Factors Affecting the Gunning Process with Key to Reference Literature

Internal Factors	References
Phase composition, chemical analysis	1, 5, 7, 8, 10-12
Granulometry	5, 7-11, 16
Binder	1, 5, 7-9, 11
Plastifier	1
Rheological properties, viscosity	16-18
Setting speed	10, 16
Bulk density, porosity	1, 7, 9, 16
Strength (cold crushing strength, hot crushing strength, HMOR, hot shear strength)	7, 8, 10, 12, 16, 17
Refractoriness (refractoriness under load, deformation under load, permanent linear change)	1, 7, 8, 10
Spalling	7, 10

the interest of most of the authors is limited to the optimization of some of the external factors. In most cases, the evaluation of gunning mixes covers the determination of the "as-received" properties as well as properties after drying and firing.[19] Regardless of how important those properties could be, they do not reveal anything about what is happening during the gunning process. To optimize the gunning process, more sophisticated cold or hot gunning tests have to be performed in the laboratory[1,3,5,7,8,10,18-20] or one has to predict the gunning behavior from simpler tests of the consistency (e.g., ball-in-hand test or flow test). The study of the rheological properties of mixes has the advantage that the viscosity, which is the most relevant value of the consistency, can be measured directly. These results can be used for formation of models which physically describe some details of the gunning process and thus allow a more theoretical understanding of them.

To the best of our knowledge, only recently are attempts being made to shed some light on this problem by investigating the relevant rheological parameters of gunning mixes. This is especially important for the manufacturers of basic gunning mixes, who have little influence on the external factors. Therefore, the products frequently have to be designed to perform even under unfavorable field conditions.

Theoretical Approach

Theoretical models seldom are perfect pictures of reality. In our case, the models cannot replace practical experience and observation; however, they allow a better understanding and analysis of the behavior of mixes during gunning, thus offering the basis for their optimization.

Let us assume that the mix consists of a high-viscous fluid and coarse particles. The fluid itself is a suspension of fine particles in water. Obviously, it is difficult to decide whether particles of intermediate sizes belong to the fluid suspension or to the coarse fraction. More or less arbitrarily, we placed the division line at 0.2 mm (65 mesh) (Fig. 1).

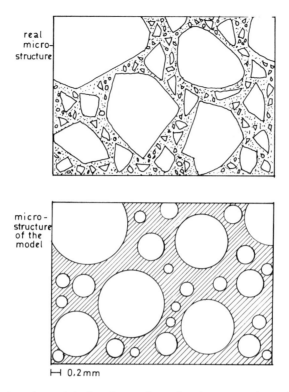

Fig. 1. Real and model structure of a refractory gunning mix.

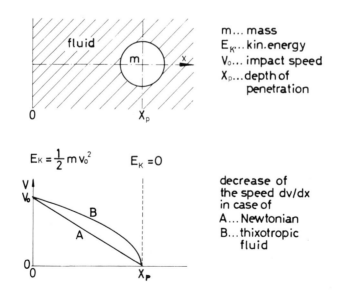

Fig. 2. Impact model of the grain in fluid.

Role of the Viscosity during Grain Impact

Impact of the Grain in Fluid: We are asking the question of how the kinetic energy $E_K = 1/2\, m \cdot v^2$ of the coarse particle dissipates when it enters the viscous suspension (Fig. 2).

When the coarse grain penetrates the distance, x_p, the kinetic energy is transformed into the work of deformation, E_D, while overcoming the resistance, W, of the fluid. Since Reynolds number $\mathrm{Re} = v \cdot d \cdot \rho / \eta$ does not exceed 1000, the stream of the fluid stays laminar. In this case, the fluid resistance W is given (according to Stokes) by:

$$W = 6\pi\eta r v \tag{1}$$

and the work of deformation E_D is:

$$E_D = \int_0^{x_p} W \cdot dx = 3\pi\eta r v_o x_p \tag{2}$$

From $E_D = E_K$, the depth of penetration, x_p, can be calculated:

$$x_p = \frac{m\, v_o}{6\pi\eta r} = \frac{2\rho_g r^2 v_o}{9\eta} \tag{3}$$

where: W is fluid resistance (N), m mass of the impacting grain (kg), η dynamic viscosity (Pa·s), r radius of the grain (m), v speed (m/s), v_o impact speed (m/s), and ρ_g density (kg/m³).

This theoretical model differs from reality for a number of reasons: (1) real particles are not spheres; (2) different conditions, immediately at the entry into the fluid, were not considered (e.g., smaller resistance); (3) the speed does not decrease linearly but progressively in a thixotropic fluid; and (4) the viscosity is not constant, but depends on the shear rate, and thus, on the speed of the particle itself.

The dissipation of the energy can be described more accurately using the formula

$$\frac{1}{2} m v_o^2 = 6\pi r \int_0^{x_p} \eta v\, dx \tag{4}$$

This equation is not very practical, however, because of the complexity of the functions $\eta(x)$ and $v(x)$.

These considerations show that low viscosity at the surrounding of the impacting grain is important for deep penetration and good packing of the grain in the structure of the gunned layer.

Impact with Grain-Grain Contact: In this case, the question is that of how the kinetic energy of the grain dissipates when it hits another particle or the hard lining surface. If there is no fluid present between the moving particle and the target, we talk about an elastic impact, and the grain will rebound (Fig. 3). If there is a liquid phase present between the grain and the target area, the kinetic energy is used for the deformation of the fluid, and we talk about the plastic impact. We describe this case by replacing the spherical particles (which were previously considered in our model), with cylindrical particles whose flat surfaces meet (Fig. 4).

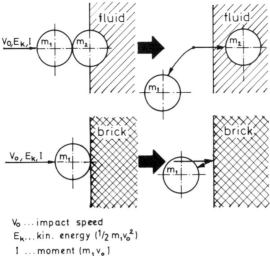

V₀ ... impact speed
E_k ... kin. energy ($1/2\, m_1 v_0^2$)
I ... moment ($m_1 v_0$)

Fig. 3. Elastic impact model (grain-grain).

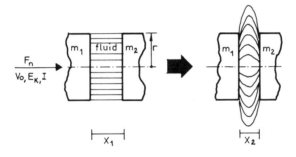

Fig. 4. Plastic impact model (grain-fluid-grain).

The force, F_n, perpendicular to the flat surfaces,[21] equals:

$$F_n = \frac{3\eta v r^4 \pi}{2x^3} \qquad (5)$$

This force is indirectly proportional to the third power of the distance between the two surfaces. The energy, E_d, required for deformation of the fluid when decreasing the distance from x_1 to x_2 can be calculated by integrating this force:

$$E_d = \int_{x_1}^{x_2} F_n dx = \frac{3\eta r^4 \pi}{2} \int_{x_1}^{x_2} \frac{v(x)}{x^3} dx \qquad (6)$$

The work required for squeezing the fluid out of the gap is directly proportional to the viscosity of the fluid and increases with decreasing distance

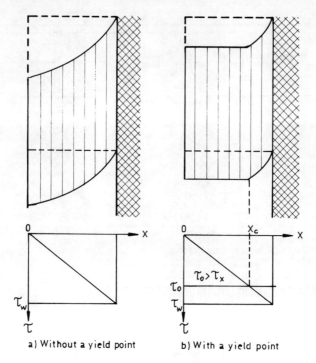

a) Without a yield point   b) With a yield point

Fig. 5. Model of a subsiding gunned layer.

between the two surfaces. A higher kinetic energy and a lower viscosity result from a closer approach of the particles. Since the kinetic energy is limited, a direct contact will never be reached. Only the contact of edges and corners can be imagined, but this is not very probable because the approaching particle tends to turn parallel to the target surface.

Role of Viscosity for Adherence of the Gunned Layer

Under certain conditions the gunned layer will slump from the vertical walls (Fig. 5). The shear stress τ_x at the distance x from the hot face is given by:

$$\tau_x = x \rho g \tag{7}$$

where x is distance from the hot face (m), ρ bulk density of the gunned layer (kg/m³), and g gravitation constant (9.81 m·s⁻²).

Without a yield value, a fluid will always flow down. The shear rate is $D_x = \tau_x/\eta$; and the flow speed at x will be

$$v_x = \tfrac{1}{2} Dx = \frac{x^2 \rho g}{2\eta} \tag{8}$$

The higher the viscosity, the slower will be the speed of the moving layer.

In most cases, however, thixotropic fluids have a yield value. A yield value exists when the viscosity η_0 is very high at a very low, almost zero, shear rate D_0. In this case, the resulting shear stress, τ_0, has a measurable value. As long as this shear stress, τ_0, is greater than the shear stress, τ_x, at the distance x, the gunned layer is stable. There exists however a critical thickness of the gunned layer, x_c, which should not be exceeded.

$$x_c = \tau_0/\rho g \qquad (9)$$

The mix in our model does not consist only of the fluid phase (suspension of magnesia fines in water), but comprises also coarse particles. The actual viscosity of this system will be higher than the viscosity of the pure fluid. Several formulas describing this correlation can be found in the literature. Some of them are listed in Table III, where the actual viscosity η is presented as a function of the viscosity of the fluid, η_f, and the volume concentration φ of the coarse particles.

Table III. Actual Viscosity of Fluids Comprising Coarse Particles

Formula	Author
$\eta = \eta_f (1 + 2.5\varphi)$	Einstein
$\eta = \eta_f (1 + 2.5\varphi + 10.6\varphi^2)$	Eirich, Riesemann
$\eta = \eta_f (1 + 2.5\varphi + 12.6\varphi^2)$	Simha
$\eta = \eta_f (1 + 2.5\varphi + 14.1\varphi^2)$	Mark
$\eta = \eta_f (1 + 2\varphi)/(1 - \varphi)$	Franck
$\eta = \eta_f \exp (2.5\varphi/(1 - k\varphi))$, $1 < k < 2$	Mooney

Table IV. Approximate Viscosity Values of Basic Mixes

Volume concentration of grain (φ)	0.4	0.5	0.6
Relative viscosity (η/η_f)	4 ± 1	5 ± 1	6,5 ± 1

At volume concentrations higher than 0.6, the fluid cannot completely fill the available interspace between the coarse particles. The formation of air-filled pores will impede further increase of the viscosity of a mix.

The Real Structure of Gunning Mixes

From the theoretical models described above, the following basic conclusions concerning the process of gunning and the structure of the gunned layer can be drawn:

(1) The rheological properties of the fluid (the thixotropic behavior of the water-fines suspension) form the basis for the adherence of the particular grains and for the stability of the gunned layer.

(2) A high viscosity is required for good stability of the gunned layer. This is accomplished by the amount of coarse grains and by the presence of a yield value of a thixotropic fluid.

(3) A low viscosity is required for a proper penetration of the impacting grains (low rebound).

(4) A low viscosity is also required for a proper packing of the approaching particles.

(5) A mosaic-type structure will be formed due to both the practical impossibility of direct contact between individual grains and their tendency to orientate parallel to each other (Figs. 6 and 7). To achieve a dense packing, a gunning mix requires more fines than a dry-pressed brick. If the gunning mix is not designed properly, the excess coarse grains will rebound during gunning.

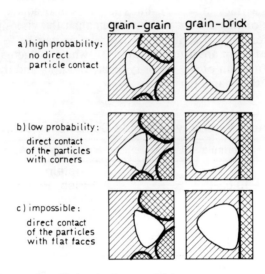

Fig. 6. Packing of particles in a gunned layer.

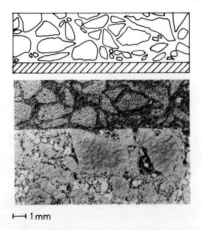

Fig. 7. "Mosaic-type" microstructure of a gunned layer.

The requirements of points (3) and (4) are met since a thixotropic fluid needs a certain time to develop a high viscosity. Therefore, the viscosity at the outside surface will always be low during gunning. Furthermore, the high shear rate in the immediate surroundings of the moving particle, as well as in the joints between two approaching grains, will result in a significant decrease in viscosity.

Experimental Procedures

Determination of the Viscosity

A rotational viscosimeter (Fig. 8) was used for our measurements. On the viscosimeter, a specific number of rpm can be set, and a torque moment detector records a value, S, which is directly proportional to the shear stress. The viscosity (mPa·s) can be expressed by the formula $\eta = G \cdot S/n$, where G is a constant of the measuring system.

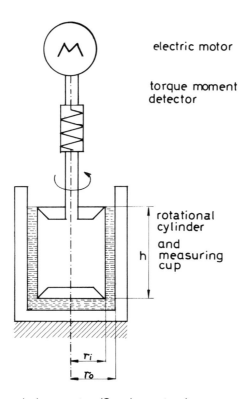

Fig. 8. Rotational viscometer (Searle system).

The test equipment consisted of the following parts:

(1) Control equipment: Selection of the rpm (30 steps from 0.01–512 rpm); digital display; and a connection for a recorder.

(2) Electric motor and torque moment detector.

(3) Different measuring devices: Rotational cylinders and cups (measuring range 10^1–10^8 mPa·s).

(4) Tempering container.

(5) Thermostat with circulating liquid (ambient temperature: $-200\,°C$ (392 °F)).

(6) X-t recorder for a time depending on the recording of the signal S.

For the determination of the flow curve, the rpm rate was increased step by step. The results can be presented in different forms (Fig. 9).

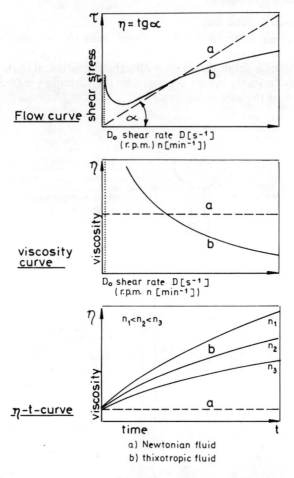

Fig. 9. Graphical presentation of viscosity measurements.

In our case, we are working with thixotropic fluids characterized by a yield value. The typical property of this kind of fluid is a decreasing viscosity with an increasing shear rate. After stirring, the fluid will regain a high viscosity.

Test Material

We carried out the experiments by using an iron-rich magnesia frequently used for the production of gunning mixes: Chemical analysis (%): MgO, 86.42; Cr_2O_3, 0.24; Fe_2O_3, 5.97; Al_2O_3, 0.51; Mn_3O_4, 0.21; CaO, 5.12; SiO_2, 1.54; and the phase composition was periclase, dicalcium ferrite, dicalcium silicate, and magnesioferrite.

The screen analyses in percents for BMF-A and BMF-B were as follows:

Table V. Screen Analyses for BMF-A and BMF-B

Screen analysis (μm)	BMF-A	BMF-B
125–200 (+115–65 mesh)	1	1
90–125 (+170–115 mesh)	2	0.5
63– 90 (+250–170 mesh)	8	1.5
32– 63	22.5	22.5
10– 32	42.5	37.7
10	24	36.8

Most trials were performed using ball-mill fines, A.

Results

The rheological properties of suspensions of this magnesia in water were studied with regard to water content, fineness, and temperature. The purpose of this investigation was the search for the explanation of the adherence of a mix on the hot wall before a chemical bond is formed. Basically, we can see at least three possible explanations of this phenomenon:

(1) The particles could be held together by the cohesion and viscosity of the liquid, but the cohesion and the viscosity of water decrease with increasing temperature.

(2) The particles could be held together by the surface tension of the liquid, but the surface tension of water is also decreasing with increasing temperature.

(3) Finally, the particles could be held together by several physical and/or chemical reactions that take place on their surfaces (hydration or colloidal surface coating), which thus lead to a strengthening of the entire structure. The rate of those reactions most probably increases at higher temperatures.

The Effect of the Shear Rate

The low-water-containing suspensions which were studied behave very thixotropically. They show more or less distinct yield values. The entire level of the viscosity increases with decreasing water content. This is more pronounced at the yield value and it is less significant at higher shear rates.

The flow curves recorded at decreasing shear rates are located below the curves recorded at increasing rates. In most cases they do not show any yield values. This is typical for thixotropic flow behavior (Fig. 10). The absolute value of the viscosity at the yield point cannot be determined exactly for two reasons: The step-by-step setting of the rpm rate on the viscosimeter does not allow registration of the accurate shear rate, D_0, where the stability of the fluid begins to break. Besides this, the viscosity at the yield point depends on the time.

Figure 11 shows very clearly that high viscosities (10^7 mPa·s) can be reached at low rotating speeds of 0.01 or 0.10 rpm, whereas the same suspensions reach only 10^3 mPa·s at high shear rates. The high viscosity at the yield point will develop relatively fast, but not immediately. That the rate of the viscosity increases with time can be observed from $\eta - t$ curves (Fig. 12).

A quick development of the yield value and the high viscosities at low shear rates are necessary requirements for the stability of a gunned layer. On

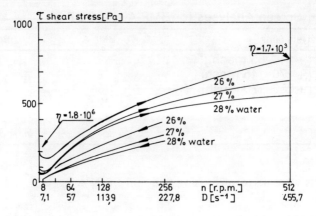

Fig. 10. Flow curves of magnesia BMF-A at ambient temperature.

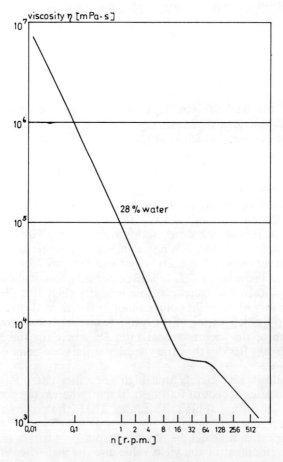

Fig. 11. Viscosity curve of magnesia BMF-A at ambient temperature.

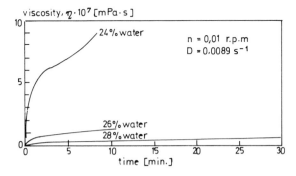

Fig. 12. Increase of the yield value of magnesia BMF-A at ambient temperature.

the other hand, the significant decrease of the viscosity at high shear rates is necessary for an easy penetration of the impacting grain and for its tight packing in the gunned layer. It is obvious that magnesia-water suspensions fulfill both requirements; thus they provide the physical basis for good gunning behavior of basic gunning mixes, even at low contents of binding and plastifying additives.

The shear strength of the wet mix is high enough that the layer will stick to the vertical wall, but it will increase further by development of the chemical bond during drying. If there are difficulties with the mix sticking during gunning, it is necessary to improve the rheological wet properties instead of the physical properties of the dry mix.

Effect of the Fineness of the Magnesia

The viscosity of magnesia-water suspensions can be increased by using finer ball-mill fines. The comparison of BMF-A and BMF-B is shown in Figs. 13 and 14. It can be seen that at the same water level BMF-B is approxi-

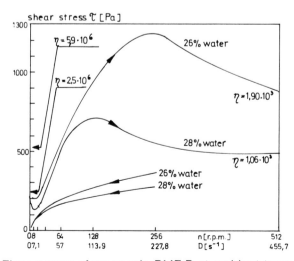

Fig. 13. Flow curves of magnesia BMF-B at ambient temperature.

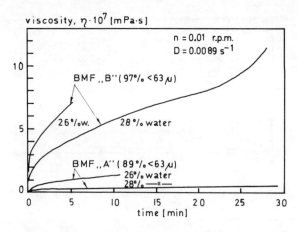

Fig. 14. Increase of the yield value of magnesia BMF-A and BMF-B at ambient temperature.

mately three times more viscous than BMF-A. Here again, the improvements of the viscosity are more pronounced at lower shear rates.

The Effect of Temperature

With increasing temperature, the viscosity of the suspensions increases at the yield values, but decreases slightly at higher shear rates (Fig. 15). The increase of the yield value is very obvious at 0.01 rpm. Here, the viscosity is so high at high temperatures that the measurements could only be carried out using more diluted suspensions. The correlations between water content and temperature at a constant viscosity of 10^7 mPa·s are illustrated in Fig. 16.

It is obvious that the same viscosity of powder-water suspensions can be obtained at 90°C with almost twice as much water than at room temperature. The following conclusions can be drawn for gunning practice: During warming of the gunned layer, setting of the mix takes place; and gunning is easier

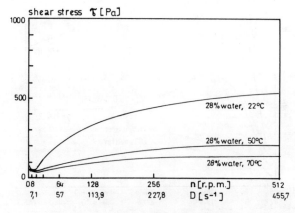

Fig. 15. Flow curves of magnesia BMF-A at 22°, 50°, and 70°C.

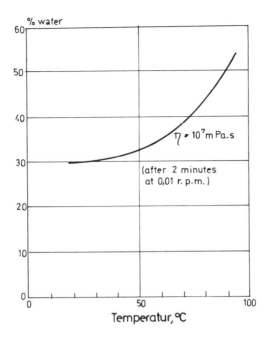

Fig. 16. Water requirements of magnesia suspensions at high temperatures.

to perform using warm water (compare Refs. 6, 8, and 9). This phenomenon has nothing to do with the dissolution and reaction of the binder; it is only a result of a behavior of the magnesia-water suspension. It can be explained neither by the cohesion and viscosity, nor by the surface tension of the added water. The setting is definitely caused by reactions on the surface of the periclase grains that leads to an immobilization of a major part of the water, which, in turn, leads to an apparent drying of the mix.

It is not clear whether water adsorption processes and subsequent hydration play a significant role. It could be assumed that the very first stages of

Fig. 17. Brucite coating on periclase.

hydration, such as chemical and physical adsorption and nucleation, form colloidal coatings on periclase surfaces.[24] This could result in the formation of a weak, three-dimensional linkage of the particles that could easily be destroyed by high shear rates, thus reversing the suspension to the liquid stage. The formation of very finely flaked brucite crystals (Fig. 17) could create thixotropic properties in much the same way as do clay minerals.[25]

Outlook

The preceding conclusions indicate that there still remain many questions to be answered. In addition to the explanation of the physical and chemical reasons for the particular rheological properties of magnesia-water suspensions, major test series are necessary to explain the different behavior of various types of magnesia sintered with different CaO-SiO_2 ratios. It is also important to study the effect of various binders and their combinations. We assume that some binders improve the rheological behavior, whereas other binders could interfere with the formation of the yield value. The viscosity determination of aqueous suspensions is obviously also a suitable method for the evaluation of plastifying agents. This would be an important addition to the conventional test methods and should be used in research and development of gunning mixes and probably of castables and vibratables as well.

References

[1] P.M. Kayworth and G.P. Carswell, "Gunning Maintenance of BOS and Arc Furnaces," *Refr. J.*, **54** [4] 11-20 (1979).

[2] R.P. Heilich, F.J. Rohr, and L.D. Hart, "Pneumatic Gunning of Refractory Concrete," *Am. Ceram. Soc. Bull.*, **46** [7] 674-78 (1967).

[3] M. Braun and A. Majdic, "Verdichtungsmechanismen beim Spritzen feuerfester Erzeugnisse," *Keram. Z.*, **34** [3] 144-48 (1982).

[4] Y. Nishikawa et al., "Some Problems with Hot Gunning," *Taikabutsu*, **33** [283] 52-56 (1981).

[5] J.G. Yount, Jr., "Hot Gunning Materials for Basic Oxygen Furnace Maintenance," *Am. Ceram. Soc. Bull.*, **47** [3] 259-63 (1968).

[6] A. Watanabe, H. Takahashi, F. Ota, M. Kondo, and O. Ebisu, "Setting and Adhesion of Gunning Mixes at High Temperatures," *Taikabutsu*, **33** [8] 436-40 (1981).

[7] A. Berthet and C. Guenard, "Contrôle des principaux paramètres de la réparation par projéction des fours d'aciére"; presented at the 10.Gemeinsames Werkstoffwissenschaftliches Kolloquium, Saarbrücken, December 12, 1981.

[8] S. Fujimoto and Y. Kawase, "Hot Gunning Repair of BOF vessels," *Taikabutsu*, **28** [1] 4-8 (1977).

[9] K. Furumi, S. Fujimoto, Y. Ochi, and Y. Kawase, "On the Methods of Gunning Repair for LD-Converters," *Taikabutsu*, **29** [3] 151-53 (1977).

[10] C. Guenard, H. Piasecki, and F. Raoult, "Quality Evaluation of Converter Gunning Materials: Industrial Measurements and Laboratory Tests," Proceedings of the 1st Int. Conf. on Refractories, Tokyo, 1983.

[11] G. Boiche, P. Steinmetz, C. Gleitzer, and C. Guenard, "Investigation of Adhesion Mechanisms of Repairing Materials for Steel Converters through High-temperature Gunning," *Rev. Hautes Temp. Refract.*, **17** [4] 339-49 (1980).

[12] L.J. Dreyling and J.H. Belding, "Advancements in AOD Gunning," AOD Intern. Conference, Geneva, May 5-8, 1975.

[13] Yu. A. Pirogov, G.D. Prokopetz, and L.A. Babkina, "Apparatus for Determining the Strength of the Adhesion of Gunite Coating to the Refractory Masonry in the Breakaway Tests," Transl. from *Ogneupory* [1] 58-60 (1977).

[14] Yu. A. Pirogov, M.I. Prokopenko, and G.D. Prokopetz, "Apparatus for Determining the Bonding Strength of Guncrete Coatings on Refractories," Transl. from *Ogneupory* [6] 14-16 (1970).

[15] H. Yamanaka, M. Ikeda, and M. Sawano, "Testing Method for BOF Gunning Materials in a High Temperature Environment," *Taikabutsu*, **34** [296] 38-42 (1982).

[16] Y. Nishikawa and H. Miyashita, "A Study of Gunning Materials in LD-Converters," *Taikabutsu*, **29** [3] 154-59 (1977).

[17]S. Yoshino et al., "Adhesive Properties of Gunning Materials for LD-Converters," *Taikabutsu*, **34** [296] 33-37 (1982).

[18]J. de Boer and M. Beelen, "Correlation between Rheological Properties and Rebound by Tests with a Viscosimeter," presented at SIPRE, WG 11 Meeting Leeuwenhorst, May 4, 1983.

[19]G. Routschka, "Die Prüfung der Bau- und Reparaturmassen im Spiegel der Literatur," *Keram. Z.*, **20** [10] 643-49 (1968).

[20]E.G. Martienssen, "Prüfung feuerfester Spritzmassen," *Keram. Z.*, **31** [3] 152-53 (1979).

[21]E. Becker, Technische Strömungslehre; pp. 98-103. Verlag B.G. Teubner, Stuttgart, 1982.

[22]J. Schurz, Einführung in die Strukturrheologie; pp. 12-21. Verlag Berliner Union, Stuttgart.

[23]M. Reiner, Rheologie in elementarer Darstellung; pp. 228-32. Carl Hanser Verlag, München, 1969.

[24]W. Feitknecht and H. Braun, "Der Mechanismus der Hydratation von Magnesiumoxid mit Wasserdampf," *Helvetica Chimica Acta*, **50** [7] 2040-53 (1967).

[25]W. Zednicek, "Über Magnesiumhydroxidbildung auf Sintermagnesia," *Radex Rundschau*, [3] 537-67 (1981).

Dryouts and Heatups of Refractory Monoliths

Norman W. Severin

Hotwork, Inc.
Lexington, KY 40523

Properties and use of castable and plastic refractory products were significantly improved by the development of high-strength, calcium aluminate and chemical bonds, better raw materials, and easier placement methods and techniques.

But whatever the refractory material selected or whoever does the professional installation job, the refractory lining cannot function and provide the expected service life until it has been properly cured and fired. Better insight into some of the refractory problems which may or may not be fully recognized in dryouts and heatups of furnaces and refractory linings would be beneficial.

Unfortunately, the heatup or dryout responsibility seems to be avoided wherever possible. The designers are usually not concerned with dryout schedules; the refractory manufacturer generally has to live with conditions existing in an operation he does not control; the contractor-installer wants to use time and resources more productively in the basic line of work; and the operator wants the process back in operation as quickly as possible.

Plans for drying and heating refractory furnaces and linings must be considered at the design stage. For example, four blast furnace stoves were built without leaving any access to preheat the combustion chamber above the main burner. This required unusual on-site expense and lost time to get them in operation.

During installation, dryout considerations (which have nothing to do with the actual dryout), should be given as they can affect the outcome of the heatup process. Primary—from our viewpoint—is the importance of having the castable, mixing water, environment, and vessel above 21°-27°C, followed with a 24-h curing temperature of at least 32°-38°C. The dryout process cannot only be considerably retarded but can also have disastrous results should much or all of the products of hydration develop into CAH_{10} instead of the more desirable C_3AH_6 or cubic phase.

Even with the best uniform heat, a shift of the refractory lining might produce the effects observed in the following areas: (1) In the case of gunned linings, whole areas have collapsed during heatup because the welds of the anchors have not been to standard; and (2) in some plastic installations, the installer may not provide sufficiently long refractory anchoring or ledges, corners, or arch transition areas. They are difficult, sometimes impossible, to install in these areas, but omission of anchors leaves unduly thick areas without proper support. This forms a weak plane which will not survive the normal heatup stresses.

Where forms are used, they should be designed to release from the refractory without rupturing or damaging the lining. After form removal, the refractory surface should be trimmed, brushed, scraped, or roughened to facilitate an easier removal of moisture during drying and firing. When ramming without forms, installations should be continuous, because air drying can cause a skin hardness that may contribute to spalling or peeling during heatup.

Any venting or scoring to control cracking should be done continuously as the job progresses, or, in the case of forms, when they are removed. Forms should be removed prior to dryout, as it is generally not good practice to burn out forms. Most forms are plywood, tightly spaced, and tend to insulate and cover the surface. The wood will superheat and finally burst into flame resulting in: (1) sudden, high, uncontrolled temperature, which can cause peeling or spalling of the hot face or ruin the total lining, and (2) a shift in the refractory as the form collapses, causing cracks in the material or damage to anchors. Where some forms must remain, provision should be made for removing alternate form boards or removable form windows must be provided. In very critical applications, diamond lath, expanded metal, or similar material can be used.

Although moisture is evaporated through the hot face, most will probably be driven to the cold face and condense at some point near the shell where it will subsequently collect at the bottom of the cladding or vessel. Where possible, drain holes should be provided at the lowest points of the vessel. Weepholes placed above these low points are relatively ineffective, although they will release some generated steam.

With today's emphasis on saving energy, fiber insulations and lightweight refractories are being placed in increasing thicknesses in furnace walls. During heatup, these linings act as a sponge and absorb the migrating moisture. Due to their low K factor and the shell heat sink, they are most difficult to dry out. A sheet of Visqueen or other moisture barrier between the refractory and the insulation will help prevent the migration of moisture into this porous material.

Although it will not affect the performance of the refractory, embedded thermocouples should be installed in the refractory, particularly at the interfaces of multiple component linings. These couples can be used effectively for heat transmission studies, to monitor changes in refractory wear, and to provide useful data during dryout.

Having completed the structure or lining and being satisfied with all of the above influences, one would want this unit in operation as quickly as possible, as pressure will certainly be brought to bear by production operations to get the product out of the door. More often than not, the refractory installation is usually the critical path, particularly in a turnaround. To accomplish the heatup, we need a heat source, the ability to measure the temperature, and of course, a schedule to follow. Let us consider some available heat sources:

(1) Drying with various combustible materials, such as wood, coal, and coke were once common practice and are still used today. Since this drying is difficult to control, it is not recommended.

(2) High temperatures can be reached with portable heaters, pipe lances, and other means. These methods can be effective if properly modulated, but they lack the ability to attain higher temperatures, and safety controls are usually nonexistent.

(3) Permanent process burner equipment has been employed by working with first one or more pilots and then sequencing the main burners. Unless multiple burners are in very close proximity and firing in the same direction, it is not a good idea to light one burner and blow combustion air through others to compensate for overheating, since damaging high-temperature differentials can develop.

(4) Heat from another process in the plant may be used to dry refractories. The problem in using such a heat source usually lies in the lack of control at low temperatures. Extreme caution and consideration should be used if such temperatures are above 107 °C.

(5) There are companies which specialize in refractory lining dryout and are recognized as effective in cases where the aforementioned methods are neither practical nor available. With proper portable burners emitting controlled heat via combustion product mass flow, typical dryout time can be considerably shortened. Further, the availability of expertise in dealing with the aforementioned and following problems, coupled with the execution of these jobs by experienced crews, makes contracting this phase worthwhile.

Good temperature measurements are necessary during heatup. Unless otherwise specified by the manufacturer, temperatures mentioned in the United States are assumed to be the combustion products in contact with the surface, rather than the refractory material itself. Placement of thermocouples within 2.54–5.08 cm of the refractory surface basically satisfies this requirement. Do affix them securely.

Thermocouples should be placed in the expected cold and hot areas of the structure both to identify a gross temperature nonuniformity and to ensure that extreme portions of the furnace basically adhere to the time-temperature plot. A record of the heatup progress should be kept, using permanent recording equipment.

A dryout or heatup schedule must be selected. The best source for this information is, of course, the manufacturer whose product is being used.

But we have found that unanimity among various suppliers for the same or equivalent product doesn't exist, as indicated by the various curves in Fig. 1. In one case, a major supplier significantly modified his heatup schedule for the very same product within three years (indicated by curves A1 and A2). These discrepancies may not be acceptable, but they are indeed understandable.

A word should be mentioned about the difference between dryout and heatup. It should be considered a dryout if the process is interrupted after a satisfactory level of dryness is attained for operation at a later time. A heatup is the continuation of this process, normally leading to the furnace or unit going directly into operation.

The mechanism by which moisture travels does not appear to be as straightforward as the scientists would have us believe. One theory is that moisture travels from the hot face to the cold face. If this is the case, how does one explain an increase and a relatively prolonged, high dew cell reading as the heatup progresses (Fig. 2)? This was a dryout of a blast furnace for winter storage. Note the increase of moisture in the exhaust stream for 3½ days. Evaporation through the hot face is certainly effective as long as the actual moisture stays above the ambient humidity line. Obviously, moisture is removed in both directions, but take as much as possible out of the lining in the combustion products' stream, and it is gone forever.

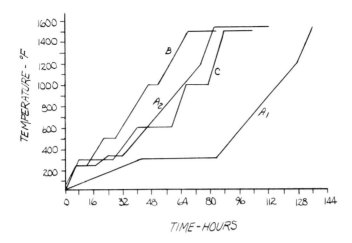

Fig. 1. Refractory manufacturers' heatup schedules.

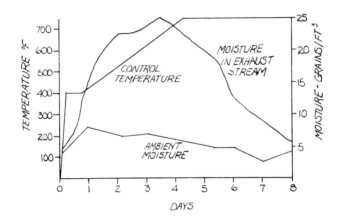

Fig. 2. Relationship between temperature and moisture removal during blast furnace dryout.

Recent tests have shown that, for a given class of refractory, the moisture removal rate is relatively fixed (Fig. 3). It would seem that a heatup rate corresponding closely to this moisture elimination rate should be chosen. Steeper heatup rates, not in proportion to the removal rate, are neither prudent nor thermally efficient, since the heat transfer rate into the lining interior is slower than the rise in temperature within the furnace. One might think that the greater the surface area, the easier the moisture removal, but this proved not to be the case. Leading edges or sills, dividing walls, or worse yet, internal piers, can be difficult to heat.

There appears to be no clear-cut relationship between density, permeability, cement purity, or percent of cement contained, but it is proper to say that the denser the product and lower the permeability, the larger the problem of moisture removal. Although there should be no reason for it, it

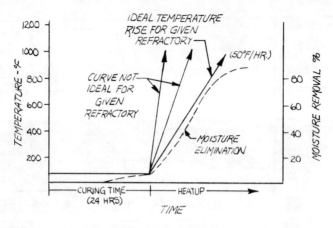

Fig. 3. Heatup rate vs moisture removal in castable linings.

appears that more time is required to dry a large furnace, such as a slab heater in a steel mill or a petrochemical, fluidized, catcracking unit, than the lining of the vestibule of a small boiler, since there may be more than a hundred times the volume of material involved.

Lining thickness is probably the major consideration for altering a given schedule. Introducing or extending hold periods by ½ h or 1 h per 2.54 cm of thickness is common practice with these kinds of masses. Further modification of a given temperature schedule should be considered, depending on the type of heat source employed. If the temperature heatup rate cannot be controlled accurately and a reasonable uniformity achieved, a slower schedule is recommended. Further, latent heat with a radiating effect, such as an oil flame, should be given consideration when working in close proximity with refractory surfaces. Gaseous fuels with little or no radiation, coupled with mass flow and good circulation, are preferred by far. More dryout time might be considered if configurations of the furnace or vessel make heat distribution difficult and uniformity cannot be achieved. Just a simple heat source into the lower manhole of a sphere would leave the refractories at the sides of this vessel cold. One method to overcome nonuniformity was to suspend a baffle inside the vessel to deflect the hot gases to the vessel sides. A simpler and better method is the placement of a mass flow burner through the manhole or upper nozzle and exhaust out of the bottom nozzle. With good uniformity, optimum dryout curves can be considered.

In the case of composite refractory linings, the slowest schedule specified by the manufacturer should be applied. Undoubtedly the manufacturer has done considerable research on his products with regard to drying and firing. But conventional laboratory firing and test conditions do not correspond with normal use conditions, mainly because in-use refractories are fired from one side only, in usually less than ideal environments. A technician is thus guided by feedback from actual installation, which he does not control. He naturally protects his product from damage during the heatup procedure by maintaining a sufficient holding time that will allow thermal equilibrium to be reached before proceeding to the next plateau. Technically, this does not make any sense. It would seem that at the least a linear profile connecting

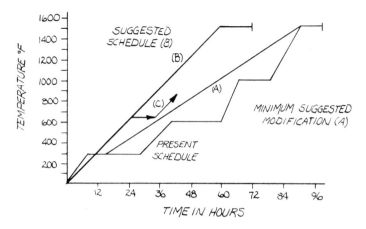

Fig. 4. Suggested modifications of heatup schedule.

starting and final temperatures, as in profile A of Fig. 4, be used. When all the other parameters are reasonably correct and good temperature control is exercised, profile B makes economic sense. One hold period of 343°–371°C might be conceded in special instances, as indicated in profile C, where curing temperatures have been less than the specified 21°–27°C. It appears that a hold in this region is most beneficial in alleviating potential explosive conditions. There is little doubt in our minds that plastic refractories should have a linear profile (Fig. 5) throughout the heatup program, as it will prevent planing and slumping due to moisture migration and accumulation. Note that some plastic refractories require firing shortly after installation, while others have a minimum use or shelf life. Regardless of the path chosen, the program should be continuous and without interruptions.

When steaming is visible, work is being done and moisture removed; however, when pressure steam is visible the heatup should be held until the steam subsides. Pressure steam is recognizable by the vapor sting-out, usually

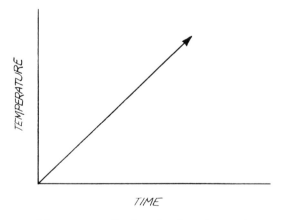

Fig. 5. Suggested heatup profile for plastic refractories.

exceeding 5-7 cm. If the furnace or vessel has been heated by auxiliary means, caution should be exercised during the takeover. Excessive temperature variations from poorly controlled process burners can cause refractory damage. Either test the main burners shortly before removing the auxiliary burners or have the auxiliary burner stand by ready to be reactivated. Starting I.D. or recirculating fans, if not under control, can unduly shock the hot refractories on startup.

Except for a few special metallurgical furnaces or reactors involving chlorine in particular, most furnaces or structures are not dry in the true sense of the word at the start of the operation. Some are never fully dry after years of operation. If it is imperative that all moisture is removed prior to operation, the unit should be taken to the highest permissible temperature and held until a satisfactory dryness is attained. If equilibrium is reached but the desired degree of dryness is not achieved, external temporary insulation, such as house insulation, can help remove the maximum moisture from the refractories. To date, we have concluded that the best means of measuring the dryness on a refractory surface is still a thermocouple at the shell. If a thermocouple hovers around 99°-107°C, you still have moisture present. Once it starts moving away from this plateau, one can assume it to be dry. No vapor pressure considerations are assumed in this case. This may appear to be a crude approach, but it is a practical one which works pretty well for us.

After all of these considerations, do not lose sight of safety during dryouts; we are operating below the ignition limits of most common fuels, and the danger of explosion from gaseous fuel exists.

Section IV
Low-Cement Castables

Effect of Microsilica on Physical Properties and Mineralogical Composition of Refractory Concretes 201
B. Monsen, A. Seltveit, B. Sandberg, and S. Bentsen

Vibrated Castables with a Thixotropic Behavior 211
R. Stieling, H.-J. Kunkel, and U. Martin

High-Technology Castables 219
E. P. Weaver, R. W. Talley, and A. J. Engel

High-Performance Castables for Severe Applications 230
C. Richmond and C. E. Chaille

Progress of Additives in Monolithic Refractories 245
Y. Naruse, S. Fujimoto, S. Kiwaki, and M. Mishima

Low-Moisture Castables: Properties and Applications 257
S. Banerjee, R. V. Kilgore, and D. A. Knowlton

A New Generation of Low-Cement Castables 274
B. Clavaud, J. P. Kiehl, and J. P. Radal

Calcium Aluminate Cements for Emerging Castable Technology 285
G. MacZura, J. E. Kopanda, and F. J. Rohr

Effect of Microsilica on Physical Properties and Mineralogical Composition of Refractory Concretes

B. Monsen and A. Seltveit

SINTEF
at the Norwegian Institute of Technology
Foundation for Scientific and Industrial Research
Trondheim-NTH, Norway

B. Sandberg

Elkem A/S Chemicals
N-4620 Vågsbygd, Norway

S. Bentsen

Norwegian Institute of Technology
N-7034 Trondheim-NTH, Norway

Pastes and concretes based on pure and microsilica-blended, high-alumina cement were studied after curing, drying, and firing. The mineralogical composition was studied by X-ray diffraction (XRD), and most of the physical properties were determined by standardized test methods. The results indicate that the mechanical strength at both moderate and high temperatures can be improved by a partial replacement of cement with microsilica.

During the last few years, there has been a rapid increase in the use of volatilized silica as a raw material in a variety of refractory products. The name microsilica was introduced by Elkem A/S as a term for the material obtained by the cleaning, classification, and homogenization of the silica-rich fume produced by smelting of ferrosilicon and silicon metal. This term will be used in this paper. Physical and chemical properties of the refractory-grade microsilica used in this investigation are given in Table I. However, it is not always possible to predict the performance of different grades of microsilica in refractory concretes from these data only. Such important processing variables as smelting furnace operation, charge composition, and filtering procedures may affect its suitability for different applications.

The present work was undertaken to study the effect of a partial replacement of cement by microsilica on a 1:1 basis in an ordinary, high-alumina cement concrete based on fireclay aggregates. Preliminary studies had indicated that optimum composition was 14% cement, 6% microsilica, and 80% aggregate when no other ingredients, such as highly reactive, fine alumina, were added.

Table I. Chemical Analyses and Physical Properties of Microsilica,* High-Alumina Cement,† and Fireclay Aggregate

Constituent	Microsilica	High-Alumina Cement	Fireclay Aggregate
SiO_2	98.0	0.2	53.2
Al_2O_3	0.21	80.5	42.9
CaO	0.10	18.0	0.1
Fe_2O_3	0.08	0.15	1.6
MgO	0.12	0.1	0.1
K_2O	0.31	0.03	0.2
Na_2O	0.08	0.25	0.1
TiO_2	< 0.01	0.03	1.7
Mn_2O_3	0.01	0.01	
SO_3	0.17	0.1	
Density (g cm^{-3})	2.25	3.2	2.71
Apparent porosity (%)			3.5–5.0
Blaine surface area (mg^2 g^{-1})		0.9	
BET surface area (m^2 g^{-1})	22		

*"Elkem MicroSilica," Elkem A/S Chemicals, Norway.
†"Secar 80," Lafarge Fondu International S.A., France.

Experimental Procedure

All test materials were made from commercial high-alumina cement and a refractory-grade microsilica. Calcined fireclay was used as the aggregate in the refractory concrete test specimens.

Preparation of Specimens

Most of the specimens used for the paste hydration studies at 40° and 65°C, respectively, were prepared by hand-mixing 10 g of powder and 5 g of water. The samples were placed in small plastic beakers and cured in a desiccator at constant temperature. Some water was added to the desiccator to keep the relative humidity close to 100%.

Test specimens of the reference concrete A were made from 20 wt% cement and 80 wt% calcined fireclay aggregate. Water addition was 11% of the weight of dry material. In the microsilica-containing mix B, the composition of the dry mix was 14 wt% cement, 6 wt% microsilica, and 80 wt% calcined fireclay aggregate. Water addition was 9% of the weight of the dry material. Sodium polyphosphate was added as a dispersing agent in an amount of 0.24% relative to the weight of dry material. By use of dispersants in microsilica-containing mixes, it is possible to maintain good flow even with very low water addition. In this work, water addition was adjusted to obtain equal flow of the two concretes. Test specimens of refractory concretes were vibration-cast in molds with dimensions of 40 by 40 by 160 mm and 100 by 100 by 500 mm, respectively.

All the dried and fired specimens were dried for 24 h at 110°C without access to CO_2 and fired in a small, electric resistance furnace in an open atmosphere. Most of the test specimens were soaked for 12 h at top tempera-

Table II. Physical Properties of Refractory Concretes with and without Microsilica

Properties	Firing Temp. (°C)	Concrete A (without Microsilica)	Concrete B (with 6% Microsilica)
Bulk density (g/cm^{-3})	110	2.20	2.18
	800	2.12	2.15
	1000	2.13	2.16
Porosity (%)	600		19.8
	800	23.9	20.8
	1000	23.5	21.2
Refractoriness under load (°C) (Rising temperature)	$T_{0.5}$	1291	1331
	T_2	1353	1412
	T_5	1422	1496
Creep in compression under load at 1190 °C between the 5th and 25th h (%)		−0.22	−0.12
Thermal conductivity (W·m^{-1}·K^{-1})	1000	1.13	1.36
Thermal conductivity after firing for 19 h at 1200 °C	1000	1.19	1.39
	300	1.16	1.29

ture. X-ray powder diffraction analyses were carried out with CuK_α radiation and Ni filter.* Scan rate was 1° 2O min^{-1}.

Hot modulus of rupture was determined according to the 1978 revised PRE/R21. Test pieces were 40 by 40 by 160 mm and the distance between supports was 122 mm. Rate of increase in bend stress was 0.15 N·mm^{-2}·s^{-1}. Heating rate was approximately 3 K·min^{-1} and test temperature was maintained 12 h before testing.

Deformation under load at high temperature was determined by the rising temperature test according to ISO/R 1893, October 1970, and the maintained temperature test according to PRE/R6, 1978 revision. The ISO recommendation describes a test designed to determine the deformation of a refractory product subject to a constant load under conditions of progressively rising temperature, 4°–5 °C/min. The test piece is a cylinder 50±0.5 mm high, with a 12.5 mm hole drilled axially through it. Temperatures corresponding to a deformation of 0.5%, 2%, and 5% of the maximum height are used as measures for refractoriness and presented as $T_{0.5}$, T_2, and T_5. Thermal conductivity was measured by the hot-wire method, PRE/R32 (1977).

---

*Philips X-ray diffractometer consisting of X-ray generator PW 1730/10, vertical goniometer PW 1050/25, and registration system PW 1360, Philips Gloeilampenfabrieken NV, Eindhoven, The Netherlands.

Results and Discussion

Pastes

The effect of curing time on phase composition of pastes during hydration at 40° and 65°C was studied using XRD. During the first days of curing, samples were analyzed at short intervals. Corundum is considered to be inert during hydration, and the peak height of the strongest diffraction line for the different phases was normalized with respect to the strongest α-alumina peak in the diagram. The variation in peak heights during hydration is shown in Figs. 1–4.

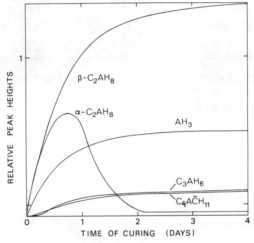

Fig. 1. Change in mineralogical composition in neat cement pastes during hydration at 40°C.

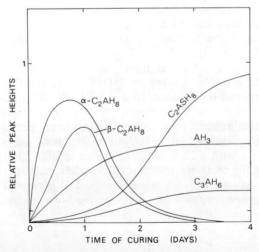

Fig. 2. Change in mineralogical composition in microsilica-blended cement pastes during hydration at 40°C.

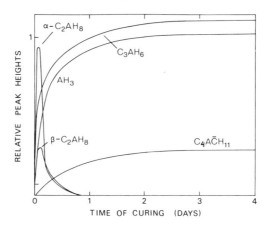

Fig. 3. Change in mineralogical composition in neat cement pastes during hydration at 65°C.

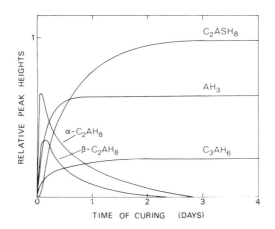

Fig. 4. Change in mineralogical composition in microsilica-blended cement pastes during hydration at 65°C.

It appears from Fig. 1[†] that considerable amounts of α-C_2AH_8 and β-C_2AH_8 were formed within the first 24 h in the neat cement paste at 40°C. However, the amount of α-C_2AH_8 seems to have reached a maximum after 15–20 h; it was later converted to β-C_2AH_8. After two days, the major phases were β-C_2AH_8 and AH_3.

The results of the X-ray analyses of microsilica-blended cement pastes are shown in Fig. 2. Initially, the same hydration products as in neat cement pastes are formed, and after 15–20 h, appreciable quantities of AH_3, α-C_2AH_8, and β-C_2AH_8 are present. In these pastes, however, the hexagonal phases α-C_2AH_8 and β-C_2AH_8 are converted to gehlenite hydrate, C_2ASH_8,

---

[†]Peak heights are normalized with respect to the peak height of the $d = 2.552$ diffraction peak of α-alumina.

by further curing. Gehlenite hydrate is observed after a 24-h cure and is the main hydration product after 3-4 days. The relatively small amounts of crystalline phases observed after two days of curing may be attributed to poor crystallinity of the newly formed hexagonal C_2ASH_8. The carboaluminate phase which is observed in the neat cement pastes do not occur in pastes with microsilica.

Hydration reactions in pastes with and without microsilica at 65 °C are illustrated in Figs. 3 and 4. The rate of reaction and phase transformation is considerably higher at the 65° than at the 40 °C cure temperature. Even at 65 °C, α-C_2AH_8 and β-C_2AH_8 are formed initially, but are rapidly converted to the stable phase C_3AH_6 in the neat cement pastes. In the microsilica-blended cement pastes, the two hexagonal C_2AH_8 phases are mainly converted to C_2ASH_8 and only minor amounts of C_3AH_6 are formed. It may also be observed that, in pastes with microsilica, carboaluminate is not found, even if some C_3AH_6 is present at 65 °C.

The effect of microsilica on the rate of hydration of CA and CA_2 is shown in Fig. 5. The hydration of these phases is retarded in the microsilica-containing pastes, especially at 65 °C. This retarding effect cannot be fully explained at present. However, it seems to be generally accepted that microcracks are formed in the bonding phase by conversion of the hexagonal phases into the stable cubic phase C_3AH_6, which has a higher density. It appears from Figs. 1–4 that C_2AH_8 in the neat cement pastes is not stable at 65 °C, whereas C_2ASH_8 is the major hydration phase in pastes with microsilica, also at 65 °C. No conversion of C_2ASH_8 seems to take place within the first few days of curing.

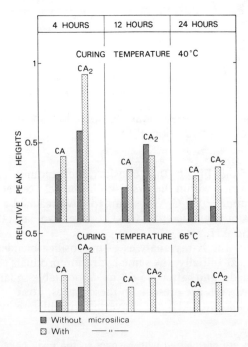

Fig. 5. Consumption of CA and CA_2 during hydration at 40° and 65 °C.

According to Turriziani,[1] gehlenite hydrate becomes unstable, and at 50°C the Al_2O_3-containing phase in pastes made from kaolin and $Ca(OH)_2$ is a hydrogarnet with the composition $C_3AS_{0.3}H_{5.3}$. Figure 4 indicates that, in a high-alumina cement that contains microsilica, C_2ASH_8 seems to be stable at even higher temperatures. This may be due to a low $Ca(OH)_2$ concentration.

When evaluating cement pastes, it is often convenient to apply a lower w/c ratio than used in the pastes described so far. The mineral composition

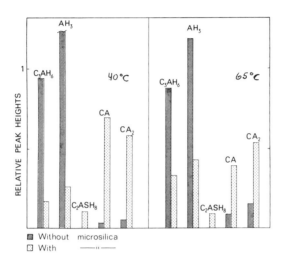

Fig. 6. Mineralogical composition of pastes hydrated for 24 h at 40° and 65°C, respectively, and dried for 24 h at 110°C.

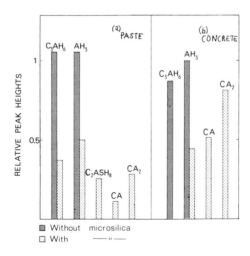

Fig. 7. Mineralogical composition of pastes and concretes after curing for 24 h and drying for 24 h at 110°C. (The water/powder ratio in pastes is 0.5. In concretes with and without microsilica the water addition is 9 and 11%, respectively.)

of vibration-cast test specimens with a w/c of 0.3 after curing and drying is shown in Fig. 6. Sodium polyphosphate was used as a dispersing agent in these mixes.

The peak heights of the different phases seem to confirm the retarding effect of microsilica, but the relative amounts of the different phases are most probably also influenced by the w/c ratio, the presence of dispersing agent, and the drying conditions. The retarding effect can also be seen in Fig. 7(a), which illustrates the phase composition in dried pastes made without use of dispersants and with a water/powder ratio of 0.5.

Concretes

Mineralogical phase compositions of concretes after curing and drying are shown in Fig. 7(b). It appears that the hydration reactions in the reference concrete A have been completed, whereas considerable amounts of CA and CA_2 are still present in the microsilica-containing concrete B. Only moderate quantities of AH_3 are found as a crystalline hydration product in concrete B. The presence of mullite, A_3S_2, quartz, and cristobalite (S) in the aggregates is not shown in the figures. Cold crushing strengths of concretes A and B were 47 and 68 MPa, respectively, which indicates that a strong bonding has been established in concrete B, despite the presence of considerable amounts of unreacted cement.

Mineralogical phase compositions after firing at 600° and 1000°C are shown in Fig. 8. It appears from Figs. 8 and 7(b) that the C_3AH_6 found in concrete A has been converted to $C_{12}A_7$ by firing at 600°C. In concrete B,

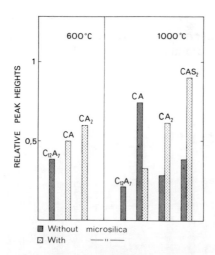

Fig. 8. Mineralogical composition in concretes after firing at 600° and 1000°C. (Soaking time at maximum temperature was 12 h.)

gibbsite has been dehydrated and is most probably present as amorphous alumina. No new crystalline phases are observed.

After firing at 1000°C, appreciable quantities of CA, CA_2, and CAS_2 were formed in concrete A. Anorthite, CAS_2, must have been formed by reaction of calcium aluminate with silica present in the aggregate. The CA_2

and CAS_2 content seems to be even higher in concrete B. Increased content of CAS_2 may be attributed to the presence of highly reactive microsilica. The high reactivity of microsilica is also illustrated by the results shown in Fig. 9. In paste specimens fired at 1000 °C, the major phases in neat cement paste are CA and CA_2. In the microsilica-blended cement paste, the major phases are CAS_2, C_2AS, and S (cristobalite). C_2AS (gehlenite) is very difficult to observe

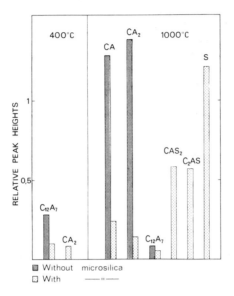

Fig. 9. Mineralogical composition of pastes after firing at 400° and 1000 °C.

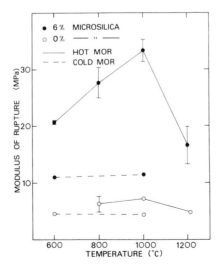

Fig. 10. Cold and hot modulus of rupture of concretes after firing for 12 h.

in the concrete test specimens because the strongest diffraction peak of C_2AS coincides with a strong diffraction peak of mullite from the aggregates.

Physical properties of concretes made with and without microsilica addition are presented in Table II and Fig. 10. The data for refractoriness under load indicate that a partial replacement of cement by silica does not result in a reduction in load-bearing capacity at higher temperatures. The observed increase in refractoriness may be ascribed to a lower porosity in concrete B, which may also account for the observed difference in heat conductivity.

The results of the modulus of rupture tests (Fig. 10) demonstrate a highly beneficial effect of microsilica on these properties. The improved strength is perhaps due to the significantly lower degree of hydration and conversion observed after curing and drying. In addition, the distribution of the microsilica particles in the voids between the coarse cement and aggregate particles may contribute to increased mechanical strength. The high reactivity of microsilica at higher temperatures must also be taken into account.

The fact that CA and CA_2 in the microsilica-containing paste and cement are only partially hydrated at all temperatures might suggest that the cement content could have been reduced. The performance of microsilica-containing, low-cement castables with approximately 5% cement supports this assumption.[2]

References

[1] R. Turriziani; p. 77 in The Chemistry of Cement, Vol. 2. Edited by H. F. W. Taylor, Academic Press, New York (1964).

[2] U.S. Pat. No. 3 802 894 and French Pat. No. 4.111.711 (Lafarge Refractaires) 1974.

Vibrated Castables with a Thixotropic Behavior

Rudolf Stieling, Hans-Joachim Kunkel, and Udo Martin

Zschimmer & Schwarz
Chemische Fabriken
Lahnstein, Federal Republic of Germany

In recent years, vibration casting has been gaining ground in both the refractories and steel industries. Compared with other procedures, vibration casting is less expensive, as it saves time and personnel. Huge and complicated building blocks can be made without risk of cracking or spalling. As a technique that combines processing ease and rapid time, vibration casting is predestined for mass production.

Early vibration-casting techniques were used with materials based on a purely hydraulic cementing medium, primarily alumina cements in proportions of 10–25%. However, since these compositions needed relatively high water ratios, they gave lower compaction and open pore results than other installation methods, for instance, ramming. Another effect that caused considerable concern with using only cement-bonded refractories was the loss of strength through the critical temperature range 600°–1000°C. Refractory managers and users have encountered these problems, and variations in performance were of immediate interest to all engaged in research.

As suppliers, we have been concerned with developing suitable bonding agents for the refractories industry. The details and results of experimentation in our laboratories indicate significant improvements through the use of silicate and aluminosilicate compounds as additives.

Object

The purpose of the development work was to offer bonding agents for varied applications with low-cement castables, cold-setting compositions without cement, and bodies containing carbon and silicon carbide. The thixotropy sought in the prepared mixes should be obtained with a minimum water proportion.

Bonding Agents

Silicate Compound (SILUBIT FB 10)

For low-cement castables, this silicate compound is being used successfully as a chemical bond in refractories. Tests have confirmed that it can be used as well with hydraulic cementing media. The important feature of this additive is to compensate for part of the cement; the cement ratio can be reduced to 5–10% (it was 8% in our laboratory). Correspondingly, 2–7% additive is required (we added 4%). The agent allows the mixing water ratio to be lowered considerably.

Aluminosilicate Compound (LITHOPIX AS 85)

For cement-free vibrated castables, our aluminosilicate compound has been successful. It is a powder—essentially aluminosilicate with a minor proportion of phosphate—used instead of alumina cement in manufacturing chemically bonded refractories that are compressed by vibration. The setting time is slow at 10°C, but may be accelerated by raising the temperature to 30°C, which allows the molds to be removed shortly after. When the mechanically held water has been evaporated, the castables can be further heated without risk of explosion or cracking. Proportions of 5–10% of this additive (normally 7%), have been satisfactory. The mixing water ratios are 4–7%, according to the nature and porosity of the aggregates involved.

Aluminosilicate with Polyelectrolytes (ZUSOSET TH 3)

In the past, refractory bodies containing carbon usually were compressed by ramming. However, it was difficult to achieve consistency with good flow properties. Therefore, an alkali- and phosphate-free bonding agent was developed. This agent allows the manufacture of bodies with good shelf life, ready-mixed for vibration casting with a minimum of mixing water. The organic component of this additive obviates the need for clay as a plasticizing medium, which means reduced shrinkage and improved flow properties. The addition is 3–5%; in the present test, we used 4%. The molds are removed after vibration, and the castables are allowed to dry slowly. They acquire their final strength at higher temperatures.

Experimental Procedures

Batches (see Table I)

The major aggregate of batches 1–3 was $\approx$ 80–88% fused alumina with a standard grading that is commonly used for refractories. This pure form of aluminum oxide was chosen to avoid certain effects on setting times due to the material. Note that batch 3 contains 10% calcined alumina to give the correct proportion of fines as compensation for cement.

The composition of batch 4 was essentially tabular alumina; other components were 15% silicon carbide, with a particle size not exceeding 1 mm, and 2% china graphite in fine flakes. A brand of cement with 80% alumina and about 18% CaO was used.

The addition of water as quoted gave optimum results for the flow qualities during vibration. The correct amount of water is essential; too much water is presumed to be the cause of trouble to manufacturers, as the effects on strength and open porosity are considerable. The mixes prepared with the above additives are comparable to a ramming body that has become a little too wet. The batch compositions also indicate that, with the addition of chemical bonds, the mixing water ratios are about 25–30% lower than that of batch 1, in which no additive is present.

Procedure

A Hobart mixer was used for the preparation. The constituents were first dry-mixed for 5 min, then mixed with the respective proportion of water. The casting was done in steel molds on a vibration table; vibrating time was 2 min, amplitude was 0.75 mm. The test prisms were 160 by 40 by 40 mm. When the setting was complete (which took between ½ and 3 h), the test pieces were dried for 24 h at 110°C and then fired at 500°, 750°, 1000°,

Table I. Batch Compositions 1-4

	Grinding (mm/mesh)	1	2	3	4
Fused alumina	1-3	40	40	40	
" "	0.5-1	15	15	15	
" "	0-0.5	15	15	15	
" "	0-0.1	10	18	13	
Tabular alumina	¼-8				25
" "	6-10				30
" "	-14				8
" "	-325				20
Calcined alumina	0-0.063			10	
SiC	0-1				15
China graphite					2
High-alumina cement		20	8		
Silicate compound			4		
Aluminosilicate compound				7	
Aluminosilicate + polyelectrolytes					4
Mixing water		7.0	5.3	5.0	4.0

1250°, and 1500°C. For firing specimen 4 (with SiC and graphite content), saggers were used in a graphite bed.

Results

The test results may be seen from Table II. The modulus of rupture (MOR), the cold crushing strengths at the firing temperatures mentioned above, the bulk densities and apparent porosities at 1500°C, and, for mixes 1, 2, and 3, the hot MOR at 1300°C were evaluated.

Modulus of Rupture (see Fig. 1)

Compared with mix 1 (high-alumina cement as the only bond), the MOR values of batches 2 and 3 prove that additions of chemical bonds result in increases in strength with rising temperatures; for the cement-bonded batch 1, the graph shows a drop in strength, starting with 6.9 N/mm² at 100°C, to 4.6 N/mm² at 1500°C. The highest initial value — 8.1 N/mm² at 110°C — is that of the low-cement castable, followed by an almost linear increase up to 1250°C with an MOR of 11.5 N/mm²; the value at 1500°C is above 15 N/mm² and off the scale of our measuring device. The initial strength of castable 3 is, of course, below that of the cement-bonded one, but it increases rapidly at higher temperatures; at 1250°C the MOR is beyond 15 N/mm².

The batch containing silicon carbide and graphite gave the lowest initial strength results, due to the organic component of this additive. It increases in strength at higher temperatures. Phosphate binders that are now available and commonly used ensure a better performance at $T = 300°-700°C$, so that here, too, the course of the graph is satisfactory.

Table II. Test Results (Physical Values)

	Firing Temperature (°C)	1	2	3	4
MOR (N/mm^2)	110	6.9	8.1	5.2	2.5
	500	6.9	9.8	6.3	n.d.*
	750	6.5	10.8	8.0	1.7
	1000	5.2	11.8	13.1	n.d.
	1250	5.1	11.5	>15	n.d.
	1500	4.6	>15	>15	>15
CCS (N/mm^2)	110	90	80	31	10
	500	87	96	31	n.d.
	750	94	92	43	8
	1000	66	81	73	n.d.
	1250	63	80	155	n.d.
	1500	52	86	180	85
HMOR at 1300°C	1500	0.4	3.8	4.4	n.d.
Water absorption (%)	1500	8.9	6.2	5.3	5.2
Bulk density (g/cm^3)		2.93	3.10	3.09	2.98
Apparent porosity (vol%)		25.9	19.1	16.3	15.5

*n.d. = not determined.

Cold Crushing Strength (CCS)

The CCS results may be read from Fig. 2. The graphs are quite similar to those of the MOR. The mix with the cement bond only loses strength—from 90-52 N/mm^2—while the low-cement castable with our silicate compound as the chemical bond maintains its CCS level through the whole temperature range—80-86 N/mm^2. Batch 3 starts with a relatively low value (31 N/mm^2) that nevertheless proved sufficient. At temperatures above 750°C, the CCS increases rapidly. At 1500°C, CCS values can exceed 200 N/mm^2. As could be expected, the CCS of castable 4 was good only at high temperatures.

Hot Modulus of Rupture at 1300°C

Figure 3 presents the hot MOR at 1300°C of mixes 1, 2, and 3. Batch 4 had to be excepted from the evaluation for lack of a kiln affording a deoxidizing atmosphere. The superiority in strength of the chemically bonded castables is evident; the results of both are about 10 times higher than those of batch 1 with a hydraulic bond only.

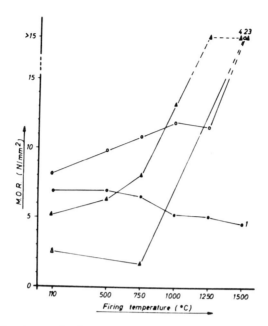

Fig. 1. MOR values of mixes 1-4.

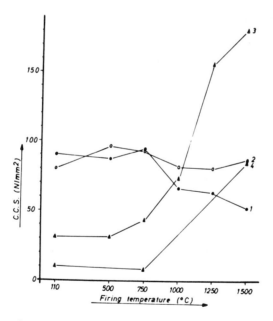

Fig. 2. CCS values of mixes 1-4.

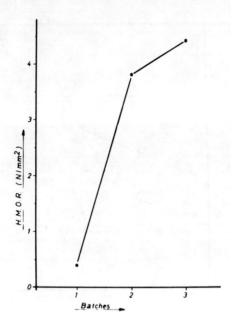

Fig. 3. HMOR values of mixes 1–3.

Bulk Density after 1500°C

Determination of the bulk densities (Fig. 4) also confirmed that additions of these bonding agents result in substantial increases in bulk density; compared with the purely cement-bonded mix 1, even the batch with SiC and graphite content gave better results.

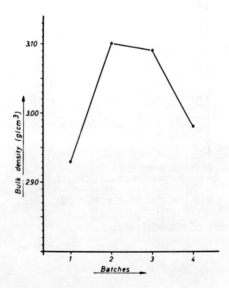

Fig. 4. Bulk density values of mixes 1–4.

Apparent Porosity after 1300°C

As may be seen from Fig. 5, the open pores can be reduced considerably by the action of the bonding agents, a reduction that is not caused by a higher rate of shrinkage, but is due to the improved flow behavior that leads to an early compaction as a result of using less water. Starting with the level of test piece 1, the percentage of apparent porosity could be lowered by 6 by the addition of the silicate compound and by about 9 with the aluminosilicate compound. Use of the aluminosilicate with polyelectrolytes results in a substantial reduction (15.5%). This is of practical importance in manufacturing runner castables, where values between 15 and 20% ensure optimum durability of the runners.

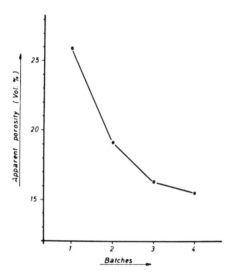

Fig. 5. Apparent porosities of mixes 1–4.

Conclusion

The bonding agents concerned have been shown to possess properties that improve the flow and thixotropic behavior of vibrated castables.

Low-Cement Castables Bonded with the Silicate Compound

The characteristics revealed through our experiments indicated very low mixing water ratios (between 4 and 8%, depending on the aggregate involved), homogeneous packing, cold crushing strengths maintained through the critical temperature range, higher hot MOR, bulk densities equal or superior to rammed refractories, and a substantial reduction in porosity.

Cement-Free, Cold-Hardening Vibrated Castables Bonded with the Aluminosilicate Compound

Experimental results in this area indicated minimum water ratios, improved cold and hot MOR and CCS values, especially through the critical and upper temperature range, increases in bulk density, very low rate of open pores, and the fact that, after drying, the cast blocks can be immediately subjected to heating and direct stress.

Vibrated Castables Containing Carbon Bonded with the Aluminosilicate with Polyelectrolytes

The experimental results for these castables revealed very low water ratios, that silicon carbide and carbon in any form could be included, a high degree of bulk density and substantial decrease in porosity, and that materials could be held in stock for longer periods with no reduction in quality.

High-Technology Castables

E.P. WEAVER, R.W. TALLEY, and A.J. ENGEL

General Refractories Company
Pittsburgh, PA 15219

Conventional castables have application limitations due to their chemical and physical properties. High-technology castables have been successfully developed by modifying the rheology of the system, resulting in a unique line of high-performance products with alumina contents of 45% to 95%. The special binding mechanism developed provides higher density, lower porosity, and improved strengths compared with conventional castables. The chemical and physical properties obtained provide these products with the ability to resist the major wear mechanisms encountered in a broad range of applications. Results of product field tests and potential application areas are also discussed.

In the past several years, refractory castable technology has advanced to a remarkable degree. The potential to install refractory linings efficiently, while maintaining the required chemical and physical properties for good performance, has been one of the major driving forces behind these advances. Improved cost effectiveness is also a major goal being addressed—virtual perpetual life through the continued patching of worn refractory linings is becoming possible. Studies on new binder systems, chemical additives, optimized grain sizing, and engineered matrix design have resulted in newly developed, high-technology refractory castables. These products rival brick in physical properties and are superior to other monolithic products. This paper compares the properties of high-technology castables to conventional products and relates these properties to improved performance against the major wear mechanisms found in service. The performance results of these products in several application areas are also covered.

Development of High-Technology Castables

Refractory castables generally fall into one or more of the following groups: conventional cement-bonded, low-cement castables, clay-bonded, ultrafine-bonded, and chemically bonded.

Each of these systems yields products with a range of physical properties. Combinations of these systems are also used to obtain the required combination of properties necessary for optimum performance.

Conventional Castables

Conventional castables typically contain 15 to 30% calcium aluminate cement as the binder. Previous improvements in this product area were largely due to improvements in cement and aggregate purities while the base technology stayed the same. The cured strength of conventional castables is due to the gel formation of the various hydrated phases of the cement binder. When heated, dehydration of the hydrated phases causes a decrease in strength between 800° and 1000°C. This decrease in strength results in a weak zone in the castable structure that is detrimental to the performance of these products. Strength at higher temperatures is attributed to ceramic bonding up to the point where liquid phases are formed and a decrease in strength occurs.

Conventional castables have lime contents of 3 to 11%, depending on the purity of the calcium-aluminate cement used and the amount present in the product. The lime present will react with available silica and alumina to form anorthite ($CaO \cdot Al_2O_3 \cdot 2SiO_2$) and/or gehlenite ($2CaO \cdot Al_2O_3 \cdot SiO_2$) liquids with relatively low melting points. These liquids will decrease the bond strength, corrosion, and erosion resistance of the product at high temperatures.

The relatively high cement content used in conventional castables also requires a relatively high water content both to achieve satisfactory flow properties for installation and to react with the cement for complete development of the hydrated phases. The required high water content limits the potential of producing lower porosity. To achieve this, new bonding systems with lower or no cement binder are required.

High-Technology Castables

High-technology castables include low-cement, clay, ultrafine, or chemically bonded products. A high-technology castable may fall into one or more of these categories, depending on the properties and characteristics required and the end-use application.

High-technology castables generally have from 0 to 5% calcium aluminate cement, which results in lime contents of up to 1.5%. The significant reduction of cement permits water contents typically in the range of 4 to 6%, which results in much lower porosity products. The low moisture requirement of these new castables also permits denser particle packing, which provides an additional improvement in density and porosity.

Due to the low lime content of these products, considerably fewer low melting point phases (anorthite and/or gehlenite) are formed at high temperatures, and strengths are dramatically improved compared with conventional castables.

Figures 1–5 compare the physical properties achieved for the new, high-technology castables with conventional monolithic and brick products of comparable alumina contents.

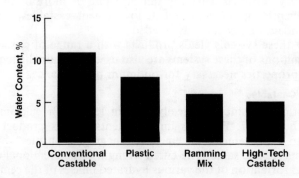

Fig. 1. Comparative water contents.

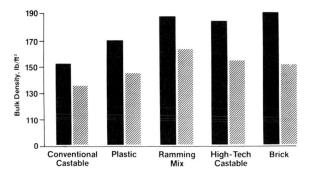

Fig. 2. Comparative bulk density; ■ 90% alumina; ▨ 60% alumina after 1370°C (2500°F) reheat.

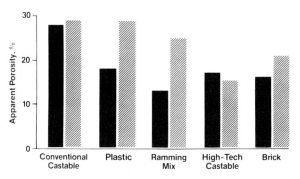

Fig. 3. Comparative porosity; ■ 90% alumina; ▨ 60% alumina after 1370°C (2500°F) reheat.

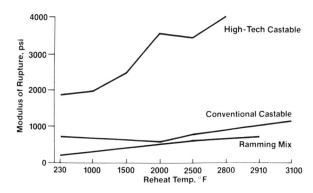

Fig. 4. Comparative cold modulus of rupture (60% alumina compositions).

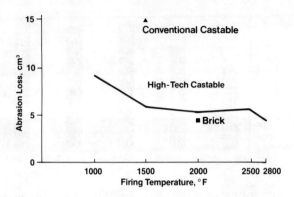

Fig. 5. Abrasion resistance (60% alumina compositions).

Installation Parameters

To achieve the potential properties of these new, high-technology castables, the proper installation is required. The key installation parameters are water content, material placement, curing, and burn-in. Figures 6 and 7 show the influence of the water content on key physical properties of a 60% alumina, high-technology castable. Excessive water contents significantly increase the porosity and decrease the strength and therefore need to be controlled within the recommended range during installation. Development tests to simulate field conditions, as well as actual field trials with this new family of products, have indicated that curing temperatures between 20° and 40°C yield optimum strengths and service life. During burn-in, heating rates of 25° to 55°C per hour are recommended to develop the maximum physical properties and prevent steam spalling of these new, dense castables.

Properties and Wear Mechanisms

The major wear mechanisms which cause refractory wear and eventual failure are corrosion, erosion, abrasion, and spalling.

The wear mechanisms present in a particular application must be analyzed when choosing the refractory product. To achieve the best performance, the key refractory properties must be reviewed in relation to the wear

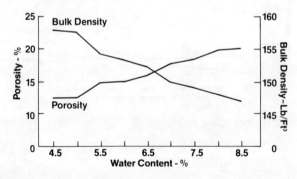

Fig. 6. Effect of water content on density and porosity after 1500°F reheat.

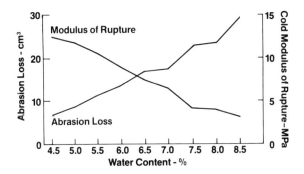

Fig. 7. Effect of water content on abrasion resistance and strength after 1500°F reheat.

mechanisms, and since there are usually several wear mechanisms present, the refractory with the best combination of key properties must be selected.

Corrosion

Corrosion of refractory products is the chemical alteration of the refractory structure due to the reaction with liquid slags or other chemical agents. Corrosion initially occurs at the refractory-slag interface, followed by penetration into the structure; further corrosion results in an altered zone near the interface, which is subsequently worn away by erosion to a penetrated zone that is subjected to further attack. The refractory-slag reaction results in an altered structure that has lower refractoriness and strength compared with the original material.

The key refractory properties required to resist wear by corrosion are chemical compatibility, density, and porosity.

Chemical Compatibility: The primary criterion in selecting the proper refractory to resist corrosion is the chemical compatibility between the refractory and the slag or other chemical agents present. Although acidic refractories are preferred for acidic slags and basic refractories for basic slags, this matching is not always possible due to other wear mechanisms present. Therefore, the resistance of refractories against a range of slag basicities must be evaluated. Since basic slags are often encountered in applications where these new, high-technology castables may be used, their slag resistance was determined vs a basic slag with a 2:1 lime-to-silica ratio. Figure 8 shows the correlation between the alumina content of these products and their resistance to corrosion and erosion by molten slag. The results show that the slag resistance increases with increasing alumina content. It should also be noted that these high-performance castables have equivalent slag resistance to a burned, 70% alumina, ladle brick at comparable alumina contents.

Density and Porosity: For products with comparable alumina contents, high-density, lower porosity materials provide improved resistance to attack by corrosion. The lower porosity reduces the penetration by molten slag, and therefore the rate of wear by corrosion and subsequent erosion is reduced. Figure 3 compares the porosity levels of the newly developed castables with other monolithic and brick products. These new products have porosity levels

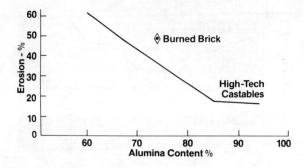

Fig. 8. Effect of alumina content on erosion resistance.

equivalent to or better than fired brick. As discussed earlier, the lower porosity is partly achieved by the significantly lower cement contents used in products that require lower water contents during installation. The low cement content also minimizes the formation of low-melting, calcium aluminosilicate liquids present at use temperature which are detrimental to corrosion resistance.

The high-technology castables have achieved the physical properties required to resist wear by corrosion. A significant improvement in physical properties has been made compared with conventional castables, and the new, high-technology castables are equal to or better than other monolithic and brick products.

Erosion

Erosion is the wear caused by the moving action of molten liquids such as iron, steel, and molten slag over the refractory. Erosion may be a primary wear mechanism in itself or a secondary mechanism following corrosion. The key refractory properties required to resist erosion are refractoriness, strength, heat stability, and density and porosity.

Refractoriness: The refractory must have acceptable refractoriness at the intended use temperature to prevent the formation of liquids with low melting points. The low lime content of the high-technology castables and the matrix design minimize the amount of low melting point phases.

Strength: The physical washing action of the liquid over the refractory surface requires the product to retain good bond strength at high temperatures to minimize the mechanical removal of the refractory hot face. Figure 9 shows the hot strength of a 60% alumina, high-technology castable compared with a burned, 60% alumina brick. The strength of the new castable is significantly higher than that of the brick in the temperature range of 540° to 1480°C. The level of strength maintained will provide excellent resistance to erosion.

Heat Stability: The positive reheat expansion exhibited by these products prevents cracks from forming under cyclic heating and cooling conditions, which would result in accelerated erosion. The relatively low thermal expansion of these products (Fig. 10) is also beneficial in maintaining a structurally sound lining.

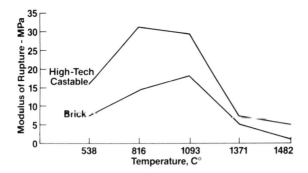

Fig. 9. Hot modulus of rupture (60% alumina).

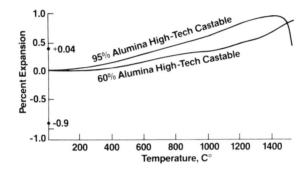

Fig. 10. Thermal expansion after drying.

Density and Porosity: The high density and low porosity obtained with these products prevents the penetration of liquids into the refractory structure, which would result in increased erosion rates. Figure 3 compares the porosity of the new castables with other conventional products.

The high density and low porosity of these castables, coupled with the excellent hot strengths, provide erosion resistance comparable to that of burned brick.

Abrasion

Abrasion of refractory products is the wearing of the refractory surface due to the surface impact by particulate matter present in the atmosphere. The key refractory properties that influence the rate of wear due to abrasion is strength.

Strength: Since many refractory aggregates will resist abrasion, the matrix of the product will determine the overall abrasion resistance of the product. The strength of the product, therefore, is an indicator of the ability of the matrix to withstand abrasion. Figure 11 shows the relationship between water content, cold modulus of rupture, and abrasion loss for a 60% alumina, high-technology castable after heating to 815°C. These results show that maximum resistance to abrasion is achieved for products with high strengths. As the water content increases, the strength and abrasion resistance are detrimentally affected. The density of the product will also influence the abrasion resistance. As previously discussed, these products have excellent densities and porosities.

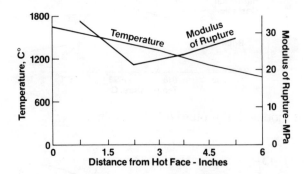

Fig. 11. Variation in temperature and strength back of hot face.

Mechanical Spalling

Mechanical spalling is the loss of refractory fragments at the working face caused by excessive stresses that develop due to the lack of sufficient expansion allowance. Resistance to mechanical spalling damage can be designed into the refractory lining; however, key refractory properties are required to minimize the potential problems.

Thermal Expansion: The lower the thermal expansion of the refractory, the less chance there is of damage due to mechanical spalling. Figure 10 shows the thermal expansion curves for 60% and 94% alumina, high-technology castables. The curves for these products show maximum thermal expansions of approximately 0.8%, which is relatively low. To resist mechanical spalling, the refractory must be strong enough to prevent cracking when exposed to the stresses present. As previously discussed, the new, high-technology castables have good strength at high temperatures.

Thermal Spalling

Thermal spalling is the loss of refractory fragments at the working face caused by excessive stresses that develop due to an excessive thermal gradient in the structure. Rapid temperature fluctuations cause severe thermal gradients. The key refractory properties which relate to good performance against thermal spalling are chemistry, porosity, strength, and thermal conductivity.

Chemistry: The susceptibility to thermal spalling depends on the rigidity of the refractory, which is influenced by the composition of the material. The range of compositions developed for the high-technology castables permits choosing the proper product to best withstand thermal shock.

Porosity: A refractory which is penetrated and reacted with molten slag densifies on the hot face and is more susceptible to thermal shock due to differences in the thermal expansion of the dense, reacted layer and the unaltered structure behind it. Since high-technology castables have lower porosity and lower lime contents than conventional castables, they are less susceptible to slag penetration and the formation of low melting point phases at the hot face that result in a densified zone.

Strength: Since strength is a function of temperature, it is possible that an excessive thermal gradient could cause cracking during burn-in of large monolithic installations in areas of low strength. A sound refractory structure is required to provide maximum resistance to further damage by thermal shock. To investigate this relationship, a 94% alumina, high-technology castable block, 45.72 by 45.72 by 15.24 cm was cast, cured at 35°C for 18 h, heated at the hot face at a rate of 55°C/h to a maximum temperature of 1650°C, held at this temperature for 5 h, and then cooled back down to room temperature. Figure 11 shows the temperature profile through the thickness of the block at the end of the hold period at 1650°C and the cold modulus of rupture strength at various distances behind the hot face. The data show that the temperature gradient through the material is essentially linear, but the strength decreases approximately 40% at a distance of 5.08 to 7.62 cm behind the hot face. These results suggest that cracking may occur behind the hot face of these monolithic linings when subjected to severe thermal shock. Additional development work is in progress that addresses this concern.

Thermal Conductivity: The composition of the product and the density and porosity will influence the thermal conductivity. A high thermal conductivity will provide better resistance to thermal shock. As indicated by the linear thermal gradients found during heatup of the block described above, the thermal conductivity is high enough to prevent excessive thermal gradients which would yield sufficient stress to cause thermal spalling.

Field Trials

Based on the results obtained in laboratory studies on 60%- and 94%-alumina, high-technology castables, field trials were conducted in various application areas to evaluate the cost effectiveness of these new prod-

Table I. Ladle Lining Results for a 60% Alumina, High-Technology Castable

Current Practice		Trial
Brick	Wet Vibratable	High-Technology Castable
High alumina	60% alumina, phosphate-bonded	Still in service since August 1983
6-9 months life	Up to 1 year life	Deskulling is much easier, with very little damage to the lining
Deskull at the end of each 16-h turn and patch with 90% alumina plastic	Deskull at end of each 16-h turn and patch with 90% alumina plastic	Less lining maintenance required "Looks as good as when it was put in service"

ucts. Table I lists the results obtained in a field trial on a small bull (teapot) ladle lined with 10.16 cm of 60% alumina, high-technology castable. The metal throughput in the ladle was 75 to 100 tons of ductile iron per shift. Heavy skulling on the ladle was periodically removed by mechanical means, usually causing damage to conventional, vibratable plastic linings. The high-technology castable proved to be extremely resistant to the mechanical abuse of deskulling, and the ladle lining is expected to have a service life of one year or more.

In a second field trial, 60% alumina, high-technology castable was used to install a 15.24 cm thick working lining in a three-strand, continuous-caster tundish. Results in Table II show that the high-technology castable gave three times the service life of the conventional vibratable plastic lining.

Table II. Tundish Lining Results for a 60% Alumina, High-Technology Castable

Standard Practice	High-Technology Castable
Wet vibratable	In service from August to November 1983.
85% alumina, phosphate-bonded	
75 heats average life	400 heats total life
100 heats shop record	300 heats without patching
Minor repairs needed every 12 to 15 heats	Minor repairs needed in the well and slagline after every 25 to 30 heats over 300 heats
Approximately 159 kg each of 85% alumina, phosphate-bonded plastic, and 85% alumina–10% chromic oxide, phosphate-bonded plastic used per patch	Approximately 159 kg each of 85% alumina phosphate-bonded plastic and 85% alumina–10% chromic oxide, phosphate-bonded plastic used per patch
Burned alumina–chrome, phosphate-bonded impact pad	After 400 heats, material below the slagline was still in good condition
	Burned alumina–chrome, phosphate-bonded impact pad

A third field trial was conducted by installing a 1.5 m long test panel in a blast furnace trough hood. A 60% alumina, high-technology castable containing 3% stainless steel fibers was vibration-cast into the hood section and tested on the main trough of the blast furnace. Results in Table III show that the high-technology castable had three times the service life of conventional gunned castables for this application.

Table III. Trough Hood Lining Results for a 60% Alumina, High-Technology Castable

Standard Practice	High-Technology Castable
15-cm thick lining consisting of 5 cm (about 2 tons) of lightweight insulating castable and 10 cm (about 8 tons) of heavyweight castable, both installed by gunning	1.5 m long test panel, 15 cm thick as part of the standard lining
Insulating castable 1201 kg/m^3, 52% alumina, 1537°C service limit	Cast rather than gunned
Heavyweight castable, 2482 kg/m^3, 75% alumina, 1760°C service limit	2563 kg/m^3, 60% alumina, 1648°C service limit
Both castables high cement and high water content	Low cement, low water content
10–12 days service life	Placed into service November 1983; removed from service January 1984
Hood sent out to be repaired as outlined above	No repairs needed for 3 campaigns
	2 hoods relined completely with high-technology castable

Conclusions

Recently developed high-technology castables have physical properties significantly better than conventional-type castables and in some areas have properties comparable to burned brick. Proper installation procedures must be followed to obtain optimum physical properties and performance of these new products. Their unique chemical and physical characteristics provide excellent resistance to the wear mechanisms present in most refractory application areas, and therefore these newly developed, high-technology castables will find a broad range of potential applications.

High-Performance Castables for Severe Applications

C. Richmond and C.E. Chaille
Babcock & Wilcox Co.
Augusta, GA 30903

Conventional castables consisting simply of graded refractory aggregates and aluminous hydraulic cement have been part of the refractories scene for many decades, but their application has been restricted to relatively less severe areas. The properties of these products are largely dependent upon the choice of refractory aggregate and hydraulic cement, and improvements generally have been restricted to improved aggregate quality and more detailed attention to the grading of the product. One of the major limitations of conventional castables has been their strength characteristics, in particular their strengths at intermediate temperatures (800°–1200°C) when the hydraulic bond breaks down and no ceramic bonding develops. Also the products tend to be porous and open textured.

Since the 1970s, much research and development has been carried out by castable refractory manufacturers to produce castables with improved characteristics to allow better performance and a much wider range of applications in the field. This work has resulted in various new ranges of castables: reduced-cement, low-cement, and cement-free types. One characteristic of these product types is that they consist of a much wider range of ingredients, sometimes present in relatively small proportions, each of which plays an important role in controlling the final product properties.

This paper describes the properties of one such range of high-performance castables, draws comparisons with equivalent-grade, conventional castables, and also reports some of the applications in which the unique properties have been successfully exploited.

Product Properties

Physical and Chemical Properties of High-Performance Castables

The high-performance castables described are based on a common bonding system that incorporates a carefully proportioned blend of fine components which, when mixed with selected aggregates, give a range of products to cover a wide range of maximum service limits and applications. The main products can be classified as follows:

40–45% alumina—based on calcined kaolin
75–85% alumina—based on bauxite
90–95% alumina—based on fused or tabular alumina

Comparisons are made with conventional castables using, in each case, the same refractory aggregates with similar alumina contents and service limits.

40–45% Alumina Castables (Table I): The maximum service limit for these products is 1550°C (2822°F). The slightly higher silica content of the high-performance castable is offset by the lower CaO and Fe_2O_3 contents, giving a lower flux content in the product at elevated temperatures.

The lower water content required for casting the high-performance castable is reflected in the density values, which are 10% higher than the conventional product. This higher density, combined with the inherent strength of the bonding system, results in cold crushing strengths significantly higher in the high-performance castable than the conventional product over the temperature range. One particular value to note is the strength at 1000°C (1832°F), where the bond in the conventional product has largely broken down, giving a very low strength.

Table I. 40-45% Alumina Castables

	Castables	
	Conventional	High-Performance
Chemical properties		
Al_2O_3	45	43.5
SiO_2	43	50
CaO	5.6	< 1.8
Fe_2O_3	2.8	1.1
Physical properties		
Max. service temp. in °C (°F)	1550 (2822)	1550 (2822)
Particle size in mm (in)	6 (0.25)	6 (0.25)
Bulk density in g/cm³ (lbs/ft³)		
110°C	2.10 (131)	2.30 (144)
1000°C	2.02 (126)	2.26 (141)
Cold crushing strengths in MPa (lbs/in²)		
110°C	23 (3300)	74 (10,500)
1000°C	13 (1850)	88 (12,500)
MST	42 (6000)	91 (13,000)
Permanent linear change (%)		
110°C	Nil	Nil
1000°C	0.2	0.3
MST	+0.2	0.8
Water required for casting (%)	11-13.5	6.0-6.5

75-80% Alumina Castables (Table II): A slightly higher alumina content and much lower CaO and Fe_2O_3 contents give the high-performance castable a much higher maximum service temperature than the standard product. However, it should be noted that the cement used in the standard product is Ciment Fondue, and not the high-alumina cement for the high-performance castable.

Again, lower water requirement for casting results in a denser product, and cold crushing strengths over the temperature range are again very much higher in the high-performance castable. Also there is no evidence of a weak intermediate zone.

Table II. 75–85% Alumina Castables

	Castables	
	Conventional	High-Performance
Chemical properties		
Al_2O_3	76	82
SiO_2	6.8	10
CaO	8.8	< 1.5
Fe_2O_3	4.8	1.5
Physical properties		
Max. service temp. in °C (°F)	1500 (2732)	1700 (3092)
Particle size in mm (in)	6 (0.25)	6 (0.25)
Bulk density in g/cm³ (lbs/ft³)		
110°C	2.56 (160)	2.76 (172)
1000°C	2.45 (153)	2.72 (170)
Cold crushing strengths in MPa (lbs/in²)		
110°C	43 (6200)	77 (11 000)
1000°C	23 (3300)	155 (16 500)
MST	60 (8500)	133 (19 000)
Permanent linear change (%)		
110°C	Nil	Nil
1000°C	0.2	0.2
MST	1.2	1.5
Water required for casting (%)	11.5–13.5	6.2–6.7

90–95% Alumina Castables (Table III): The refractory aggregate in both products is the same, and both products use high-alumina cements as part of the bonding systems. Following the pattern shown in the two previous grades of castables, the high-performance castable uses less water to cast, has higher densities, and very high cold crushing values.

From these typical values, some general differences between the conventional and high-performance product ranges can be stated. The new products are characterized by a lower water requirement to cast, higher densities, exceptionally high crushing strengths, and high-temperature strength.

One characteristic not defined in the tables is that of texture. Visual examination of the products clearly shows that the new castables have a much denser, tighter texture. Again, this is largely related to the lower water requirement to cast and also to the use of carefully selected gradings.

Relationship between Cold Crushing Strength (CCS) and Prefiring Temperature of High-Performance Castables

The results in this section have been obtained on 7.6-cm cubes prefired at temperatures up to the maximum service temperature and then crushed at room temperature. Comparisons are made with equivalent conventional castables and fired refractory brick in each grade.

Table III. 90-95% Alumina Castables

	Castables	
	Conventional	High-Performance
Chemical properties		
Al_2O_3	96	93
SiO_2	< 0.3	5
CaO	3.6	< 1.5
Fe_2O_3	< 0.2	< 0.2
Physical properties		
Max. service temp. in °C (°F)	1800 (3272)	1800 (3272)
Particle size in mm (in)	6 (0.25)	6 (0.25)
Bulk density in g/cm³ (lbs/ft³)		
110°C	2.87 (179)	3.05 (190)
1000°C	2.77 (173)	3.00 (187)
Cold crushing strengths in MPa (lbs/in²)		
110°C	42 (6000)	95 (13 500)
1000°C	31 (4500)	122 (17 500)
MST	42 (6000)	175 (25 000)
Permanent linear change (%)		
110°C	Nil	Nil
1000°C	0.1	0.2
MST	+0.5	1.2
Water required for casting (%)	7-9	4.2-5

40-45% Alumina Castables (Fig. 1): The conventional castable, cast and dried, has a CCS of around 24.5 MPa (3500 psi). This value falls slightly as the hydraulic bond breaks down to a minimum in the range of 800°-1200°C (1472°-2192°F). At higher temperatures, the strength increases significantly due to the development of ceramic bonding.

The high-performance castable strength, cast and dried, is just over 70 MPa (10 000 psi) and progressively increases to around 91 MPa (13 000 psi) at 1500°C (2822°F) with no indication of any intermediate weak zone. For comparison, the CCS of a typical, equivalent-alumina-content, fired brick is shown as being just under 70 MPa (10 000 psi) and is shown to be constant on the graph since it has already been fired to a high temperature.

75-85% Alumina Castables (Fig. 2): The conventional product, cast and dried, has a CCS of just over 42 MPa (6000 psi), but with increasing temperature, this falls to a minimum of 21 MPa (3000 psi) in the temperature range 1000°-1200°C (1832°-2182°F); a sharper decrease starts at 700°-800°C (1291°-1472°F), when the hydraulic bond starts to break down. Again, at higher temperatures, a ceramic bond appears to develop with an associated increase in strength.

The high-performance castable again shows a progressive increase from around 77 MPa (11 000 psi) as dried to nearly 140 MPa (20 000 psi) with no

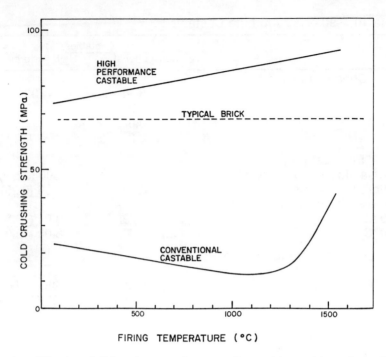

Fig. 1. Effects of firing temperature on the cold crushing strength of 40–45% alumina castables.

evidence of a weak zone. A typical fired brick of similar alumina content would have a CCS of around 91 MPa (13 000 psi).

90–95% Alumina Castables (Fig. 3): With this grade of castable, the overall picture is the same as for the two previous grades, with the conventional product having much lower CCS values and exhibiting a weaker zone at 900°–1200°C (1652°–2182°F). The high-performance castable shows a progressive increase in strength with temperature to extremely high values far in excess of those found in equivalent-alumina content bricks.

In general it can be seen from these figures that high-performance castables have extremely high strength, maintained strength at all temperatures, no weak zone, and equivalent or better strengths than similar brick.

Some caution should be used in the interpretation of CCS figures since it is possible for high cold strengths to be achieved at temperatures approaching the maximum service limit due to the formation of liquid phase within the body. On testing such compositions at room temperature, high strength values would result due to bonding contributed by the solidified liquid phase. However, in the case of the high-performance castables, the strength builds up progressively over the temperature range with no dramatic increase at elevated temperatures, which indicates that high strengths are not associated with the formation of liquid phase. Further information on the high temperature characteristics can be gained by examination of the hot modulus of rupture over the temperature range.

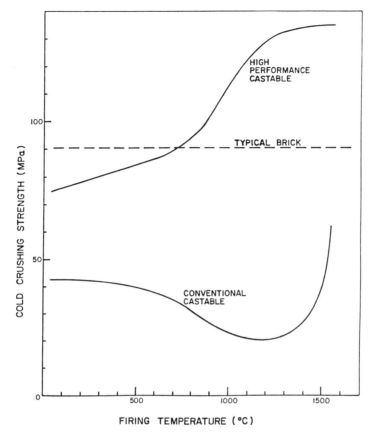

Fig. 2. Effects of firing temperature on the cold crushing strength of 75–85% alumina castables.

Hot Modulus of Rupture (MOR) of High-Performance Castables

Hot MOR values of the different grades of castables have been measured by three-point loading 152 by 25 by 25 mm (6 by 1 by 1 in) test bars at test temperature. The test bars were not prefired, but were taken up to test temperature and held for 10 min before the load was applied. Again for comparison purposes, results have been included in each case on fired bricks of equivalent alumina content.

40–45% Alumina Castables (Fig. 4): The conventional castable exhibits low hot MOR values falling from 4.6 MPa (650 psi) at room temperature to 1.4 MPa (200 psi) at 1400°C (2552°F). The high-performance castable MOR increases from 16.1 MPa (2300 psi) to a peak at 700°–800°C (1292°–1472°F) of 22.8 MPa (3250 psi), and the strength then falls progressively to about 10 MPa (700 psi) at 1400°C (2552°F). Up to approximately 950°C (1742°F), the castable is stronger than the equivalent brick. However, the strength of the brick reaches a peak at 1100°C (2012°F) and then falls dramatically.

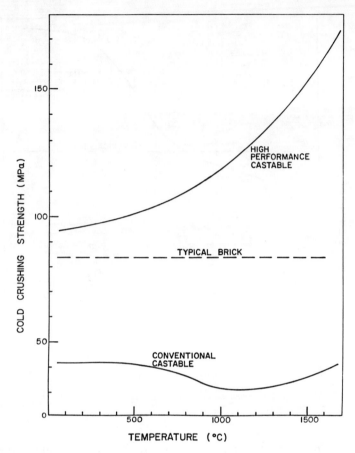

Fig. 3. Effects of firing temperature on cold crushing strength of 90–95% alumina castables.

75–85% Alumina Castables (Fig. 5): The highest values are shown by the high-performance castable, which reaches a peak value of 21.4 MPa (3050 psi) at around 800°C (1472°F) followed by a sharp decrease in strength. The hot MOR of the brick does not reach the high values of the castable; but the strength is maintained to a higher temperature before decreasing rapidly at temperatures approaching 1400°C (2552°F). The hot MOR of the conventional castable tends to be more or less constant to 1000°C (1832°F), followed by a progressive decrease.

90–95% Alumina Castables (Fig. 6): Extremely high hot MOR values are shown by the high-performance castable, starting at 23.1 MPa (3300 psi) and rising to 28 MPa (4000 psi) at 800°–900°C (1472°–1652°F). The strength then falls steeply over 1000°C (1832°F). A similar trend is shown by the brick, but it does not reach such high values below 900°C (1652°F).

Again, the hot MOR of the conventional castable remains more or less constant at 11.9 MPa (1700 psi) up to 1000°C (1832°F); then it falls progressively but maintains a higher value at the higher test temperature.

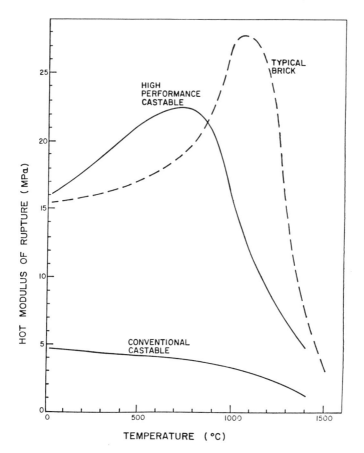

Fig. 4. Effect of temperature on hot modulus of rupture of 40–45% alumina castables.

Some general points can be made from examination of these results:

(1) At temperatures below 800°–900°C (1472°–1652°F) the high-performance castables are very strong and exceed the hot strength of fired products.

(2) At progressively higher temperatures (about 1000°C); the hot MOR value of the high-performance castable decreases more rapidly then brick. This may, to some extent, be due to the short time at temperature before the load has been applied, allowing insufficient time for ceramic bonding to develop.

(3) Conventional castables generally show much lower hot MOR values.

(4) The rapid decrease in strengths at elevated temperatures are indicative of liquid-phase formation in the products. This is supported by the results on the conventional 90–95% alumina product which has very few liquid-forming constituents; its hot strength, although generally lower than the other products, is maintained at the higher test temperatures.

Abrasion Resistance of High-Performance Castables

Good resistance to abrasion is usually associated with high cold crushing

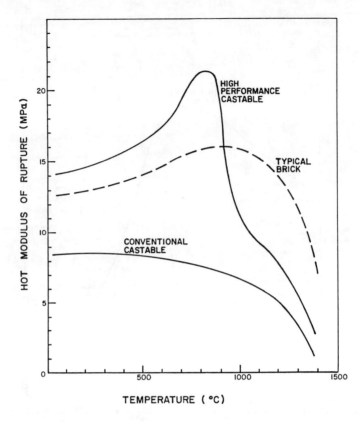

Fig. 5. Effect of temperature on hot modulus of rupture of 75–85% alumina castables.

strengths so it would not be unreasonable to expect high-performance castables to exhibit good abrasion resistance.

The results below indicate the abrasion resistance of high-alumina products as measured by the Morgan Marshall Abrasion Index (MMAI). The test measures the amount of material abraded from a test piece under standardized conditions, so that a low number indicates a good abrasion resistance:

	MMAI
Conventional 90% Alumina Castable	80
Conventional 80% Alumina Castable	90
86% Alumina Brick	80
Phosphate-Bonded 97% Alumina Castable	65
High-Performance 90% Alumina Castable	60

The high-performance castables all generally show good abrasion resistance on this test and particularly the 90% alumina grade. This value compares favorably with that of a two-part, phosphate-bonded castable, which has a good performance record in abrasive environments.

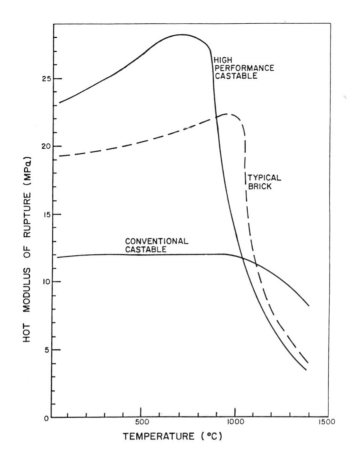

Fig. 6. Effect of temperature on hot modulus of rupture of 90–95% alumina castables.

Installation and Initial Heat

The high-performance products have to be installed under closely controlled conditions using mechanical mixers and vibration (hand-mixing is not satisfactory). It is important to adhere to the recommended water addition to achieve optimum properties.

In the early stages of development, several setbacks were encountered with most types of improved castables during the initial heating stage. Due to the density and tight texture of the products, they were very prone to explosive spalling. However, this disadvantage has been overcome by the use of such suitable additives as organic fibers, and the improved castables can now be treated in much the same way as conventional castable products.

Product Applications

Summarizing the product properties described above, the high-performance products have the following features compared with conventional products: improved strength at all temperatures; lower porosity; closer, dense texture; better hot-strength properties; and lower flux content. These features result in improved performance in conditions of severe hot abrasion, mechanical impact, and slag and metal contact.

The following brief descriptions of various application areas illustrate the success of high-performance products under these conditions.

Iron and Steel

Torpedo Ladle Throats (see Fig. 7): In a large U.K. steelworks, torpedo ladles transporting iron some distance from the blast furnace to the steelmaking unit generally showed excessive wear in the throat area after about 110 000 metric tons throughput, with additional damage to adjacent steelwork. Also severe wear was encountered during Polysius lance injection, which created heavy splash buildup on the throat.

Fig. 7. Installation of castable in the throat of a 350-ton torpedo ladle.

To counter the problem, the throat was lined with the 40–45% alumina high-performance castable in these ladles, and the first 2 installations have completed their campaigns. Both ladles operated extremely well, with no intermediate repair or any steelwork maintenance. One ladle achieved 146 141 metric tons and the other 146 914 metric tons (U.K. record).

Concast Tundishes: The high strength and mechanical abrasion resistance of the high-performance castables have given good results in the area of concast tundishes, where the impact damage in the strike area has been a limiting factor in performance.

Reheat Furnace Hearths (see Fig. 8): The high wear areas in two large, walking hearth reheat furnaces have been lined with the 75–85% alumina high-performance castable, replacing existing conventional monolithic products. Both have been successful to date, the new product giving more than double the previous performance.

Fig. 8. High-performance castables in the hearth of a walking hearth reheat furnace.

Pelletizing Kilns: On a large pelletizing kiln (iron ore), the nose ring at the discharge end gave excessive wear problems. Conventional castables and bricks were tried without success, all giving very short lives. The nose ring was installed using the 80–85% alumina, high-performance castable. This was very satisfactory, surviving over 1.25 million metric tons of pellets throughput until, unfortunately, the plant was "mothballed."

Direct-Reduction Kilns (see Figs. 9 and 10): Severe problems due mainly to extremely abrasive conditions resulted in poor lining performance in large, direct-reduction kilns producing prereduced pellets. Several conventional

Fig. 9. Large, rotary ore, direct-reduction kiln.

Fig. 10. Installing high-performance castable lining in direct-reduction kiln.

refractory materials were tried without success; all bricks and castables showed unacceptably high wear rates.

A small, controlled trial of high-performance product gave very promising results, so a large part of the working lining of one kiln was installed during mid-1983. The performance of this lining quickly demonstrated that the product was giving a much lower wear rate, and subsequently 5 out of the 11 kilns on-site have now been lined with the material.

Cement Kilns

Nose Rings: Several installations of high-performance castables have been

Fig. 11. Precast, high-performance castable lining in planetary cooler bends.

carried out in the nose rings of rotary cement kilns, all giving satisfactory performance, and generally out-performing previous materials.

Planetary Cooler Bends (see Fig. 11): The cooler bends in planetary coolers are an area of very high abrasive conditions and have been a problem area for kiln operators. At a plant using planetary coolers, some initial repairs that used the 75-85% alumina castable indicated the potential of this type of material for the application. In January 1983, a complete set of 10 cooler bends was lined with high-performance castables. The 40-45% grade was used in the less abrasive areas and the 75-85% grade in the more abrasive areas. Ten months later, after continuous operation, the linings were inspected and revealed negligible wear.

Lifter Blocks: To improve the efficiency of the cement kiln operation, several kilns have been installed using precast lifter blocks. The environment to which the blocks were subjected is extremely abrasive, and little protection is offered since they stand above the main lining. Under these conditions, high-performance castables have proved very satisfactory.

Retaining Rings: In this application in the cement kiln, abrasion resistance and high mechanical strength are required to give good lining performance. The high-strength characteristics of the high-performance castable (75-85% grade) have enabled the product to give satisfactory performance in this area.

Oil and Petrochemical Industry

Catalytic Crackers: In such areas of highest abrasion as the feed bayonet, dip leg shields, and areas inside the reactor, it has long been the practice to line these areas with high-alumina castable materials. To give the strengths and abrasion resistances required in these areas at the relatively low operating temperatures, it has been necessary to use phosphate-bonded products that are usually provided as two-part mixes.

Special grades of the 90-95% alumina, high-performance castable have been developed for this application and have been very successfully introduced into the U.K. oil and petrochemical industry over the past two years. The product performs well; it is easy to install and is less wasteful than existing materials. Installers find the single-part mix, which requires only the addition of water, simplifies the mixing, and its ability to remain pliable for up to ¾ hour gives greater flexibility. The product is now proving successful in other high wear areas in this industry.

Miscellaneous Applications

Ferrosilicon Furnaces (see Fig. 12): One of the first successful applications of the 90-95% alumina high-performance castable was in very large precast blocks for lining the upper rotating ring in a large ferrosilicon furnace (electric arc furnace operation). The lining gave a very satisfactory performance and easily out-performed the previous 90% alumina conventional castable due to its high strength and mechanical abrasion characteristics.

Kiln Cars: The 40-45% alumina, high-performance castable has been used very successfully in a kiln car application where severe mechanical damage to bricks on the car deck was giving problems. Large, precast blocks were used, which gave a quicker installation than brick, a flatter deck, as well as improvement in impact resistance.

Fig. 12. Large, precast, 90–95% alumina castable blocks for rotary ring in ferrosilicon furance.

Aluminum Ladles: At a large, primary aluminum producer, the aluminum anodes are remelted in a reverberatory furnace. Attached to this furnace are large, semifixed syphon ladles which allow the cryolite slag to be cleaned off the metal before ingot casting. Two problems were associated with these ladles: In the first case, penetration of the molten aluminum into the lining promoted rapid wear; in the second, mechanical damage occurred when cryolite buildup was removed from the ladle by pneumatic tools. The ladles were lined with a single-thickness lining of the high-strength, high-density and low-porosity, 75–85% castable to give a good resistance to penetration and mechanical damage. The lining was successful in achieving at least a 50% improvement in life and in lowering maintenance and cleaning requirements, giving greater ladle availability.

Conclusions

A range of high-performance castables have been developed having the following features: improved strength at all temperatures; lower porosity; close, dense texture; better creep resistance; and lower flux content. The above features result in improved performance in conditions of severe hot abrasion, mechanical impact, and slag and metal contact.

Progress of Additives in Monolithic Refractories

Yoichi Naruse, Shoichiro Fujimoto,
Sukekazu Kiwaki, and Masaaki Mishima

Kurosaki Refractories Co., Ltd.
1-1, Higashihama, Yawata-nishi,
Kitakyushu, Japan

On the basis of knowledge of the dispersion properties of aluminous cement, clay, and other refractory particles in an aqueous system, new types of monolithic refractories were developed. Additives such as deflocculants, superfine particles, and other active powders are indispensable to these monolithic refractories.

Due to recent technological progress in pig iron and steelmaking, refractories used in these areas are increasingly subjected to severe conditions. This situation also applies to monolithic refractories, so that highly refractive, corrosion-resistant, etc. refractories are required.

Castable refractories are mixed with a quantity of water appropriate to the method of installation (e.g., casting troweling, gunning, or injection) and contain considerable amounts of aluminous cement as a binder. The refractoriness and the corrosion resistance of installations are governed by either the density and homogeneity of installations or the quantity of compounds which have low refractoriness, i.e., are produced by the reaction between aluminous cement and aggregates at high temperatures. The most important problem is how the quantity of compounds having low refractoriness can be decreased without affecting other properties of castable installations.

In this report, some functions of the recently developed additives presently used in monolithic refractories are described as well as some of their advantages over conventional castables. Additives used in a clay-bonded castable due to the flocculation-deflocculation phenomenon of clay suspensions are also described.

In addition, some information is given concerning the addition of the active metal powders frequently used to shorten curing and drying times for monolithic installations.

Utilization of Superfine Powders, Deflocculants, and Other Additives

Since conventional castables contain a considerable quantity of aluminous cement as a binder, they possess some resultant defects: lower refractoriness and lower mechanical strength.

Refractoriness

Aluminous cement reacts with aggregates in a castable at high temperatures, which lowers its refractoriness. The principal mineral component of aluminous cement is calcium aluminate ($CaO \cdot Al_2O_3$). This component and its hydrates react with silica in the castable aggregates to produce anorthite

($CaO \cdot Al_2O_3 \cdot 2SiO_2$) or gehlenite ($2CaO \cdot Al_2O_3 \cdot SiO_2$); hence, the refractoriness and corrosion resistance of the resulting castable is low. These phenomena are remarkable in the matrix of a castable. The relation between the quantity of aluminous cement and the refractoriness of the castable is represented in Table I.

Table I. Effect of Aluminous Cement Content on Refractoriness of Castable Refractories.

	(I)	(II)	(III)	(IV)
Aggregate (%)				
Chamotte (coarse)	60	60	60	60
(fine)	10	15	20	25
Alumina (fine)	10	10	10	10
Aluminous cement	20	15	10	5
Refractoriness (Segel Kegel)				
Castable	31.5	32	33.5	35.5
Matrix	26	27	28.5	33

Mechanical Strength

Mechanical strength of a castable installation is lowered due to the conversion and dehydration of aluminous cement hydrate. The hardening of a castable installation results from the hydration of aluminous cement. Decrease in castable strength occurs after conversion to the metastable hydrate phase ($3CaO \cdot Al_2O_3 \cdot 6H_2O$, $Al_2O_3 \cdot 3H_2O$) at room temperature.[1,2] It has been reported that the castable installations gradually carbonated into dusting and peeling in the air.[3] Moreover, the conversion and dehydration of aluminous cement hydrates at temperatures between 300° and 1000°C result in decreased strength of a castable installation.

Aluminous cement is detrimental to the physical properties of castable installations at high temperatures or after heating; therefore, the most important problem is how to lower the quantity of aluminous cement in a castable while maintaining the strength of the castable installation.

Observing the matrix of a conventional castable, it can be seen that aluminous cement grains are nonuniformly distributed, thus rendering aluminous cement imperfect in filling the role of a binder.

Since the strength of a concrete mass is increased by the decrease in the water/cement ratio, a high-strength castable installation should be produced through the uniform distribution of aluminous cement grains and a decrease in the quantity of mixing water, notwithstanding the considerably lowered quantity of aluminous cement in a conventional castable.

The zeta potential of an aluminous cement grain was determined by means of electrophoresis to clarify the effect of some defloculants on an aluminous cement grain. The results are shown in Fig. 1. An aluminous cement grain originally has a positive zeta potential. The Ca^{2+} and Al^{3+} dissolved from a cement grain contribute to the sign. When defloculants are added to the cement suspension, the zeta potential of an aluminous cement

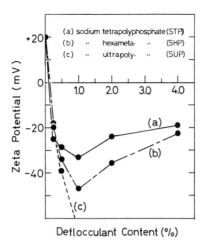

Fig. 1. Zeta potential as a function of defloculant content for aluminous cement at 20°C.

grain changes to a negative sign, and the minimum point in the zeta potential appears at the specific value of the deflocculant content. The magnitude of the zeta potential of an aluminous cement grain is close to zero in the region where the deflocculant content is high, a phenomenon ascribed to the screening effect of a charged particle.

Therefore, the flocculation among aluminous cement grains in a castable mix is prevented by the use of an appropriate quantity of these deflocculants. These deflocculants also act to increase the zeta potential of oxide grains, so a decrease in the quantity of mixing water is also accomplished.

The setting of a castable is due to the hydration of aluminous cement. Various theories of the hydration mechanism of aluminous cement have been reported. On the supposition that dissolution of Ca^{2+} and Al^{3+} from a cement grain is closely related to the hydration, the influence of these deflocculants on the change of the concentration of Ca^{2+} in an aluminous cement suspension was studied. The results, shown in Fig. 2, reveal that the action of a deflocculant on suppression of increased Ca^{2+} concentration in the aluminous cement suspension is the strongest from sodium ultrapolyphosphate (SUP) and less strong from sodium hexametaphosphate (SHP) or sodium tetrapolyphosphate (STP). Suppression from SUP increases with its content in the suspension.

These results have the following implications:

(1) A uniform distribution of aluminous cement grains in a castable mix and a decrease in the quantity of mixing water is obtained by adding 1% SHP or SUP to the aluminous cement of a castable.

(2) Regarding the pot life of a castable mix, it is desirable to add more than 1.5% SUP to the aluminous cement of a castable. A sufficient pot life is not gained by the addition of SHP or SUP alone, so it is necessary to use an appropriate retarder with these deflocculants.

(3) The density of a castable installation can be improved by addition of superfine powders along with these deflocculants. The deflocculants affect

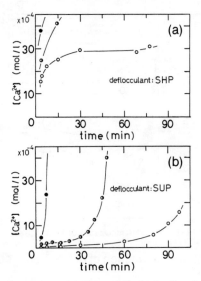

Fig. 2(a). Change in concentration of Ca^{2+} in the liquid phase of aluminous cement–deflocculant–water system. (b) Change in concentration of Ca^{2+} in the liquid phase of aluminous cement–SUP deflocculant–water system. Deflocculant content: 1.0 (——●——), 1.5 (——◐——), and 2.0% (——○——) of the aluminous cement.

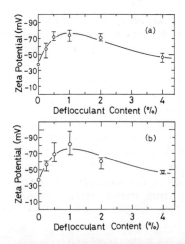

Fig. 3(a). Zeta potential as a function of STP deflocculant content for alumina at 20°C; (b) zeta potential as a function of SHP deflocculant content for alumina at 20°C. Solid lines are the prediction from a simple model (Ref. 4).

superfine powders and other fine aggregates as well. Their effect on the zeta potential of alumina grain is shown in Fig. 3.

Using the additives described above, a high-strength, dense castable, superior in density, strength, and refractoriness, was developed. A comparison of microscopic distribution of aluminous cement grains in a high-strength, dense castable and a conventional one is shown in Fig. 4. Flocculated grains of aluminous cement are seen in a conventional castable but not in a high-strength, dense castable.

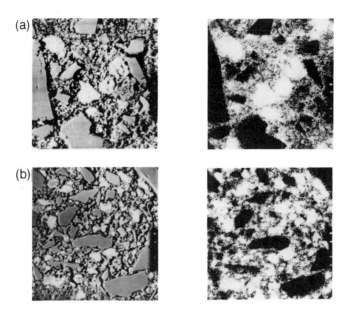

Fig. 4. Secondary electron images (×150) and characteristic X-ray images by calcium of conventional castable (a) and high-strength, dense castable (b).

The physical properties of a high-strength, dense castable compared with a conventional castable are shown in Table II. The results of refractoriness under load for these castables are shown in Fig. 5. The apparent initial softening temperature in the high-strength, dense castable is higher than that in the conventional castable. The degree of deformation in the high-strength, dense castable is less than that in the conventional castable. The X-ray diffraction patterns of these castables after heating at 1400°C are shown in Fig. 6. The diffraction intensity from the anorthite produced by the reaction between aluminous cement and chamotte aggregate in the high-strength, dense castable is weaker than that in the conventional castable.

Utilization of Clays

Collodial properties are observed in clay suspensions. Clays swell to form a gel as they absorb a large quantity of water. Gel clays are changed into the suspension by the addition of more water. When an electrolyte is added to the suspension, the clay particles cause the aggregated structure to precipitate because clay particles have a property of cation substitution. Studies of these properties yielded the following results:

Table II. Properties of a Typical High-Strength, Dense Castable and Conventional Castable

	H.S.D.C.*		C.C.†
	Casting	Vibrating	Casting
Chemical analysis (%)			
Al_2O_3		54	56
SiO_2		36	35
CaO		2.9	5.3
other		7.1	3.7
Mixing water (%)	8.5	6.7	13.0
Mixing time (min)	6	3	3
Bulk density			
110°C/24 h	2.42	2.56	2.18
1500°C/3 h	2.40	2.51	2.12
Apparent porosity (%)			
110°C/24 h	15.2	10.7	22.1
1500°C/3 h	15.0	11.4	19.8
Crushing strength (MPa)			
110°C/24 h	53.4	58.8	16.7
1500°C/3 h	96.1	106.2	36.3
Modulus of rupture (MPa)			
110°C/24 h	9.3	13.5	3.7
1500°C/3 h	20.2	27.2	10.3
Hot modulus of rupture (MPa)			
1400°C/1 h	1.5	2.2	0.6
Permanent linear change (%)			
1500°C/3 h	0.08	0.10	−0.20
Index of abrasion resistance			
sandblast method	46	38	100

*High-strength, dense castable.
†Conventional castable.

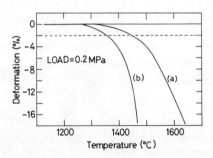

Fig. 5. Characteristics of refractoriness under load of a high-strength, dense, 8% aluminous cement castable (a) and a conventional 15% aluminous cement castable (b).

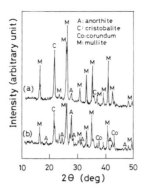

Fig. 6. X-ray diffraction patterns of a high-strength, dense castable (a) and a conventional castable (b) after heating at 1400°C.

The presence of montmorillonite in clay is a major factor affecting the rheological properties of clay suspensions. Water and ions are absorbed into the interlayers of montmorillonite crystals. The dependence of the viscosity of a clay suspension on water content is shown in Fig. 7. The clay slurry with

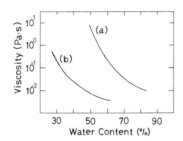

Fig. 7. Variation in viscosity with water content for two clay slurries containing a deflocculant. Clay in slurry (a) contains montmorillonite; clay in slurry (b) does not contain montmorillonite.

montmorillonite shows a higher viscosity value than that without montmorillonite. The influence of additives on the flocculation of clay slurries can be seen in Fig. 8. When sodium pyrophosphate is added to a clay slurry containing 20% calcium aluminate of clay, the slurry sets after 10 or 15 min. The Ca^{2+} potential of the slurry is in the region of -40 and -30 mV at this time, as shown in Fig. 8(a).

When SHP is added to the same slurry, the slurry sets after 30 or 35 min. The Ca^{2+} potential of the slurry is in the region of -10 and -5 mV at this time (Fig. 8(b)). When sodium pyrophosphate is added to the clay slurry containing 20% calcium silicate of clay, the slurry sets after 25 min (Fig. 8(c)).

The Ca^{2+} potential and precipitation of the polyphosphate solution were studied after adding 0.1 $mol^{-1} \cdot L$ calcium chloride to examine the occurrence

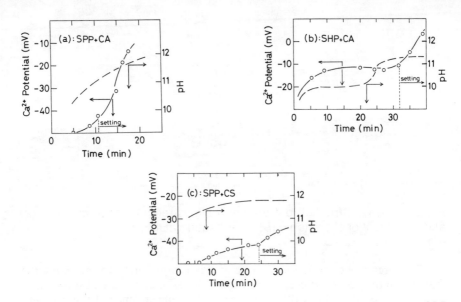

Fig. 8(a). Changes in Ca^{2+} potential, pH, and setting time of clay–SPP deflocculant–CA lime compound system. (b) Changes in Ca^{2+} potential, pH, and setting time of clay–SHP deflocculant–CA lime compound system. (c) Changes in Ca^{2+} potential, pH, and setting time of clay–SPP deflocculant–CS lime compound system.

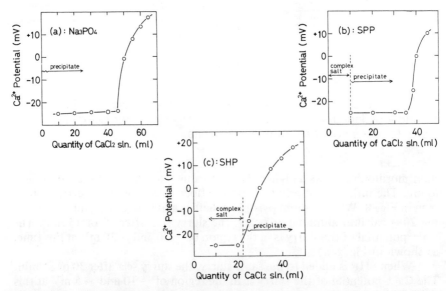

Fig. 9(a). Variation of Ca^{2+} potential in the tribasic sodium phosphate solutions with additional quantity of 0.1 mol·L^{-1}. (b) Variation of Ca^{2+} potential in the sodium pyrophosphate solutions with additional quantity of 0.1 mol·L^{-1}. (c) Variation of Ca^{2+} potential in the sodium hexametaphosphate solutions with additional quantity of 0.1 mol·L^{-1}.

of precipitation from the reaction between these polyphosphates and Ca^{2+} ions. The results are shown in Fig. 9. When the calcium chloride is added to tribasic sodium phosphate for reference, tribasic calcium phosphate is produced immediately. An increase in Ca^{2+} potential of the solution is observed when more than equivalent calcium chloride quantities are added (Fig. 9(a)). When the calcium chloride is added to the SHP, precipitation is not observed before the increase in Ca^{2+} potential of the solution; thus appears that there is a calcium complex in the solution (Fig. 9(b)). When the calcium chloride is added to the sodium pyrophosphate, the intermediate behavior between the tribasic phosphate and hexametaphosphate is seen as in Fig. 9(c). It is conjectured that the decreased fluidity of clay slurries resulting from precipitation is chiefly related to the setting of clay slurries.

It is possible to use clay as a binder for monolithic refractories utilizing the properties described above: A clay-bonded castable was developed which

Table III. Properties of Clay-Bonded Castable vs Conventional Castable

	Clay-Bonded	Conventional
Chemical analysis (%)		
Al_2O_3	52	50
SiO_2	43	41
Mixing water (%)	9.7	12.0
Bulk density		
110°C/24 h	2.31	2.15
800°C/3 h	2.28	2.13
1500°C/3 h	2.20	2.10
Apparent porosity (%)		
110°C/24 h	19.4	19.0
800°C/3 h	20.9	17.0
1500°C/3 h	20.4	14.5
Crushing strength (MPa)		
110°C/24 h	5.7	15.7
800°C/3 h	16.7	13.7
1500°C/3 h	37.5	31.4
Modulus of rupture (MPa)		
110°C/24 h	1.2	3.9
800°C/3 h	2.1	2.9
1500°C/3 h	9.2	8.3
Hot modulus of rupture (MPa)		
1400°C/1 h	1.5	0.5
Permanent linear change (%)		
110°C/24 h	−0.24	−0.05
800°C/3 h	+0.08	−0.10
1500°C/3 h	+0.86	−0.84
Thermal conductivity ($W \cdot m^{-1} \cdot K^{-1}$)		
350°C	0.93	0.78
500°C	0.95	0.79
Refractoriness (S.K.)	36	33+

had less decrease in refractoriness based on the reaction between an aggregate and a binder and had superior properties at high temperature. A comparison of the physical properties of the clay-bonded castable with those of conventional castables is shown in Table III. Hot modulus of rupture and refractoriness of the clay-bonded castable are superior to those of a conventional castable. The temperature dependence on setting time of the clay-bonded castable is shown in Fig. 10. The clay-bonded castable has a very long setting time at less than 10°C.

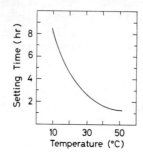

Fig. 10. Temperature dependence of setting time for a clay-bonded castable.

Utilization of Metal Powder

Explosive spalling sometimes happens to recent monolithic refractories on a conventional drying schedule, whereas denseness, strength, and corrosion resistance of these monolithic refractories are noticeably improved by making use of appropriate superfine powders and deflocculants. There are some reports on a method of adding a metal powder to clear the explosive spalling.[5,6]

It is expected that the exothermic reaction of aluminum powder is effective in decreasing the setting time and strengthening a clay-bonded castable. The heat evolution of a clay-bonded castable containing aluminum powder is shown in Fig. 11. A plain exothermic peak appeared in a clay-bonded castable containing aluminum powder, whereas the peak did not appear in a clay-bonded castable. Temperature dependence of the strength of a clay-bonded

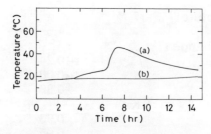

Fig. 11. Heat evolution curves for Al-containing clay-bonded castable (a) and clay-bonded castable (b).

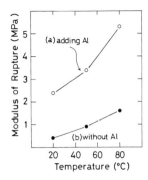

Fig. 12. Variation of modulus of rupture with curing temperature for Al-containing clay-bonded castable (a) and clay-bonded castable (b).

castable is shown in Fig. 12. It can be seen that addition of aluminum powder to a clay-bonded castable is effective in hastening the generation of strength. It is necessary to make a further study to determine the dominant factor affecting the generation of strength of a clay-bonded castable containing aluminum powder.

Summary

Utilization of the deflocculants and superfine powders in the high-strength, dense castable; the rheological and colloidal properties of clay suspensions in the clay-bonded castable; and the metal powder in the high-strength clay-bonded castable are described. These castables are developed by using appropriate properties of the additives. The use of monolithic refractories will undoubtedly be extended because of their jointless installations. Therefore, monolithic refractories with such various properties as insulation, thermal shock resistance, corrosion resistance, and high-temperature strength, not possessed by conventional monolithic refractories, are required.

It will be possible to develop these monolithic refractories by using various kinds of additives according to their characteristics. Studies of these monolithic refractories are now in progress.

Acknowledgments

We would like to express our appreciation to the president, Mr. T. Shibayama, for permitting us to present this report. This report was made possible due to the pioneering efforts of I. Takita and J. Yoshitomi.

References

[1] A.C.C. Tseung and T.G. Carruthers, Refractory Concretes Based on Pure Calcium Aluminate Cement," *Trans. Br. Ceram.Soc.*, **62** [4] 305-20 (1963).

[2] P.K. Mehta and G. Lesnikoff, "Conversion of $CaO \cdot Al_2O_3 \cdot 10H_2O$ to $3CaO \cdot Al_2O_3 \cdot 6H_2O$," *J. Am. Ceram. Soc.*, **54** [4] 210-12 (1971).

[3] Y. Hongo, "Carbonation of Castable Refractory," *Refractories* (in Japanese), **30** [12] 690-95 (1978).

[4]Y. Naruse, K. Semba, S. Kiwaki, and M. Mishima, "Influence of Deflocculant on the Isoelectric Point of Refractory Powders," *Refractories* (in Japanese), **33** [2] 33-39 (1981).

[5]T.Taniguchi and K. Watanabe, "Some Results of Testing on Prepared Unshaped Refractories Containing Aluminum Powder, *Refractories* (in Japanese), **29** [11] 591-95 (1977).

[6]T. Taniguchi, M. Isikawa, and J. Ohba, "Drying Character of Castable Refractories Containing Al Powder," *Refractories* (in Japanese), **33** [8] 440-45 (1981).

Low-Moisture Castables: Properties and Applications

SUBRATA BANERJEE, RONALD V. KILGORE, AND DAVID A. KNOWLTON

Gunning Refractories, Inc.
Subsidiary of BMI, Inc.
South Webster, OH 45682

A new generation of low-moisture castables has been developed with high density, low porosity, high reheat and hot strengths often exceeding values from brick with comparable alumina contents. Their applications range from blast furnace troughs, iron and slag runners, blow pipes, reheat furnaces, desulfurizing lances, ladles, tundishes for concast, and coke oven door plugs to petroleum refineries and aluminum and other nonferrous industries. The ease of installation, short preparation time before use, superior performance, and nonhazardous nature make them highly attractive for the consumer.

Over the past decade, more and more monolithic refractories have been used in place of brick, since brickmaking is an extremely energy-intensive process. Moreover, given increased labor costs, the monoliths require less labor for both installation and maintenance, an obvious advantage. Among the monoliths, castables and gunning mixes containing calcium aluminate cements rather than other binders are the most used because of their refractoriness, easy installation, and longer shelf life.

The major disadvantage of conventional castables that contain a 10-30% cement has been their need for high water content, which causes higher porosity, lower strength, and slow, controlled curing and drying processes. Another major drawback is their loss of strength during the 538°–982°C dehydration process. Moreover, the upper limit to their hot strength is 1371°C, above which there is a sharp drop due to the fluxing action of CaO in the cement.

The present work demonstrates some of the reasons why this new generation of low-moisture castables has emerged as the most viable type of product for a wide variety of applications.

Previous work using lower amounts of cements in conjunction with fine-grain-sized material with high surface area and a dispersing aid was reported by Prost and Pauillac,[1] Crookston and Fitzpatrick,[2] Havranek and Thornbald,[3] and Havranek.[4] Kiehl reported that use of Cr_2O_3 powder in place of amorphous silica resulted in much higher hot strength.[5] Highlights of these publications are shown in Figs. 1–5.

Experimental Procedure

The total amount of alumina content dictated the choice of main aggregates in the mix compositions. The grain sizings of the aggregates with respect to their bulk density were carefully selected to obtain the highest density and lowest porosity in the composition. The dry ingredients were

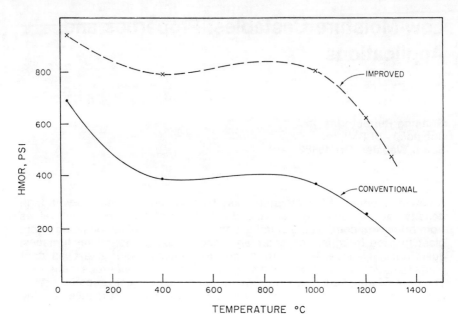

Fig. 1. Hot modulus of rupture at different temperatures of conventional and improved castables (Ref. 1).

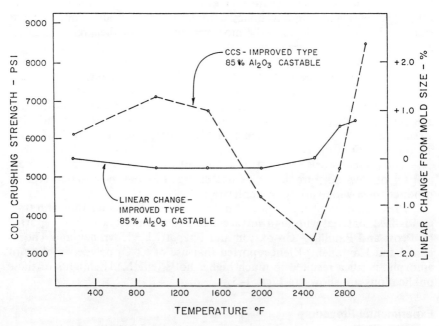

Fig. 2. Cold crushing strength and linear change of improved 85% alumina castable fired at different temperature (Ref. 2).

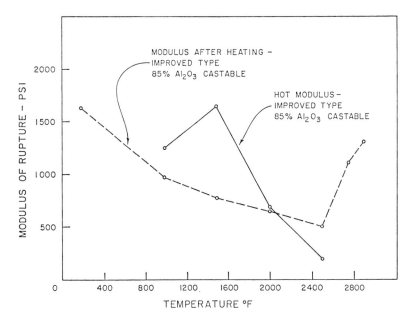

Fig. 3. Cold and hot moduli of rupture of 85% alumina castable at different temperatures (Ref. 2).

mixed for 1 min in a Hobart mixer; the appropriate amount of water was then added and mixed for another 3 to 4 min. After wet mixing, the mix looked drier than conventional castables. These mixes were then cast into 3.8 by 3.8 by 20.3 cm bars on a vibrating table. Vibration (3600 vpm) was applied for about 1 min until the material flowed in the mold and most of the trapped air was removed. The vibratility was determined by the degree of flow of a special volume of material on the vibrating table.

The bars formed in the molds were then covered by plastic and cured for 24 h at 21°–32°C. The cured bars were then stripped from the mold and dried at 104°C for 24 h. Subsequently, bars were heated to 954°C for 4 h, 1371°C for 5 h, and 1593°C for 5 h. Bulk density, cold modulus of rupture (MOR) (three-point bending load), cold compressive strength, and linear changes were measured. Hot modulus of rupture for 90% alumina was determined at 1371°C and also at 1482°C. All tests were conducted according to ASTM standard procedures.

Thermal shock tests were performed in the ribbon furnace on 22.8 by 6.4 by 3.8 cm vibracast bars that were cured for 24 h, dried for 24 h at 104°C, and then fired at 1371°C for 5 h. The material was thermally shocked for 10 cycles by heating the bars at 982°C for 15 min and then cooling with forced air for 15 min. Sonic velocity was measured before and after the shock test.

Slag erosion testing was done in a 3 kW induction furnace. The 3.8 by 3.8 by 20.3 cm bars were cast, cured 24 h, and dried for 24 h before they were subjected to slag erosion testing. Twenty-three kg of iron and 2.7 kg of blast furnace slag were used. The temperatures were measured from the top by an optical pyrometer and ranged from 1500°C (2732°F) to 1530°C

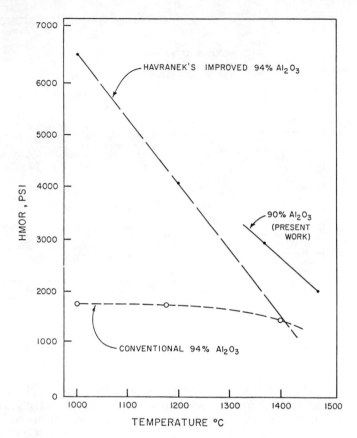

Fig. 4. Hot modulus of rupture of 94% alumina, low-moisture and conventional castable at different temperatures (Ref. 3).

(2786°F). The set of 4 bars was rotated at 12 rpm for 2 h. After each test, visual examination was made first. The bars were then sawed through the worst eroded faces to measure with a planimeter the area lost by slag erosion. Abrasion resistance was run according to ASTM C-704 on 11.4 by 11.4 by 6.4 cm brick samples fired to 815°C for 4 h.

Experiments were conducted to determine whether the material would adhere to itself and how strong the bond would be. Bars were fired to 1538°C for 5 h and broken for MOR values; fresh material was then vibrated against one portion of the bar, followed by curing, drying, and refiring at different temperatures. The bars were then broken for MOR after reheating at 954°C for 4 h, 1371°C for 5 h, and 1593°C for 5 h.

Results and Discussion

The properties (cold compressive strength, cold MOR, and porosity) of some of the high-alumina castables are shown in Figs. 6–8 on the basis of the alumina content of the products. Figure 9 shows the hot MOR run at 1371°C, where the strength increased with alumina content and peaked at 75% alumina and decreased at 90% alumina, probably because in 75%

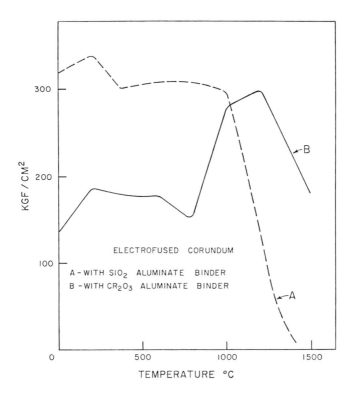

Fig. 5. Hot modulus of rupture of castables with silica and chromium oxide at different temperatures (Ref. 5).

alumina, the matrix and the aggregate approached a condition of all mullite. Chemical analysis and physical properties are shown in Table I.

There are some highly noticeable characteristics of these products. The dried strengths of all the products are very high and continue to increase throughout the temperature range, quite unlike conventional castables. The porosities are also low, better than brick with equivalent alumina content.

Since the low-moisture castables need small quantities of water, attempts have been made to determine the effect of water content on their properties. Figure 10 shows the relationship of the different water contents to the bulk densities and reheat strengths of one particular mix composition. The properties deteriorate linearly with water content until the temperature is reached at which ceramic bonding starts. Strength due to ceramic bonding is not affected until a very high water content is reached, whereupon the bulk density drops off noticeably.

Since calcium aluminate cement is the binder in these products, the effects of curing time and conditions on their strengths were also studied (Fig. 11), and revealed that these low-moisture castables can be cured and dried in a reasonably short time without sacrifice of strength.

Figure 12 shows the thermal expansion data run on the 60, 65, and 90% Al_2O_3 castables. One of each sample was run after drying at 104°C, and the

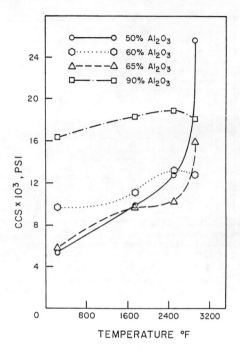

Fig. 6. Cold compressive strengths of low-moisture castables with varying alumina contents after firing at different temperatures.

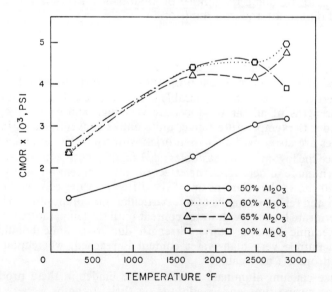

Fig. 7. Cold modulus of rupture of low-moisture castables with varying alumina content after firing at different temperatures.

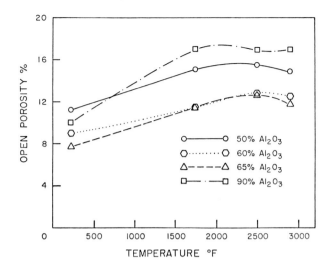

Fig. 8. Open porosity of low-moisture castables with varying alumina contents after firing at different temperatures.

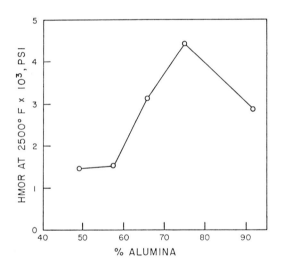

Fig. 9. Hot modulus of rupture at 2500°F of low-moisture castables with varying alumina contents.

other after prefiring at 1538°C for 1 h. The differences among the three castables are noticeable; the 65% alumina castable seems to be the most volume-stable.

Creep deformation was tested on some of these materials to compare with published data on brick of similar alumina contents. Figure 13 shows the creep run for 100 h at 1425°C and 172.4×10^2 Pa load. The 60% alumina castable has a maximum creep of 1.5%, and the 65% alumina castable has a

Table I. Properties of Some Low-Moisture Castables

Chemical analysis	1	2	3	4	5	6
SiO_2	47.8	37.7	39.3	30.0	22.0	7.2
Al_2O_3	48.2	57.8	57.0	65.9	74.8	91.6
Fe_2O_3	0.9	0.8	0.7	0.9	0.7	0.04
TiO_2	1.2	1.8	1.8	1.6	1.2	
CaO	1.2	1.2	1.2	1.2	1.2	1.2
Percent water used for vibracast	5.2	4.1	5.3	4.4	4.0	4.5
Physical properties						
Bulk density (lb/ft³)						
after 220°F/24 h	150	160	159	163	169	178
after 1750°F/ 4 h	148	160	158	163	168	179
after 2500°F/ 5 h	145	155	155	163	167	178
after 2900°F/ 5 h	151	155	154	160	166	177
Porosity, open (%)						
after 220°F/24 h	11.2	9.0	8.7	7.7	9.2	10.0
after 1750°F/ 4 h	15.1	11.7	11.8	11.3	14.3	15.0
after 2500°F/ 5 h	15.5	12.8	12.1	12.7	13.7	15.0
after 2900°F/ 5 h	14.9	12.5	12.0	11.9	13.9	15.0

CMOR (psi)					
after 220 °F/24 h	1 475	2 400	2 400	2 350	2 200	2 600
after 1750 °F/ 4 h	2 250	4 400	4 300	4 200	3 775	4 400
after 2500 °F/ 5 h	3 000	4 500	4 500	4 150	3 561	4 550
after 2900 °F/ 5 h	3 150	4 950	5 000	4 750	4 180	3 900
CCS (psi)					
after 220 °F/24 h	5 250	9 700	9 700	5 850	9 985	16 205
after 1750 °F/ 4 h	9 800	11 100	15 400	9 800	13 580	18 300
after 2500 °F/ 5 h	12 750	13 350	17 300	10 100	10 900	18 800
after 2900 °F/ 5 h	25 500	12 700	17 200	15 900	13 000	18 000
Linear change (%)					
after 1750 °F/ 4 h	0.00	−0.2	−0.3	−0.3	−0.2	−0.1
after 2500 °F/ 5 h	+0.05	+0.7	+0.6	−0.3	−0.4	−0.3
after 2900 °F/ 5 h	−0.05	+0.8	+0.7	+0.3	−0.1	−0.4
HMOR (psi)					
at 2500 °F	1 470	1 510	1 650	3 150	4 400	2 900
at 2700 °F						2 000
Abrasion loss (cm³) (ASTM C-704)	9.0	8.0	8.0			4.0

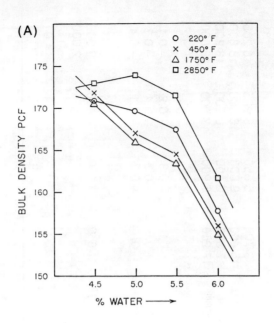

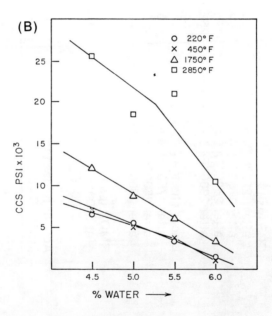

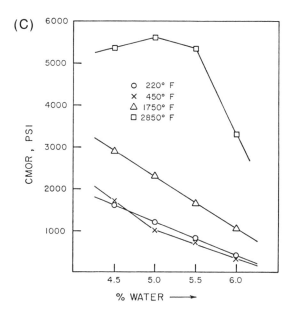

Fig. 10. (A) Variation of bulk density with water used for casting; (B) variation of cold modulus of rupture with water used for casting; (C) variation of cold compressive strengths with water used for casting.

deformation of only 0.8%. Brick of similar alumina contents could have creep in the range of 5 to 10%.

The thermal shock resistance was run in a ribbon furnace; Table II shows the relative percent loss of modulus of elasticity (MOE): The MOE retained on relative thermal shock resistance increases with the alumina content, probably because of the higher thermal conductivity of alumina.

Table III shows the MOR values of patching to previously fired surfaces and indicates that the material patches to itself very effectively and that in all cases the break during MOR determinations took place through lines other than joints.

There are some important points to note concerning these low-moisture castables:

(1) The water content used for casting is very important, as it will dictate the ultimate performance properties of the product.

(2) Since the binder is calcium aluminate cement, the temperature of the aggregates and water must be kept above 15°C, and the temperature of the mix in place must not be allowed to fall below 15°C; it has been observed from field application of these materials that higher hydrates will cause explosive spalling during dryout, and the setting time will be prolonged. If the above temperatures are maintained, the curing and drying processes can be considerably shortened.

(3) These products are cast by vibration, so the selection of proper vibrating equipment is very important. The flow and resulting densification will depend on the vibrating parameters.

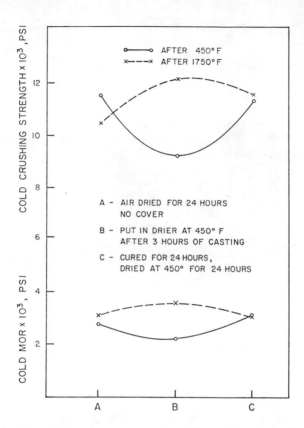

Fig. 11. Cold modulus of rupture and cold crushing strengths under different curing and drying conditions after 450°F and 1750°F heats.

Reasons for the exceptional properties of the low-moisture castables are not entirely clear, and nobody has given an in-depth explanation. We think that the preliminary high strength is developed from the compaction of the material by vibration due to the pseudothixopropic nature of the mix. Because of lower water content, the particle-to-particle distance is small, and hence the dried strength is higher. During dehydration of a small amount of cement, recrystallization of CA phases occurs on the submicrometer particles and not in void spaces, resulting in strength increases during the dehydration process. At higher temperatures, the formation of mullite makes it stronger. Although CA cement tends to form anorthite which, in combination with mullite and corundum, forms a low-melting eutectic at around 1380°C (2515°F) and causes a loss of strength above 1399°C (Fig. 14), we have been able to eliminate the problem by inducing mullite formation at the expense of anorthite; CaO is then converted to CA phases, as has been substantiated from X-ray diffraction studies.

Applications

Because of their superior properties, ease of installation, and rapid preparation for use, low-moisture castables have had a large impact in several

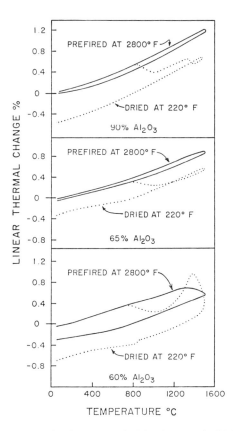

Fig. 12. Linear thermal change of 60, 65, and 90% alumina, low-moisture castable with temperature of dried (220°F/24 h) and prefired (2800°F/1 h) material.

areas of application where brick or other conventional refractories have generally been used. The following are a few samples.

Blast Furnaces

Perhaps the most low-moisture castables used today are in the blast furnace casthouse in trough, iron, and slag runners. These components have compositions which are basically similar to a ramming mix, differing primarily in the binder system. Unlike ramming mixes, the linings with low-moisture castables can be patched with the same material repeatedly, resulting in less cost per kg of metal. The relative slag erosion resistance is equivalent to or better than commonly used ramming mixes (Fig. 15).

In the blast furnace casthouse, a recent trend has been to use precast pipes in the iron and slag runners, particularly for the small- and medium-sized blast furnaces. These pipes are expendable and have the advantages of reducing pollution, preventing losses of hot metal temperature, and being maintenance-free. Less temperature fluctuation of the refractory results in longer life.

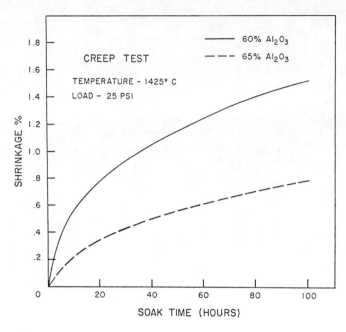

Fig. 13. Creep deformation of 60 and 65% alumina, low-moisture castables at 1425°C (2600°F) under 25 psi load with time.

Table II. Thermal Shock Resistance (Ribbon Furnace)

Type of Material	Modulus of Elasticity Retained (From Sonic Velocity Measurement) (%)
48% Al_2O_3	63.0
57% Al_2O_3	68.0
66% Al_2O_3	73.0
91% Al_2O_3	80.0
67% Al_2O_3 + 25% SiC%3C	93.0
68% Al_2O_3 + 24%SiC + C	89.0

Table III. Modulus of Rupture Measurements on Patching Fired Bars With Same Type of Trough Castable

Treatment	Cold MOR (psi) Patching of Type A to	
	Type A	Type B
Initial reheat (2800°F/5 h)	1260	1250
Patched bars (220°F/24 h)	900	430
Patched bars (1750°F/4 h)	270	690
Patched bars (2500°F/5 h)	1260	1130
Patched bars (2800°F/5 h)	1220	1080

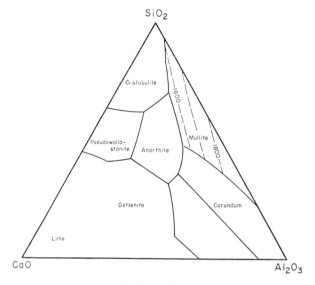

Fig. 14. Selected area of CaO-Al_2O_3-SiO_2 phase diagram.

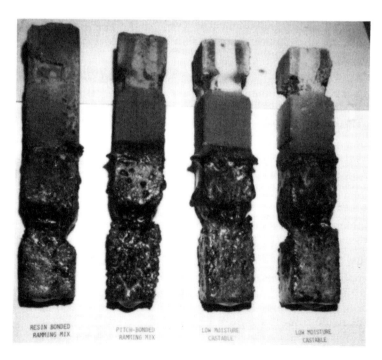

Fig. 15. Bars of ramming mixes (resin-bonded and pitch-bonded) and low-moisture castables after slag test.

Conventionally, SiC and C compositions are used in iron runners, similar to those materials used in troughs. Nishi reported that, for iron runners, the Al_2O_3–SiO_2 castables (low-moisture) are better than castables containing SiC and C, because the latter becomes porous and brittle due to oxidation.[6]

Skimmers and skimmer blocks have shown longer life than conventional refractories (so far over 2.7×10^8 kg of metal flow has been achieved).

Steelmaking

The use of low-moisture, vibratable castables in Concast tundishes has been fairly successful for continuous casting. Usually a 60–65% alumina, low-moisture castable can be used directly as the working lining or beneath the expendable SiO_2 or MgO boards. A basic parting material can also be used on the low-moisture castable.

Alloy steel ladles, steel ladle pouring pads, desulfurizing and argon stirring lances, and electric furnace delta areas have also found susccessful uses for low-moisture castables.

Coke Oven

Low-moisture castables have been used in coke oven door plugs in place of fused silica or conventional castables. Because of their high density and low porosity, carbon adhesion seems to be minimal. Their structural stability also appears to be far superior to that of conventional castables. The improved performance far outweighs the initial cost of conventional castables.

Coke oven cast doors, end flue walls, stand pipe linings, and charging port rings are also potential application areas for low-moisture castables.

Reheat Furnaces

A very effective and successful use has been in the skid rails where high strength is needed at higher temperatures, along with better thermal stability. These materials are also applied in soaking pits, slab reheat and walking beam, stationary hearth furnaces, as well as burner blocks.

Foundries

Low-moisture castables are applied in iron and steel ladles, ladle pocket blocks, mechanical iron stirrers, and ladle drying walls in the foundry industry.

Petroleum Refinery

Low-moisture castables formed by vibracasting have performed well in high-abrasion areas, CAT regenerator and return lines, and slide valves.

Aluminum Industry

Application in aluminum industries has been very broad-based. From subhearth to metal contact areas and superstructures, low-moisture castables have been very effective. Trough runners and filter boxes also have been successfully made from low-moisture castables.

The future seems to be very promising for low-moisture castables, where a wide variety of aggregates can be applied with particular applications in mind to produce a much broader range of products.

References

[1] L. Prost and A. Pauillac, "Hydraulically Setting Refractory Compositions," U.S. Pat. No. 3 802 894, April 9, 1974.

[2] J.A. Crookston and W.D. Fitzpatrick, "Castables with Improved Properties"; private communication.

[3] P.H. Havranek and L.O. Thornbald, High-Strength Refractory Casting Compound for the Manufacture of Monolithic Linings," U.S. Pat. No. 4 244 745 Jan. 13, 1981.

[4] P.A. Havranek, "Recent Developments in Abrasion and Explosion Resistant Castables," *Am. Ceram. Soc. Bull.*, **62** [2] 234–35 (1983).

[5] J.P. Kiehl and Y.L. Mat, "Use of Chrome-Alumina Castables and Shapes in Blast Furnace, Carbon Black Reactors and Coal Gasifiers"; unpublished work (March 1982).

[6] M. Nishi, et al., "Application of Al_2O_3–SiO_2 Casting Materials for Blast Furnace," *Taikabutsu*, **34** [289] 26–29 (1982).

A New Generation of Low-Cement Castables

B. Clavaud, J.P. Kiehl, and J.P. Radal

Research Center
LaFarge Refractories
Vénissieux, France

Castable refractories, during the past 15 years, have undergone a considerable technical evolution, leading to greatly increased mechanical resistance and compactness and reduced cement content. These low-cement castables, giving high performance at intermediate temperatures, are complemented by new ultralow-cement castables for high-temperature applications.

At the First International Conference on Refractories in Tokyo, Japan, in November 1983, we presented a paper "15 Years of Low Cement Castables in Steelmaking."[1] This paper showed our company's interest in the research of a new range of products: the ultralow-cement castables.

Six months later we can make a presentation of this new generation of materials which is to outdate traditional products in applications at high temperature. Until now, low-cement castables at holding temperatures above 1500°C could not compete with traditional mixes, which consequently were still employed. This new temperature capability implies an open competition between low-cement, castable specialties and bricks. For the first time, based on equivalent aggregates, specialties have more favorable characteristics.

Furthermore, the excellent properties of the ultralow-cement castables allow them to be used as support elements at high temperature, which is not the case for traditional and low-cement, castable specialties. This fact will help specialties take a new share of the market and contribute to the progression of the general refractory technology in terms of increased reliability, performance, substantial savings, reduced energy consumption for firing, and shorter delivery time due to the absence of special shapes characteristic of monolithic products.

Traditional Castable Refractories: Evolution toward Low-Cement Castables

Traditional Castables and Bricks

As already mentioned several times,[2] the years 1970–1980 will be a watershed in the history of refractories. During these years, specialties, and mainly castables, proved their real value. At the beginning, castables were dependent on bricks and used as a necessary complement; later, despite their wider applications, little research was devoted to them. Up to the 1960s, they had a relatively simple composition: refractory aggregates with a refractory cement, which acted principally as the only fine ingredient (mainly 10–100μm), in enough quantity to serve both as a lubricating medium and as a bond to give mechanical strength at cold temperatures. But they presented three defects compared with bricks: (1) lack of compactness due to the high percentage of water required for casting; (2) mechanical strength was very dependent on firing temperatures for the conversion of the hydraulic bond into a ceramic bond; and (3) a very high CaO content that had obvious deleterious effects (see Table I). At this stage, with the exception of nonsilica-containing products, the castables were only a partial substitute for fired brick.

Table I. Properties of a Dense, Fired Brick and a Traditional Castable, Based on Fireclay Aggregates

Properties	Traditional Castable	Dense, Fired Brick
Refractory aggregate	Fireclay	Fireclay
Al_2O_3 (%)	48.6	44.2
CaO (%)	6.2	0.5
Open porosity after 110°C (%)	9.0	16.5
Open porosity after 1000°C (%)	26.0	16.5
Hot modulus of rupture at 1000°C (N/mm^{-2})	3.5	6.0

Progressive Reduction of Cement Content: Low-Cement Castables

At the end of the 1960s, general evolution in the refractory industry predicted these developments. Increasing knowledge and the use of high-quality synthetic aggregates were applied to specialties and led to the creation of new, improved ramming mixes, plastics, insulating concretes, and jointing cements. Research work in France was concerned with improving the characteristics of castables: Part of the cement content was replaced by fine particles (1–10 μm) or by additives to reduce the percentage of casting water.

Table II. Properties of Low-Cement Castables

Properties	Traditional Castable	Low-Cement Castable
Aggregate	Fused alumina	Fused alumina
Al_2O_3 (%)	90.0	90.0
CaO (%)	6.5	1.1
Open porosity after 110°C (%)	8.0	5.0
Open porosity after 1000°C (%)	22.0	11.0
Cold crushing strength after drying at 110°C (N/mm^{-2})	80.0	110.0
Cold crushing strength after firing at 1000°C (N/mm^{-2})	40.0	120.0

The decisive step was the use of ultrafine particles from different origins (natural or synthetic), the size of which was ≤1 μm. Several patents were filed.[8,9] (Subsequent research has led to even more advanced materials: the ultralow-cement castables.)

Without entering into technical details,[1,2] the low-cement castables very soon appeared as materials with very interesting properties, as shown in Table II.

(1) Improved compactness: After firing at 1000°, $d/D \times 100$ was about 84–92%, instead of 70–80% as for traditional castables. Their permeability and porometry provide a high resistance to penetration by fluidized metal in working conditions.

(2) Mechanical strengths improved in two ways: Constant progression (Fig. 1) (strength anisotropy through a lining can create spalling parallel to the hot face); and possible development of exceptional mechanical strengths of up to 250 N/mm^{-2}.

(3) Low lime content: The use of a low content of a high-alumina cement, less than 3–8%, gives percentages of CaO of less than 2.5% and more often of about 1.2%, with an exceptionally good slag resistance compared with traditional castables (Fig. 2). There is less corrosion at the metal-slag interface and less impregnation.

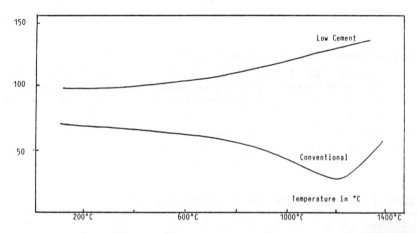

Fig. 1. Cold crushing strength of a conventional and a low-cement castable (corundum aggregates) in N/mm^2.

Owing to these improvements, a new range of materials was produced for the following industries:

Cement industry[3]: Materials based on hard aggregates with very high crushing strength were produced for antiabrasion applications; and special bonding agents were produced for increased resistance to corrosion by alkalis.

Nonferrous industry[4]: Highly impermeable materials, materials with a reacting agent creating a total impermeability, and materials with a special aggregate were developed.

Ferrous industry[5-7]: Materials for blast furnace injections, blast furnace run-

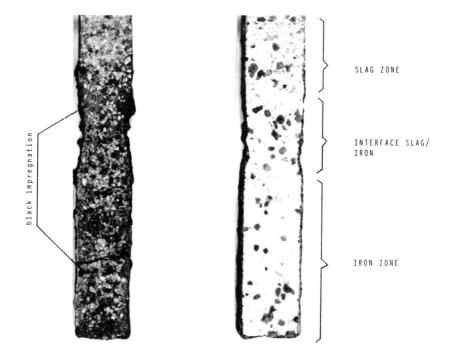

Fig. 2. Comparison of corrosion between conventional and low-cement castable (blast furnace slag and iron).

ners, torpedo ladles, open ladles, desulfurization lances, casting accessories, and tundishes were produced.

Petrochemical industry[7]: Carbon black was produced for civil engineering works.

Miscellaneous industries: For the glass industry, materials for furnace floors and feeders were developed, and for the ceramic industry, materials were produced for use in car decks, furnaces, burner gathering holes, and saggers.

Owing to other important improvements occurring at the same time, the strict control of heatup can now be avoided. Initially, installation was slowed down because of the risks associated with explosion or partial demolition due to the water present in the product. Two methods have been developed and applied: The first ensures total safety from explosion (drying schedule between 5 and 10 h), and the second, for more sophisticated linings, uses a drying schedule that is halved.

The special properties of the low-cement castables have been clearly established and, together with the new drying techniques, the ultralow-cement castables provide those same characteristics with an even greater refractoriness.

Ultralow-Cement Castables: Composition and Properties

Limitation of the Low-Cement Castables

Given their cement content of less than 3–8%, these castables still have too high a percentage of CaO for developing a good hot strength, which causes a rapid decrease of the hot modulus of rupture at 1200°–1400°C. At 1500°C, the Al_2O_3-SiO_2-CaO low-cement castables have a hot modulus of rupture near zero. Figure 3 shows the variation as a function of temperature of the hot modulus of rupture of a fused-alumina-based, low-cement castable, prefired to the testing temperature. The falling off above 1400°C occurs no matter which aggregate is used—fireclay, bauxite, or fused alumina. Figure 4 gives the curves for refractoriness under load under 2 bars, measured on prefired test bars of low-cement castables based on different aggregates. It shows that, whatever the aggregate, these low-cement castables start falling at 1300°–1400°C. Considering the SiO_2-Al_2O_3-CaO ternary diagram, the formation of a liquid phase can be observed at these temperatures.

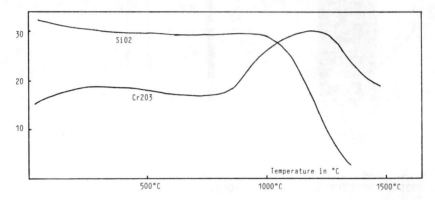

Fig. 3. Hot modulus of rupture of low-cement castables with superfine silica and chromic oxide powders in N/mm² (fused corundum aggregates).

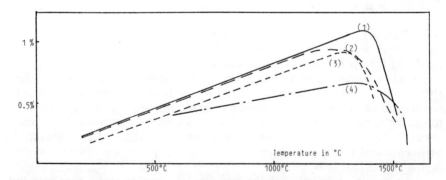

Fig. 4. Refractoriness under load of prefired, low-cement castables with different aggregates: (1) fused corundum, (2) tabular alumina, (3) bauxite, and (4) andalusite.

Attempts to Solve the Problem of Hot Strength

Suppression of the silica content to maintain the binary system Al_2O_3-CaO is one of the solutions to improve the behavior of the low-cement castables at hot temperature. Development of these castables without silica is continuing although their performances do not compete with those of ultra-low-cement castables.

The use of other fine powders (chrome oxide) allows a very rigid high-temperature bond to develop, as shown in Fig. 3 and Table III. However, for economic and pollution reasons, the use of chrome oxide limits the application of these castables.

Table III. Properties of Low-Cement Castables with Superfine Silica and Chrome Oxide Powder

	Low-Cement Castables	
Properties	SiO_2-type	Cr_2O_3-type
Aggregate	Fused alumina	Fused alumina
Al_2O_3 (%)	90.0	85.0
Cr_2O_3 (%)		9.5
CaO (%)	1.1	1.1
Cold crushing strength after 110°C (N/mm^{-2})	110.0	80.0
Cold crushing strength after 1000°C (N/mm^{-2})	120.0	120.0
Hot modulus of rupture at 1500°C (N/mm^{-2})	1.0	18.0

Possible Solutions

Advantage of the System SiO_2-Al_2O_3: The introduction of fine reactive particles of silica and alumina gives, at temperatures as low as 1300°C, a very strong mullite reaction. The hot modulus of rupture at 1500°C varies between 10 and 15 N/mm^{-2}; with an increased CaO content, this modulus drops rapidly. Figure 5 shows that, with more than 0.6% CaO, the hot modulus of rupture at 1500°C is close to zero.

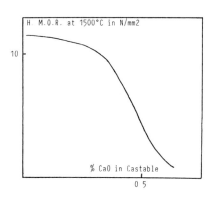

Fig. 5. Influence of CaO content (%) on hot modulus of rupture of pre-fired, ultralow-cement castables.

Mullite Bond with CaO: The only advantage in introducing CaO into the above compound is solving of the setting and hardening problems. The thermomechanical strengths of ultralow-cement castables will be dependent first on their CaO content (Fig. 5), and second on the resulting mullite content after firing. Figure 6 shows the evolution of the hot modulus of rupture at 1500°C, of an ultralow-cement castable, depending on the percentage of mullite reaction. These two factors are obviously not independent: The bond must therefore be carefully selected from the Al_2O_3-SiO_2-CaO phase diagram to develop the maximum mullite at hot temperature, while keeping good cold setting and hardening properties.

Properties of Ultralow-Cement Castables

Despite their low cement content, the ultralow-cement castables have a high crushing strength after drying at 110°C. Their hot modulus of rupture is

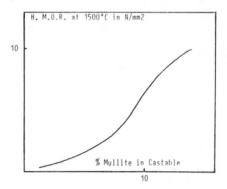

Fig. 6. Influence of mullite content (%) on hot modulus of rupture of prefied, ultralow-cement castables.

Table IV. Properties of Low-Cement and Ultralow-Cement Castables with Superfine Silica

Properties	Low-Cement Castables	Ultralow-Cement Castables
Aggregate	Fused alumina	Fused alumina
Al_2O_3 (%)	90.0	91.0
CaO (%)	1.1	0.3
Cold crushing strength after 110°C (N/mm^{-2})	110.0	80.0
Cold crushing strength after 1000°C (N/mm^{-2})	120.0	150.0
Open porosity after 1000° (%)	11%	8.5%
Hot modulus of rupture at 1500°C (N/mm^{-2})	1.0	10.0

also very high when mullite develops, as mentioned in the above paragraph and shown by the curves of refractoriness under load (Fig. 7).

These new materials are destined to compete with bricks of equivalent composition. Figure 8 compares the evolution of the hot modulus of rupture, according to temperature, of low-cement castables with ultralow-cement castables, and with equivalent mullite-bonded bricks. The better behavior of ultralow-cement castables is due both to their strong bonding and high degree of compactness.

Depending on the desired application, different ultralow-cement castables can be supplied; three examples are given in Table V.

Application of the Ultralow-Cement Castables

Before commercialization, these products were used in several applications and others are being studied.

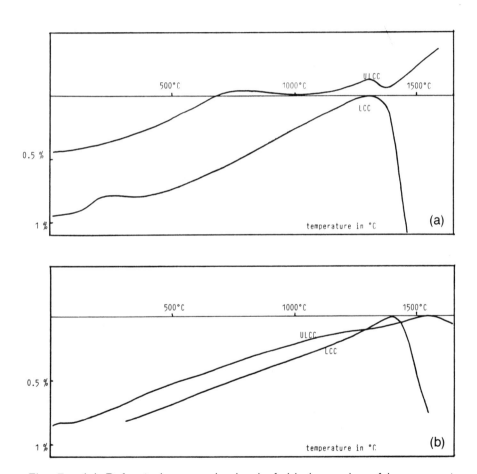

Fig. 7. (a) Refractoriness under load of dried samples of low-cement and ultralow-cement castables (fused corundum aggregates); (b) refractoriness under load of prefired samples of low-cement and ultralow-cement castables (fused corundum aggregates).

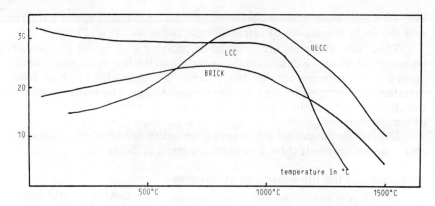

Fig. 8. Hot modulus of rupture in N/mm² of low-cement, ultralow-cement, and dense fired brick (fused corundum aggregates).

Table V. Properties of Three Types of Ultralow-Cement Castables

Properties	Ultralow-Cement Castables		
Aggregate	Andalusite	Bauxite	Fused alumina
Al_2O_3 (%)	60.0	81.0	88.0
SiO_2 (%)	38.0	14.0	7.5
Fe_2O_3 (%)	1.4	1.2	0.1
Bulk density after 110°C (g/cm³)	2.80	2.95	3.45
Open porosity after 1000°C (%)	10.0	14.0	8.5
Cold crushing strength after 110°C (N/mm³)	60.0	70.0	80.0
Permanent linear change after 1600°C at 100 h (%)	+2.4	+0.8	+1.3

Blast Furnace Runners

In blast furnaces, producing 5500 tons/day, with 2 tapholes and a main runner length of 11.50 m (without syphon), maintenance with an SiC-containing graphitic castable is providing partial repair of the impact area by cast-vibration, after 40–50 000 tons of iron. Complete repair after 30–40 000 tons of iron is predicted. Other applications, not discussed, show that even more savings are possible.

Steel Ladles

First trials were with test panels of about 1 ton, either in the slag area or in the metal bath area. Results with a basic slag ($CaO/SiO_2 = 5$) at 1600°–1650°C show that 50–70 heats can be done without repair and more than 100 heats with some repair. Zoning of the ladle lining should noticeably improve their performances.

Table VI. Comparison of Specific Consumption Savings between Low-Cement and Ultralow-Cement Castables

	Specific Consumption*
Low-cement castables	389 g/ton
Ultralow-cement castables	306 g/ton
Savings	83 g/ton = 21%

*Approximate consumption figures vary from 170 to 450 g/ton of iron from one blast furnace to another.

Steel ladles are, for us, the most important immediate outlet (20% of the refractory consumption for steel production) for the ultralow-cement castables, since traditional and low-cement castables cannot withstand this high-temperature environment.

Reheating Furnaces and Impact Area of Tundishes

The application of low-cement castables is now well developed in full hearth walking beam furnaces, providing a good service life of 1-4 years, depending on service conditions. In practice, the low-cement castables are often applied in the hottest zones: corrosion, creep, and abrasion of the castables. The average service life of a lining should be increased by the ultralow-cement castables, for which properties are better and permeability lower above 1300°C.

The behavior of the impact zone of the tundishes became a real problem when sequences were raised to 4-5 ladles to increase productivity. First trials of low-cement castables with mild steel proved they had restricted applications. With ultralow-cement castables, sequences of 4-5 ladles are now possible, with a low wear of about 7-10 mm per ladle of 85 tons.

Future Developments

The ultralow-cement castables will permit the original concept of low-cement castables to be used at high temperatures. The low CaO content helps to develop a strong ceramic bond at medium temperatures; thus the production of economic compositions based on a mullite-type bond is now possible. This new range of materials appears very progressive compared with low-cement castables; their characteristics compete favorably with those of bricks based on similar aggregates, and a more dominant role of specialties can be expected.

The applications of low- and ultralow-cement castables are clearly delineated; low-cement castables should generally be used for applications at temperatures lower than 1500°C, and ultralow-cement castables for temperatures over 1500°C. This is not an absolute rule since different corrosion resistances must be taken into account. The new technology is being extended to special castables, i.e., selection of the type of chemical ultrafine particles depends on the aggregate—Cr_2O_3, SiC, C, Si, ZrO_2, TiO_2—and to basic castables, i.e., Mg-Al_2O_4, MgO-CaO, and MgO.

The Research Department is anticipating the next step: a lime-free castable, but the production difficulties associated with ultralow-cement castable products become even more stringent.

References

[1] B. Clavaud, J.P. Kiehl, and R.D. Schmidt Whitley, "15 Years of Low Cement Castables in Steelmaking," First International Conference on Refractories, November 15-18, 1983, Tokyo, Japan.

[2(a)] M. Durand and J.P. Kiehl, "Ram Castables and their Applications," VI Colloquium on Refractory Castables, May 24-26, 1977, Karlory-Vary, Czechoslovakia.

[(b)] B. Clavaud, "Special Monolithics," Société Française de Ceramique, Bulletin No. 129, pp. 13-27 (1980).

[3] M. Robert, "Refractory Materials Alkali Resistant to Cement Industry Slag," XXVI International Colloquium on Refractories, October 6-7, 1983, Aachen.

[4] B. Clavaud and V. Jost, "Refractories used in Melting Furnaces For Aluminum Alloys," XXIII International Colloquium, September 25-26, 1980, Aachen.

[5] B. Clavaud and M. Van Zynghel, "Materiali Non Formati Speciali per Iniezione in Alto Forno; Tecniche Utilizzate, Risultati Altuali," SIPRE Meeting, October 20, 1980 TARANTO-REFRATARI, Vol. VI N.10, 1981.

[6] B. Clavaud, G. Landman, and R.D. Schmidt Whitley, "Modern Lining of Blast Furnace Runners," *Fachberichte Hüttenpraxis Metallweiterverarbeitung*, **20** [10] 914-19 (1982).

[7] J.P. Kiehl and Y. Le Mat, "Use of Chrome-Alumina Castables and Shapes in Blast Furnace, Carbon Black Reactors and Coal Gasifiers"; for abstract see *Am. Ceram. Soc. Bull.*, **61** [3] 396 (1982).

[8] Fr. Pat. No. 69.34405.

[9] Fr. Pat. No. 77.14717.

Calcium Aluminate Cements For Emerging Castable Technology

George MacZura, Joseph E. Kopanda, and Frank J. Rohr

Alcoa Laboratories
Alcoa Center, PA 15069

Paul T. Rothenbuehler

Alcoa International, Inc.
Lausanne, Switzerland

Monolithic refractory technology has recently advanced to accommodate increasingly severe processing environments posed by new iron- and steel-making techniques. Low-cement (LC) castables aided this advance, but not without placement problems. A new 80% Al_2O_3 cement, having extended working times in a "generic" LC tabular Al_2O_3 castable containing fumed-SiO_2 dust and/or superground submicrometer Al_2O_3, is characterized in comparison with commercial calcium aluminate cements (CAC). Alternatives to LC castables are presented that use a new 90% Al_2O_3 and commercial CAC in conventional castables designed for maximum density. Hot MOR is supplemented by a sag-type deformation test that differentiates refractories at higher temperatures.

The successful development of new technical refractories has enabled significant advances to be made in processing technology by the iron and steel, petrochemical, ferrous and nonferrous, energy, and environmental industries. Contributing significantly to this success has been the development of low-cement (LC) castable technology.[1-13] These LC castables generally contain 4–8% calcium aluminate cement (CAC) and typically require only 3–7% water for placement by vibration casting. The low water requirement is accomplished by carefully grading the particle-size distribution so that successively finer pores are filled with increasingly smaller particles, down to the submicrometer range below 0.1 μm diameter. Minimizing pore-water requirements thus limits the mixing water for good fluidity to only that required to coat the surfaces of all the particles. The surface water requirements can be further reduced by the proper selection of deflocculants or dispersive additives. Consequently, with packing density maximized, the LC castables are characterized by their low porosity, high density, and exceptional hot strength, which enhance erosion, corrosion, and spalling resistance. These characteristics provide superior performance over such other monolithic refractories as ramming mixes, plastics, slinging materials, or gunning refractories, and durability is equal and sometimes superior to equivalent conventionally fired, dense refractories.

Submicrometer powders commonly used in LC castables, ultralow-cement (1–2%), or low-moisture castables include fumed-SiO_2 dust, Al_2O_3,

Cr_2O_3, ZrO_2, TiO_2, SiC, clay minerals, and carbon. The ease of dispersion, availability, and cost effectiveness of fumed SiO_2 has resulted in its use in many commercial LC castables. However, the sensitivity of the fumed-SiO_2 LC castables to water requirement, and the unpredictably short working times, have caused mixed reactions regarding their total acceptance and full utilization. One problem seems to be that the various CAC types do not perform similarly in LC castables that contain submicrometer fumed SiO_2.

Another limiting factor is refractoriness. When LC castables with fumed SiO_2 are heated in the range of 1350°–1400°C (2460°–2550°F) or above, the CaO and Al_2O_3 from the CAC react with the fumed SiO_2 to form anorthite ($CaO \cdot Al_2O_3 \cdot 2SiO_2$) and gehlenite ($2CaO \cdot Al_2O_3 \cdot SiO_2$). These are chemical compounds with low melting points, 1553°C (2827°F) and 1593°C (2900°F), respectively (see Fig. 1). A low-melting eutectic between these two compounds exists at 1380°C (2515°F) on the Al_2O_3 side of the triaxial CaO-Al_2O_3-SiO_2 phase diagram. An even lower eutectic can occur at 1335°C

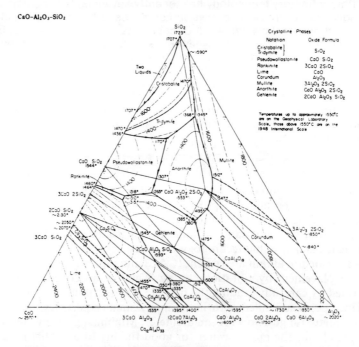

Fig. 1. System CaO-Al_2O_3-SiO_2; Composite.

(2435°F) if $2CaO \cdot SiO_2$ is developed in the presence of $12CaO \cdot 7Al_2O_3$ and $CaO \cdot Al_2O_3$. Thus, hot strength is limited by the formation of a glass at temperatures above this eutectic. Although special LC castables that contain submicrometer Al_2O_3 or Cr_2O_3 have excellent hot strengths at 1500°C (2730°F) and higher, only the submicrometer Al_2O_3 LC castable can be used safely without such potential toxic complications as hexavalent chrome, which can form from Cr_2O_3 in CAC castable refractories.[14]

This paper gives properties of a "generic" LC tabular alumina castable derived by substituting a fully ground calcined Al_2O_3 and either fumed SiO_2 or submicrometer Al_2O_3 for equivalent fractions of CAC, in a conventional test castable. A new CAC, having extended working times in fumed-SiO_2 LC castables, is characterized and compared with conventional castables designed for maximum density using an experimental higher Al_2O_3-lower CaO cement. Hot strength properties are supplemented by a sag-type deformation test to differentiate refractoriness at higher temperatures.

Tabular Alumina, Test Castable Mixes

Only the high-purity CAC binders containing 70–90% Al_2O_3 have sufficient purity for use in LC castables containing fumed SiO_2 (Table I). The compositions of the high-purity cements used in this study are given in Table II, and Table III lists the chemical analyses for tabular alumina, ground calcined aluminas, and fumed silica used in this work. Figure 2 shows the particle-size distributions for the subsieve components.[15]

Table I. Composition of Calcium Aluminate Cement Binders

Type	Oxide Composition Range (%)			
	Al_2O_3	Fe_2O_3*	CaO	SiO_2
Low-purity	39–50	7–16	35–42	4.5–9.0
Intermediate-purity	55–66	1–3	26–36	3.5–6.0
High-purity	70–90	0.0–0.4	9–28	0.0–0.3

*Total iron as Fe_2O_3.

The 15% CAC tabular castable mix was designed for maximum density, using the principles of particle packing according to Furnas[16] and Anderegg.[17] Substitution of 5% each of the 3-μm Al_2O_3 and fumed SiO_2 for an equivalent weight of cement in the 15% CAC conventional castable provides a 5% CAC "generic" tabular castable that closely approximates the "ideal" continuous particle-size distribution curve for maximum density illustrated in Fig. 3. The dense conventional and LC tabular mixes are given in Table IV. The LC castable mixes are differentiated by their submicrometer SiO_2 and Al_2O_3 components.

Specimen Preparation and Testing

The working times were determined on the castable compositions using the Vicat needle test,[18] after mixing the batches according to the ASTM standard practices described below.

Twenty-five mm (1 in.) prism specimens for bend-strength testing were prepared according to ASTM Standard Practice C862, "Preparing Refractory Concrete Specimens by Casting." No vibration was used during forming of the 25 by 25 by 178 mm (1 by 1 by 7 in.) specimens. The water required for casting was determined by the ball-in-hand test described in ASTM Standard Practices C860, "Determining and Measuring Consistency of Refractory Concretes." Ambient conditions were nominally 24°C (75°F) and 40–60% rh.

Table II. Composition of High-Purity CA Cements

Oxides %	Commercial Types				New Products*	
	70%Al$_2$O$_3$[†]	80%Al$_2$O$_3$[‡]	90%Al$_2$O$_3$[§]	80% Al$_2$O$_3$	90% Al$_2$O$_3$
Al$_2$O$_3$	72.00	80.40	88.10	80.40	87.70
CaO	28.00	17.60	9.70	17.60	10.60
Fe$_2$O$_3$	0.10	0.26	0.18	0.26	0.26
SiO$_2$	0.45	0.20	0.08	0.20	0.11
TiO$_2$	0.00	0.02	0.03	0.02	0.02
MgO	0.25	0.10	0.00	0.10	0.10
MnO	0.00	0.06	0.01	0.06	0.03
Na$_2$O	0.38	0.60	1.00	0.60	0.40
K$_2$O	0.00	0.01	0.00	0.01	0.00
P$_2$O$_5$	0.00	0.06	0.06	0.06	0.03
Moisture (110°C)	0.10	0.28	0.48	0.30	0.21
LOI (110°–1100°C)	0.39	0.92	1.20	1.30	0.69
XRD Phases (%)					
α-Al$_2$O$_3$	<1	36	69	35	40
CaO·Al$_2$O$_3$	48	43	18	43	24
CaO·2Al$_2$O$_3$	46	5	2	5	2
12CaO·7Al$_2$O$_3$	<1	6	5	5	4
Other	4	10	6	12	30
Surface area (m^2/g)	2	6	8	6	6
Specific gravity	2.93	3.35	3.5	3.35	3.5

*Experimental Alcoa Products.
[†]Alcoa CA-14.
[‡]Alcoa CA-25.
[§]Denka S-90.

Table III. Properties of Tabular Al$_2$O$_3$ and Fine Powders

	Crushed Tabular Al$_2$O$_3$*	Reactive Al$_2$O$_3$		Fumed-SiO Dust [†,§]
		Fully Ground 3 μm[†]	Superground 0.5 μm[‡]	
Oxides (%)				
Al$_2$O$_3$	>99.50	99.80	99.80	0.13
SiO$_2$	0.20	0.030	0.040	95.30
Fe$_2$O$_3$	0.08	0.012	0.020	0.02
TiO$_2$	0.001	0.001	0.001	0.06
CaO	0.07	0.015	0.015	0.21
MgO	0.00	0.003	0.05	0.00
Na$_2$O	0.08	0.07	0.08	0.13
K$_2$O	0.00	0.01	0.01	0.40
Carbon				1.50
Sulfur				0.05
LOI	< 0.10	0.30	0.80	1.90
Surface area (m²/g)	< 0.1	2.6	9.0	24.0
Specific gravity	3.60	3.98	3.98	2.26

*Alcoa T-61.
[†]Alcoa A-17.
[‡]Alcoa A-16SG.
[§]Reynolds RS-1.

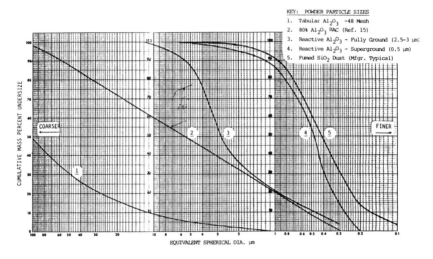

Fig. 2. Castable component subsieve particle-size distributions.

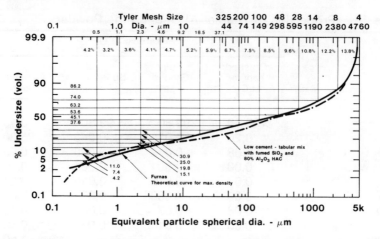

Fig. 3. Low-cement tabular castable particle-size distribution curve for maximum density (after Furnas).

The dried 25 mm (1 in.) bars were electrically heated to 815°C (1500°F) and held for 5 h, according to ASTM Standard Practice C865, "Firing Refractory Concrete Specimens." Hot modulus of rupture (MOR) was determined on the 815°C prefired specimens using ASTM Standard Test Method C583, "Modulus of Rupture of Refractory Materials at Elevated Temperatures," except that specimens were held at the test temperature for only 3 h instead of the 12 h normally required for unheated castables.

The sag deformation test was used to establish a relative indication of hot strength by the bar specimen's ability to resist gravity deformation or sagging when supported unrestrained on a 127 mm (5 in.) span. The percent sag deformation was calculated by dividing the amount of deformation, as measured between the center of the specimen and the 127-mm cord, and is shown by the schematic and equation in Fig. 4. The 1% sag-deformation temperature indicates the maximum use temperature for the test castable at which sufficient hot strength exists (about 1.4 MPa (200 psi) MOR) to support its own weight, based on the limited data reported herein.

Low-Cement (LC) Test Castables

Table V shows varied water requirements and working time responses for commercial 70 and 80% Al_2O_3 cements in the LC test castable, formulated by substituting 5% each of a 3-μm Al_2O_3 and fumed SiO_2 for an equivalent weight of cement in the conventional 15% CAC castable mix. Replacement of the finer components for CAC in the conventional castable without sodium polyphosphate (SPP) electrolyte addition causes a marked reduction in working time to an unacceptable level of about 10 min for both 70 and 80% Al_2O_3 cement types. Also, the 80% Al_2O_3 cement water requirement is reduced much more than the 70% Al_2O_3 cement in the LC castable.

The reverse is true when using small additions of SPP. The lower alumina cement responded to the electrolyte addition by further reducing the water requirement and extending the working time to acceptable levels; the 80% Al_2O_3 cement fails to show a significant decrease in water requirement,

Table IV. CAC*—Tabular Alumina Dense Castable Mixes

Component	Screen Size		Castable Compositions (wt%)		
	Tyler Mesh	μm	Conventional	Low-Cement	
				SiO$_2$	Al$_2$O$_3$
Tabular Al$_2$O$_3$	4–10	4760–1680	22.0	22.0	22.0
Tabular Al$_2$O$_3$	8–14	2380–1190	9.8	9.8	9.8
Tabular Al$_2$O$_3$	14–28	1190–595	18.7	18.7	18.7
Tabular Al$_2$O$_3$	–48	298–2	34.5	34.5	34.5
CAC	–100	149–0.3	15.0	5.0	5.0
Al$_2$O$_3$ (3 μm median)	FG†	10–0.3		5.0	5.0
SiO$_2$	Fume	2–0.1		5.0	
Al$_2$O$_3$ (0.5 μm median)	SG‡	2–0.1			5.0

*CAC = Calcium aluminate cement.
†Fully ground.
‡Superground.

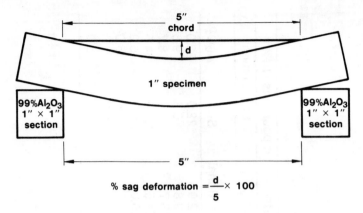

Fig. 4. Sag-deformation test.

and working times remain unacceptably short. The high $12CaO \cdot 7Al_2O_3$ (Table II) content of the 80% Al_2O_3 CAC is suspected of causing the shorter working times, because it is known for its fast hydration rate and early strength development.[19-21] Surprisingly, both cements responded the same to SPP additions in the 15% CAC castable mixes by becoming faster setting with only minor reductions in water requirements.

It is apparent from these results that each CAC-type LC castable must be characterized individually or with additives to establish product performance characteristics.

New 80% Al_2O_3 Cement for LC Castables

LC Castable Properties

The unacceptably short working times obtained with commercial 80% Al_2O_3 cement with or without the addition of SPP electrolyte spurred the development of an experimental 80% Al_2O_3 cement specifically for use with LC castables formulated with fumed SiO_2. The properties of this new product are shown in Table VI. An acceptable working time of 52 min is obtained with only 4.8% water and no additional electrolyte. Working times are extended still further with small additions of SPP for tests conducted at 24°C (75°F).

Excellent hot strength properties are obtained with this new 80% Al_2O_3 cement, which exhibit hot MOR values at 1370°C (2500°F) equivalent to the cold MOR after firing at 815°C (1500°F). Typical of fumed-SiO_2 LC castables, the hot MOR drops very significantly from 8.55 to 1.52 MPa (1240 to 220 psi) as the test temperature is increased from 1371° to 1510°C (2500° to 2750°F) with no SPP addition. The latter strength approximates the 1.4 MPa (200 psi) strength requirement thought necessary for a refractory to support its own weight.

Very small SPP additions seem to aid hot strength at 1370°C (2500°F) but the 1510°C (2750°F) hot MOR shows a decreasing trend with increasing SPP addition. The higher lime, 70% Al_2O_3 cement hot strength in the fumed-SiO_2 LC castable was barely measurable, being 0.3 MPa (45 psi) at 1510°C (Table VII). This is significantly lower than that obtained with the new 80% Al_2O_3 cement, even though the 1370°C higher MOR values were equivalent.

Table V. Effect of CAC-Type and Sodium Polyphosphate Additions on Working Time of Conventional and Low-Cement Castables

Castable Type	Composition* (% on total weight basis)				Tabular Alumina Castable			
					CAC-Type			
					70% Al_2O_3		80% Al_2O_3	
	CAC	Reactive Al_2O_3 (3 μm)	SiO_2 Fume	Sodium Poly-phosphate[†]	GBIH[‡] H_2O (%)	Working Time[§] (min)	GBIH H_2O (%)	Working Time[§] (min)
Low-cement	5	5	5	0	8.5	9	5.5	13
	5	5	5	1.0	5.0	44	6.0	14
	5	5	5	2.0	5.0	136	5.3	12
Conventional	15	0	0	0	10.5	294	9.0	44
	15	0	0	0.2	10.0	15	8.5	18
	15	0	0	1.0	10.0	19	8.5	16
	15	0	0	2.0	10.0	11	8.5	14

*T-61 tabular aggregate: 1680–4760 μm = 22%; 1190–2380 μm = 9.8%; 595–1190 μm = 18.7%; 2–298 μm = 34.5%.
[†]A dry powder distributed by Alcoa under the name Agilu.
[‡]GBIH = Good ball-in-hand consistency by ASTM C860.
[§]Vicat test on castable conducted at 24°C (75°F).

Table VI. Properties of New 80% Al_2O_3 CAC for Low-Cement Castables (LC (SiO_2))* Tabular Castable)

Additive SPP[†] (% twb)	GBIH H_2O (%)	Working Time (min)	Modulus of Rupture (MPa/psi)				
			Cold			Hot	
			Moist Cured 24 h at 32°C/90°F	Dried at 110°C/230°F	Fired at 815°C/1500°F	1370°C/2500°F	1510°C/2750°F
0	4.8	52	2.24/325	4.38/635	9.17/1330	8.55/1240	1.52/220
0.005	4.8[‡]	65	2.10/305	3.65/530	10.50/1520	ND[§]	1.21/175
0.01	4.8	74	2.28/330	4.10/595	12.30/1780	11.90/1730	0.97/140

*LC (SiO_2) = Low-cement castable with SiO_2 fume as submicrometer fines: 5% CAC, 5% Al_2O_3 (3 μm), 5% SiO_2 fume.
[†]Sodium polyphosphate (AGILU).
[‡]Slightly dry.
[§]Not determined.

Table VII. Hot Strength of Tabular Al_2O_3 Castables

Castable Type	Al_2O_3 (wt%)	CAC Amount (wt%)	SPP (% twb)	GBIH H_2O (%)	Hot MOR, (MPa/psi)* 1370°C/2500°F	1510°C/2750°F
Low-cement[†]	70	5	0	6.5	8.69/1260	0.31/45
Low-cement[†]	70	5	1.0	6.5	2.76/400	0 /0
Low-cement[†]	70	5	2.0	6.5	0.83/120	0 /0
Conventional	80	25	0	7.8	15.2 /2205	15.6 /2265
					18.7 /2715[§]	16.9 /2455[¶]
Conventional	90	20	0	8.3	ND**	8.31/1205
						11.7 /1695[¶]
Conventional	90[‡]	20	0	6.9	ND**	12.4 /1800
						14.7 /2130[¶]

*25 mm (1-in.) bars prefired at 815°C (1500°F), except where noted, bars held 3 h before loading.
[†]Contains 5% fumed-SiO_2 dust and 5% Al_2O_3 (fully ground 3 μm median).
[‡]New experimental cement.
[§]Prefired 5 h at 1650°C (3000°F), cooled, and reheated for testing.
[¶]Prefired 5 h at 1510°C (2750°F), cooled, and reheated for testing.
**Not determined.

The fluxing effect of the SPP is very noticeable for both the 1.0% and 2.0% (total weight basis) SPP additions to the 70% Al_2O_3 CAC mix.

Sag Deformation in 5% Fumed-SiO$_2$ LC Tabular Castables

The data in Table VII suggest that the higher lime, 70% Al_2O_3 cement has lower refractoriness in the fumed-SiO$_2$ LC castable than the new 80% Al_2O_3 cement, as indicated by the much lower strength at 1510°C (2750°F). In an effort to verify this observation, sag deformation tests were conducted at 1425°C (2600°F), 1540°C (2800°F), 1650°C (3000°F), and 1760°C (3200°F). Figure 5 shows the deformed bars and the percent sag deformation after heating for 5 h at test temperatures. Sag deformations with the 70% Al_2O_3 cement are approximated by each succeedingly higher 110°C (200°F) test result obtained with the new 80% Al_2O_3 cement. This suggests that the 80% Al_2O_3 cement has about 110°C higher refractoriness than the 70% Al_2O_3 cement in the LC tabular castable containing 5% fumed SiO$_2$.

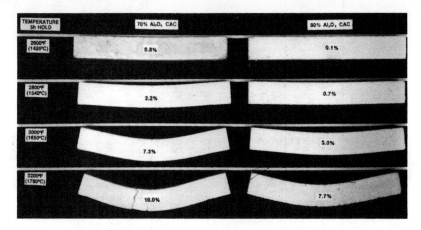

Fig. 5. Sag deformation of low-cement tabular castables containing 5% fumed-SiO$_2$ dust (percent sag deformation noted on bars).

Reduced Fumed-SiO$_2$ LC Castable Mixes Substituting Superground 0.5 μm Al$_2$O$_3$

The superground 0.5 μm median-reactive, calcined Al_2O_3 was substituted for fumed SiO$_2$ in the percentages shown in Table VIII, using the LC tabular castable mixes listed in Table IV, in an effort to establish the effect of fumed SiO$_2$ on the refractoriness of LC castables. Table VIII also shows an increased casting water requirement for both the 70 and 80% Al_2O_3 cements with increasing substitution of the 0.5 μm Al_2O_3 for the 5% fumed SiO$_2$. This nominal 30% increase in water requirement for the two castables, after replacing the 5% fumed SiO$_2$ totally with 0.5 μm superground Al_2O_3, is phenomenally large, considering the reasonably close approximation of the particle-size distributions of these two fine materials when fully dispersed (as shown in Fig. 2). There are several physical characteristics of powders that can upset the packing characteristics of continuous particle-size distributions designed for maximum packing and cause an increase in castable water requirement. These will be discussed briefly.

Table VIII. Effect of Substituting Superground Alumina for Fumed-SiO_2 Dust on Low-Cement Castable Tempering Water

Low-Cement Tabular Castable		Tempering Water (%)	
Submicrometer Fines (%)		GBIH Consistency (CAC Type)	
Al_2O_3 0.5 μm Median	Fumed SiO_2	70% Al_2O_3	80% Al_2O_3
0	5	5.5	5.1
2	3	5.5	5.5
4	1	6.5	6.2
5	0	7.0	6.8

Particle shape can significantly affect particle packing if the diameter-thickness aspect ratios differ greatly. As might be expected, as shown in Fig. 6 (A), the shape of the spherical fumed-SiO_2 dust shown in the transmission electron photomicrograph is markedly different from the sharp-cornered, irregularly shaped, fractured α-Al_2O_3 crystal shown for comparison in Fig. 6 (B). Although the superground alumina has sharply angled corners, the original α-Al_2O_3 crystal plates appear to have been sufficiently fragmented to approximate spheres with regard to aspect ratio and, therefore, should have somewhat similar packing characteristics in a wet mix.

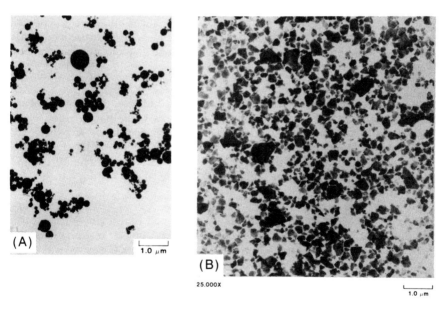

Fig. 6 (A) TEM — fumed SiO_2 dust. (B) TEM — superground submicrometer Al_2O_3.

Assuming complete dispersion of the submicrometer fines, another variable that could account for the increase in water requirement is the variance of the particle-size distribution from the ideal. The particle-size distribution for the SiO_2 LC castable, as calculated on a volume basis from the various components contained in the mix, approximates the Furnas theoretical curve for maximum density (illustrated in Fig. 2), except for the intermediate sizes (10–150 μm) and insufficient fines below 0.3 μm, the latter of which so greatly affects rheology. The low, nominal 5–5.5% H_2O requirement confirms this close approximation. The deficiency of ultrafines below 0.3 μm becomes more apparent when substituting the lower surface area 0.5 μm median Al_2O_3 (Table III, 9 m^2/g) for the 24 m^2/g fumed SiO_2 in the LC castables. This reduction in fines on the volume basis is attributed in part to the difference in densities of the two materials. The 0.5 μm Al_2O_3 has only 58% of the submicrometer particle volume contributed by the fumed SiO_2 on an equivalent mass basis. The conventional castable containing 15% CAC is even more deficient in ultrafine particles, because no submicrometer grains have been added except those occurring in the CAC as manufactured. This probably accounts for the higher water requirements seen at the bottom of Table VII.

The third, and possibly most important factor, causing an increase in water requirement when substituting fine alumina for fumed SiO_2 is the inability to obtain complete dispersion of the alumina fines that pelletize during supergrinding when mixed for only 4 min in a Hobart mixer. Incomplete dispersion of the fine alumina pellets would cause an even greater variance from the ideal Furnas distribution, both in the subsieve and submicrometer ranges. The inability to obtain complete dispersion of dry, superground, submicrometer particles during short mixing times in low shear mixers is a problem that requires additional investigation. Further studies on changing rheological surface charge effects with surface active agents should provide fruitful results in this area.

Sag-Deformation Refractoriness of Fumed-SiO_2/Al_2O_3 LC Castables

The refractoriness of LC tabular castables is indicated by the sag-deformation characteristics shown in Fig. 7 for mixes containing 70 and 80% Al_2O_3 cements and fumed-SiO_2 additions to 5%, in accordance with the compositions shown in Table VIII. The percent sag deformation represents the total deformation of unrestrained 25 mm (1 in.) bars, that remain unsupported over a 127 mm (5 in.) span during heatup and a 5 h hold at the test temperature.

Besides the predominant sag deformation resulting from the anorthite/gehlenite glass formation, it may be assumed that a small portion of the deformation occurs as a result of CAC phase transformations: $CaO \cdot Al_2O_3 \rightarrow CaO \cdot 2Al_2O_3 \rightarrow CaO \cdot 6Al_2O_3$. The Al_2O_3 LC castables, represented by the 0% SiO_2 curves, are assumed to represent the amount of deformation resulting from these phase transformations because the occurrence of glass phases should be essentially nil. The 0.5 μm Al_2O_3 LC tabular alumina castable with 0% fumed-SiO_2 addition shows a maximum deformation of about 1% after heating 5 h at 1760°C (3200°F). Since this composition functions readily as a 1760°C castable, it may be assumed that a 1% sag deformation by this test still provides a functional castable at the 1% deformation temperature.

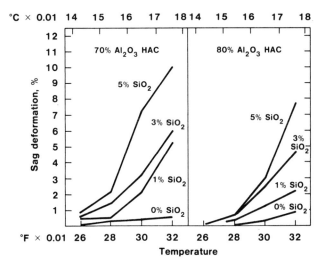

Fig. 7. Effect of fumed-SiO_2 dust/superground 0.5 μm Al_2O_3 on sag deformation of low–cement tabular castables.

Accepting the 1% sag-deformation temperature as a measure of refractoriness is further supported by comparing the 1510°C (2750°F) hot strengths for the 70 and 80% Al_2O_3 cements containing 5% fumed SiO_2 with the sag deformation at 1510°C taken from the Fig. 7 plots. The 0.3 and 1.5 MPa (45 and 220 psi) hot strengths at 1510°C correspond to 1.8% and 0.6% sag deformations, respectively. The sag-deformation limit for acceptability falls somewhere between these values. Until further data are obtained, it will be assumed that 1% sag deformation equates to an acceptable refractory performance.

Using the 1% sag-deformation limit for refractoriness, it is apparent from Fig. 7 that the 80% Al_2O_3 CAC provides 55° to 110°C (100° to 200°F) higher refractoriness than that obtained with the 70% Al_2O_3 cement, and that increasing the fumed SiO_2 in LC tabular castables severely reduces the refractoriness.

Hot Strength of Conventional Tabular Alumina Castables

Table VII lists the hot strength properties of conventional CAC castables designed for maximum density when using a casting-grade 80% cement, a 90% Al_2O_3 commercial cement, and an improved experimental 90% Al_2O_3 cement. The hot strengths for these castables at 1510°C (2750°F) far surpass the best fumed-SiO_2 LC castables, even though the water requirement for casting is considerably higher. Further experimentation on castable mix designs with 10–20% additions of these CAC types should permit additional water requirement reductions to further enhance hot-strength properties and reduce pore size to improve slag resistance. The use of organic fibers with these more refractory tabular castables should also permit faster heatup and greater use in new iron and steel processes as a result of improved performance over fumed-SiO_2 LC castables.

The new experimental 90% Al_2O_3 CAC requires only 6.9% H_2O when used in a 20% CAC tabular castable. Hot strengths are exceptional, and 1815°C (3300°F) sag deformation is only 1% compared with 1.5% for the 25% CAC castable bonded with the 80% Al_2O_3 CAC. The lower CaO content afforded by the 90% Al_2O_3 cement offers potential slag resistance advantages over the lower Al_2O_3 CAC when used in equivalent amounts.

Regarding hot MOR testing (Table VII), the higher strengths exhibited by fully supported bars preheated at test temperatures or higher prior to hot MOR testing emphasizes the need for reporting the specific thermal treatment of specimens along with hot MOR values so proper judgment can be used in comparing results.

Conclusions

The "generic" LC tabular alumina castable, obtained by substituting 5% each of 3 μm Al_2O_3 and fumed SiO_2 for an equivalent weight of cement in a conventional 15% CAC castable mix designed for maximum density, exhibits the low water requirement typified by commercial LC castables. It can be used effectively for characterizing CAC types in LC castables using various submicrometer powders and for establishing the effect of dispersant and set modifying additives on castable properties. The following conclusions can be drawn from results obtained by use of this "generic" LC test castable:

1. LC castables containing fumed SiO_2 as the submicrometer component exhibit varied water requirements and working time responses with 70 and 80% Al_2O_3 cements and additions of SPP electrolyte.

2. A new 80% Al_2O_3 cement containing significant amounts of the $12CaO \cdot 7Al_2O_3$ phase, known for its fast hydration and early strength development, exhibits extended working times in fumed-SiO_2 LC castables with low water requirements and good hot-strength properties at 1510°C (2750°F).

3. Hot strength properties can be supplemented by a sag deformation test to differentiate refractoriness at higher temperatures.

4. The 80% Al_2O_3 cement exhibits 55°–110°C (100°–200°F) higher refractoriness than a 70% Al_2O_3 cement in fumed-SiO_2 LC castables.

5. Exceptional hot strengths are exhibited by conventional tabular alumina castables designed for maximum density using 80% Al_2O_3 and 90% Al_2O_3 cements. These higher cement castables merit further mix design development to reduce water requirements and obtain lower porosities for improvement in corrosion- and erosion-resistant characteristics.

6. A new 90% Al_2O_3 cement offers reduced water requirements and the potential of reducing the CaO content of castables for improved slag resistance and greater refractoriness when using normal cement additions.

References

[1] "Special Topics: Recent Developments in Japanese Refractories Technology," *Taikabutsu Overseas*, **1** [1] (1981).

[2] "Special Topics: Refractories for Steelmaking and Casting," *Taikabutsu Overseas*, **1** [2] (1981).

[3] "Special Topics: Refractories for Ironmaking," *Taikabutsu Overseas*, **2** [1] (1982).

[4] "Special Topics: Evaluation and Testing Methods for Refractories," *Taikabutsu Overseas*, **2** [2] (1982).

[5]Papers from the Proceedings of the XXVth International Colloquium on Refractories (October 14-15, 1982), Aachen, Germany.

[6]D. H. Hubble and K. K. Kappmeyer, "Future Raw Material Requirements for Steel Plant Refractories," Workshop on Critical Materials, Vanderbilt University, Nashville, TN (October 4-7, 1982); to be published in *U.S. Bureau of Standards Bulletin*.

[7]B. Cooper, "Continuous Casting of BSC Port Talbot Works," *Steel Times*, 235-40 (May 1983).

[8]B. Cooper, "Conquest Keeps BSC Stocksbridge at the Forefront of Technology," *Steel Times*, 242-47 (May 1983).

[9]J. Mitchell, "Scunthorpe Completes Concast Trio," *Steel Times*, 249-53 (May 1983).

[10]W. Welburn, "Developments in Continuous Casting of Special Steels," *Steel Times*, 254 (May 1983).

[11]K. Yamamoto, "Some Properties and Applications of Low-Moisture Castables"; reviewed at the ACI Refractory Concrete Committee 547 Workshop-Seminar IV on Refractory Concrete at Carnegie-Mellon University, Pittsburgh, PA (June 7-9, 1983).

[12]G. MacZura, V. Gnauck, and P. T. Rothenbuehler, "Fine Aluminas for High Performance Refractories," presented at the First International Conference on Refractories, Tokyo, Japan (November 15-18, 1983).

[13]B. Clavaud, J. P. Kiehl, and R. D. Schmidt-Whitley, "Fifteen Years of Low Cement Castables and Steelmaking," presented at the First International Conference on Refractories, Tokyo, Japan (November 15-18, 1983).

[14]D. J. Bray, "Chromium Compounds, Refractories and Toxicity"; for abstract see *Am. Ceram. Soc. Bull.*, **62** [3] 431 (1983).

[15]R. P. Heilich, G. MacZura, and F. J. Rohr, "Precision Cast 92-97% Alumina Ceramics Bonded with Calcium Aluminate Cement," *Am. Ceram. Soc. Bull.* **50** [6] 548-54 (1971).

[16]C. C. Furnas, "Grading Aggregates: I," *Ind. Eng. Chem.*, **23** [9] 1052-58 (1931).

[17]F. O. Anderegg, "Grading Aggregates: II," *Ind. Eng. Chem.*, **23** [9] 1058-64 (1931).

[18]W. H. Gitzen, L. D. Hart, and G. MacZura, "Properties of Some Calcium Aluminate Cement Compositions," *J. Am. Ceram. Soc.*, **40** [5] 158-67 (1957).

[19]T. W. Parker, p. 485 in Proceedings of the Third International Symposium on Chemistry of Cement, London, England, 1952.

[20]T. D. Robson, High Alumina Cements and Concretes. Wiley & Sons, New York, 1962.

[21]W. S. Treffner and R. M. Williams, "Heat Evolution Tests with Calcium Aluminate Binders and Castables," *J. Am. Ceram. Soc.*, **46** [8] 399-406 (1963).

Section V
Monolithics for Blast Furnace Usage

Designing a Casthouse for Preformed Shapes 305
    R. A. Howe, J. W. Kelley, and T. A. Dannemiller

Application of Dry-Forming Method to Blast Furnace Troughs 313
    S. Nishizawa and A. Kondo

The Use of Monolithic Refractories in Blast Furnaces 323
    L. Krietz, R. Woodhead, S. Chadhuri, and A. Egami

Wear Mechanisms in Alumina-Silicon Carbide-Carbon Blast Furnace Trough Refractories 331
    S. B. Bonsall and D. K. Henry

Progress in Casting Trough Materials and Installing Techniques for Large Blast Furnaces 341
    Y. Toritani, T. Yamane, S. Yamasaki, I. Nishijima, T. Kawakami, and Y. Kadota

Comparison of Monolithic Refractories for Blast Furnace Troughs and Runners 355
    C. M. Jones

Designing a Casthouse for Preformed Shapes

Robert A. Howe, James W. Kelley, and Thomas A. Dannemiller

North American Refractories Company
Cleveland, OH 44115

North American Refractories has been supplying preformed shapes for casthouse floors since July 1980. This experience has shown that preformed shapes can extend campaign life, which is necessary to maximize production of small furnaces and meet EPA requirements. To realize the full benefit of preformed shapes, a comprehensive casthouse design approach is necessary.

Advantages of Preformed Shapes

When properly installed in a well-designed floor, preformed shapes have many advantages over a field-installed monolithic material in terms of properties, installation time, labor requirements, and campaign life.

Properties

Preformed shapes are manufactured under more controlled conditions than field-installed refractories, which provides better uniformity in forming and improved properties of the cured refractories. The properties of the material are not dependent on the skill of the installation crew or the limitations of field curing.

Oven curing gives a superior product to the same product cured under ideal conditions in the field. Table I shows the properties of an alumina, silicon carbide, carbon ramming mix as a preformed shape and after an ideal field curing. To simulate field curing, the samples were placed in the door of an electric furnace, and the furnace was cycled through the prescribed field curing schedule. The oven-cured samples have a higher strength and a lower porosity partly due to the one-sided heating. Another factor is that the hot face of the field-cured sample has been heated to a higher temperature, which altered the organic bonding.

Table I. Properties of Mix A

	Modulus of Rupture MPa (psi)	Porosity (%)
Preformed shape*	16.50 (2400)	11
Simulated field-cured sample†	3.45 (500)	15

*Pressed shape, heated in oven from all sides for 5 h to 150°C (300°F), 5-h soak at 150°C (300°F).
†Pressed shape, heated in oven door from one side for 2 h to 700°C (1300°F), 6-h soak at 700°C (1300°F).

Properties after coking at 700°C (1300°F) were measured on oven-cured and field-cured samples to provide a more realistic comparison and eliminate temperature effects on cured samples. Properties on coked samples are shown in Table II. The oven-dried samples again have a higher strength, which is attributed in part to the better polymerization of the binder when the material is slowly heated in the oven. The increase in porosity of the field-cured sample partially results from less polymerization and partially from oxidation. Of the samples heated from one side, the hot-face portion which is exposed to the most oxidation shows higher porosity than the cold face. Both portions had higher porosities than the oven-cured samples.

Table II. Coked Properties of Mix A

	Modulus of Rupture MPa (psi)	Porosity (%)
Preformed shape*	2.41 (350)	13
Simulated field-cured sample† (hot face)	1.45 (210)	16
Simulated field-cured sample†(cold face)	1.79 (260)	14

*Pressed shape, heated in oven from all sides for 5 h to 150°C (300°F), 5-hr soak at 150°C (300°F).
†Pressed shape, heated in oven door from one side for 2 h to 700°C (1300°F), 6-hr soak at 700°C (1300°F).

Installation Time

Preformed shapes have been placed at the rate of 1 every 5 min. The typical installation proceeds at about 6 pieces per hour of working time when the working drawings accurately represent the casthouse floor. Table III shows comparative times for a main trough installation using preformed shapes and a ramming mix. The 15-18-h turnaround is done routinely in one shop on 4 different troughs, and they hope to be able to lower this to a 12-h turnaround.

Over one-half of the 33 h required for field installation is for dryout. In the press of getting a furnace back on line, this step is often shortened signifi-

Table III. Time Estimated for Main Trough Installation

	Inserts (h)	Ramming Mix (h)
Tear out	4	4
Bottom installation	3	4
Form installation	None	2
Sidewall installation	3	4
Dryout/preheat	5-8	19
Total time	15-18	33
Total labor	30 man-hours	60 man-hours

cantly. Poorer material properties and reduced life result when drying time is sacrificed to achieve faster installation times.

Labor

Preformed shapes can be installed with a four-man crew, if storage logistics and labor jurisdictional disputes do not interfere, as compared with the six-man crew required for a ramming installation. Also, the work involved in placing preformed pieces is much less physically demanding than a ramming installation, which prevents quality deterioration as the day progresses.

Campaign Life

Longer runs between patches have been achieved with preformed shapes over field ramming due to the reduced wear rates resulting from improved material properties. With periodic maintenance, preformed shapes in a main trough have routinely obtained 400 000 tons of iron produced per campaign with one record campaign of 500 000 tons of iron being produced in one campaign.

Disadvantages of Preformed Shapes

Preformed shapes for use on the casthouse floor have some disadvantages, including cost, fabrication time, and installation.

Cost

Preformed shapes are expensive. The customer is not only paying for the ramming mix, but also is paying for forming by skilled workers as well as the equipment and energy required for forming and curing.

Fabrication Time

Preformed shapes take a minimum of two weeks to manufacture after the mold is made. Curing and ramming equipment limit the total output possible. Due to these factors, the normal lead time on preformed shapes is six to eight weeks. These lead times may require larger inventories.

Installation

A potential disadvantage is difficulty of installation due to inaccurate drawings of the casthouse floor. If a shape does not fit, field cutting is necessary, and the superior material properties make field cutting a difficult task requiring diamond cutting equipment. Accurate drawings of the casthouse floor are essential for successful application of preformed shapes.

Casthouse Design

The superior properties of preformed shapes will not be adequately utilized to extend campaign life unless the casthouse system is designed to accommodate the unique features of a preformed shape.

Backup Lining

The thermal conductivity of the backup materials is the single most important factor affecting successful use of preformed shapes. The thermal conductivity should be low enough to provide insulation and hold heat in the iron. At the same time, the thermal conductivity should be high enough to ensure that the metal freeze plane stays in the working or safety lining. The effects of different trough configurations are illustrated below using a one-dimensional, equilibrium model to calculate thermal gradients. Actual field results may be significantly lower and should be monitored by embedding thermocouples at selected locations.

Figure 1 shows a temperature gradient in an air-cooled trough wall. The safety and backup linings are both 11.4-cm superduty fireclay brick. The iron freeze plane of 1205°C (2200°F) is shown on the figure as a horizontal line. Above the line, metal will flow in a crack; below the line, metal in a crack will be frozen and seal off further penetration. Figure 2 shows the same wall near the end of its useful life. The freeze plane has shifted into the safety lining, but the system is still relatively secure from breakouts since freezing will occur within the safety lining.

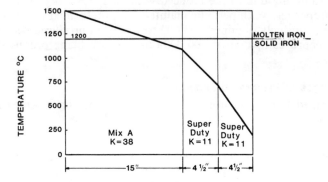

Fig. 1. Temperature gradient for normal backup.

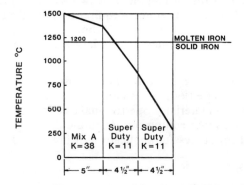

Fig. 2. Temperature gradient for normal backup near the end of its useful life.

The use of insulating firebrick for backup significantly alters the thermal gradient. Figure 3 shows the installation. The iron freeze plane starts over halfway through the safety lining of superduty firebrick. As the working lining wears to a thickness of 12.7 cm, as shown in Fig. 4, the freeze plane has shifted into the insulating firebrick. If a crack develops under these conditions, molten iron will flow to the backup brick and quickly deteriorate the insulation, resulting in a breakout. The possibility of a breakout should be

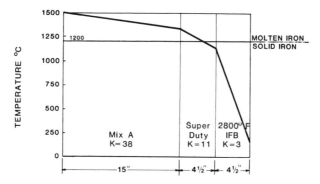

Fig. 3. Temperature gradient for insulating backup.

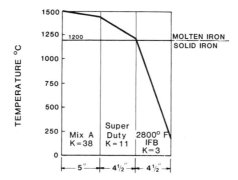

Fig. 4. Temperature gradient for insulating backup with working lining worn to 12.7 cm.

sufficient to demonstrate the importance of considering the thermal conductivity of the lining configuration when designing the backup.

Support Configuration

Another important factor in casthouse design is the configuration of the metal box or supports. For a cooled trough, a steel box supports the refractories. It must be strong and rigid to provide proper support for the refractories. If heat transfer to the steel is too high, it must be force-cooled to prevent warping.

If the main trough is buried in sand, the heat cannot escape to the surroundings at a fast enough rate to cool a steel box. Figure 5 shows the temperature gradient when a trough is buried in 0.3 m of sand. Even with insulating firebrick in the backup and this minimum amount of sand, the temperature of the steelwork could be about 538°C (1000°F). Larger amounts of sand will increase the temperature of the steelwork. Any temperature above 427°C (800°F) will cause warping of the steelwork. When the steel box warps, it either pushes the preformed shapes out of alignment or creates a void. A void leaves the preformed shape unsupported, which could lead to cracking when the molten iron presses against the preformed shape.

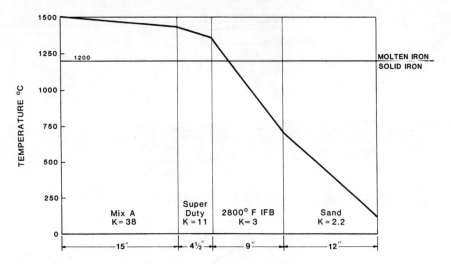

Fig. 5. Temperature gradient for sand backing.

No, or very little, buried steelwork should be used, since a total refractory backup can be designed for buried troughs.

The last important feature of the backup is its strength or structural integrity. Preformed shapes will not bridge a large gap under the pressure of the iron. Therefore, the backup lining should provide support that does not shift, shrink, soften, or crush over the entire temperature use range. Improper support will cause cracks in the working lining, which leads to metal penetration and turbulence, resulting in accelerated wear.

Working Lining

An important factor in designing a working lining with preformed shapes is to reduce the turbulence in the flowing iron or slag. Turbulence is undesirable because it not only causes mechanical erosion, but also increases chemical attack by reducing boundry layers and removing corroded refractories.

Figure 6 shows two runner shapes. By going from the design on the left to the design on the right with the flat bottom, the amount of wall contact and turbulence was reduced. This shape change resulted in a 20% increase in service life. A second example of reducing turbulence is shown in Fig. 7. The original elbow design is shown on the left, and the current elbow design is on the right. The original design with the flow impacting directly into a wall resulted in high wear on the sidewall and a premature failure of the impact wall. The new design with the gradual sweeping curve results in much longer life. Because there is no way to avoid wear on the outside wall of a curve, the outside wall is thicker, which allows longer life before maintenance begins. Reducing turbulence is particularly important with preformed shapes because of the long life expected before the beginning of a maintenance program.

The material used to manufacture the preformed shapes must be chosen based on service conditions. In most of the main trough, a dense, strong, slag-resistant material is required. The bond phase of the material must be

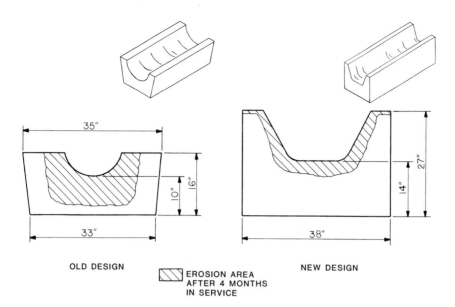

Fig. 6. Design evolution for the shape of runners; erosion area after 4 months in service.

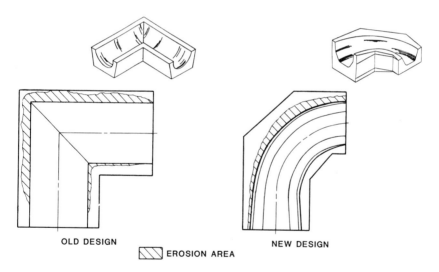

Fig. 7. Turbulence reduction in two elbow designs; erosion area.

selected based on the hot metal temperature in the trough. As shown in Table IV, when iron temperatures are above 1510°C (2750°F), mix A with a high-temperature bond is preferred. Mix B with an intermediate-temperature bond is necessary for lower temperature applications. In the skimmer, however, where thermal stress causes cracking in these materials, a very thermal-shock-resistant and volume-stable, clay-bonded castable, mix C, is used.

Table IV. Properties of Mixes

	Mix A	Mix B	Mix C
Chemical Analysis (%)			
Al_2O_3	68	74	68
SiO_2	5	7	5
SiC + C	24	15	22
After drying at 150°C (300°F)			
Porosity (%)	10	11	16
Bulk density (Mg/m^3)	2.98	2.94	2.88
(psi)	(186)	(184)	(180)
Modulus of rupture (MPa)	16.5	14.5	4.83
(psi)	(2400)	(2100)	(700)
Bond formation temperature	High	Intermediate	High

Operational Practices

As in most steel mill applications, the maintenance program and operational variables significantly influence refractory performance.

Maintenance

A scheduled maintenance program is important for successful and long campaigns with preformed shapes. While main trough campaigns of 150 000 tons of iron can be achieved with no maintenance, 400 000 tons is a realistic goal with a program of periodic repair. Theoretically, unlimited campaign lives are possible with the new low-moisture castables and an aggressive patching program. The most economical campaign life is still being determined.

Operational Factors

The following operational factors affect refractories: tap temperature, slag/iron ratio, lime/silica ratio in the slag, tap stream velocity, taphole angle, and taphole misalignment.

Briefly, the first three factors affect chemical wear, whereas the last three affect mechanical wear. For maximum life, the operational factors causing the wear must be reduced. The operator's control over these factors may not always permit the desired adjustment, but any effort at reducing these factors will usually be rewarded with longer refractory life.

Summary

Present steel mill operating conditions require extended campaign lives for casthouse refractories. The required campaign lives have been obtained with preformed shapes when the following requirements have been met: proper thermal gradient in lining; adequate structural support of preformed shape; proper working lining material and configuration; scheduled maintenance program; and consistent control of operating variables.

Application of Dry-Forming Method to Blast Furnace Troughs

Shozo Nishizawa and Atsushi Kondo

Wakayama Steel Works
Sumitomo Metals
Wakayama 640, Japan

The iron tapping trough of the blast furnace in the Wakayama Works is of the fixed type from its main trough to iron trough. As there are only two iron tapping troughs at each blast furnace, installation work is required to be completed within the shortest time possible.

The lining method using ramming material, which is presently used for its ease of installation, is apt to provoke some partial failures, due to uneven packing and the impossibility of using the patching technique for lining repair. Therefore, it is difficult to achieve a reduction in the unit consumption of the trough material, and much manual work at high temperature is required during repairs.

To solve those problems, in collaboration with the Tsurumi Synthetic Refractories Co., Ltd., the Wakayama Works has developed an installation method using dry trough material, improved the installation equipment, and the trough material quality, and put it to practical use for the main trough of blast furnaces in the Wakayama Works. We have obtained good results.

Sumitomo-Tsurumi-Dry-Forming Method (STD Method)

Features of the STD Method

(1) Dry trough material with a water content of 0–0.5% STD material is used.

(2) A method that employs packing equipment with pressure and vibration (Hi-pack) is used to pack material, and pressure vibration is applied to increase packing density.

(3) Compared with a ramming method, uniform installation without uneven packing is possible, and the risk of premature failure is reduced.

(4) During repair, only a little of the retained material is removed, and unlimited repair by patching is possible, so that the trough material unit consumption is reduced.

(5) Required curing time after installation is short, and curing energy is also saved. (Even sudden curing does not provoke any trouble such as spalling.)

(6) Heavy manual work at high temperature may be alleviated by mechanization of the packing device and the extension of trough service life.

Dry-Forming Method

The procedure for installing a trough by the dry-forming method is given in Fig. 1.

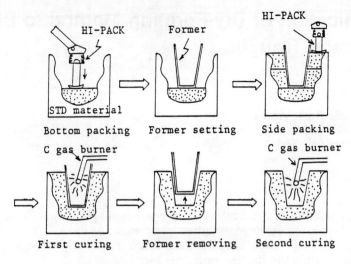

Fig. 1. Working procedure of STD material.

Dismantling of Retained Material: After water-spray cooling, dismantle and remove the retained material using a backhoe loader. (As unlimited repair by patching is possible, remove the minimum quantity of the retained material.)

Bottom Packing: Put the dry material onto the bottom, install Hi-pack on the backhoe loader, and make bottom packing by pressure vibration.

Form Setting: After completion of bottom packing, set form in designated position.

Side Packing: Put dry material to the side area, and perform packing with Hi-pack in same manner as for the bottom area.

First Curing: Cure trough with form, set for one hour using coke oven gas burner until the shape-holding strength of the dry material is achieved.

Form Removing: Remove form after first curing is completed.

Second Curing: After form removal, cure trough again for one hour before casting molten iron.

Packing Device

Generally the life of monolithic refractories may be extended by increasing packing density. In the case of dry material, the packing density varies greatly depending on the packing device and packing method used.

When dry-forming materials were first used, the rod vibrator or fork was employed; however, this method gave a low packing density which caused abnormal wear. Those problems were eliminated by the adoption of the fork vibrator, or Hi-pack, which increased packing density. Figure 2 shows the relation between various packing devices and packing density.

Actually, the backhoe loader equipped with Hi-pack is used for packing in the ordinary installation work. If Hi-pack cannot be used due to a narrow

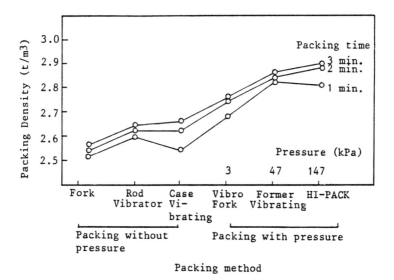

Fig. 2. Relation between packing method and packing density of dry material.

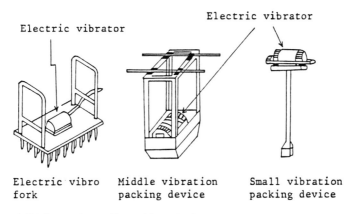

Fig. 3. Middle and small packing devices.

working place, the electric fork vibrator or a middle or small vibration packing device is used (Fig. 3).

Dry Material (STD Material)

Electrically fused, Al_2O_3-SiC is used as dry material for the main trough, and its grain distribution is regulated to give the densest packing. The dry material is solidified by heating, due to the addition of small amounts of organic and inorganic binders, which fuse and solidify at low temperature. Figure 4 shows the cold crushing strength at low temperature.

As Fig. 5 shows, the liquid phase of the dry material appears at 150°–200°C and solid bonding occurs at around 300°C, the temperature at

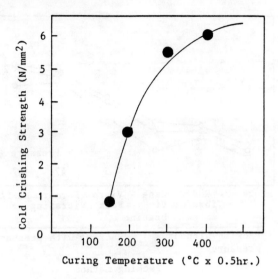

Fig. 4. Cold crushing strength at low temperature.

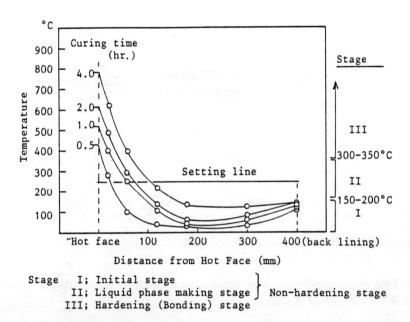

Stage I; Initial stage
       II; Liquid phase making stage } Non-hardening stage
       III; Hardening (Bonding) stage

Fig. 5. Relation between curing time and internal temperature of installed STD material.

which the shape-holding strength is achieved. Form removal is determined when the shape-holding strength is developed 50 mm from the hot face. Actually, one hour is established as the curing time, and after this period the form is removed.

Quality Improvement

When first used, cracks in the dry material or exfoliation occurred during curing. The exfoliation problem has been solved by changes in the main aggregate (from 80–98% Al_2O_3 material), changes in the binder quality, and alteration of blending practices. Table I shows the present dry material and ramming material used for main troughs.

Table I. Specification of STD Material

		STD Material	Ramming Material
Chemical composition (%)	SiO_2	3	4
	Al_2O_3	82	67
	SiC	10	22
	FC	1	5
Bulk density (ton/m³)	400°C·2 h	2.61	2.57
	800°C·2 h	2.60	2.58
	1450°C·2 h	2.61	2.58
Apparent porosity (%)	800°C·2 h	27.2	28.3
	1450°C·2 h	21.0	22.3
Cold crushing strength (N/m²)	400°C·2 h	5.9	29.0
	800°C·2 h	14.4	34.0
	1450°C·2 h	15.7	39.9

STD specimens are arranged by hand-packing with fork.
Ramming specimens are arranged by pressing (9.8MPa).

Table II. Comparison of Working Schedule (Material used 20 ton)

	STD Method	Ramming Material	*Casting Method
Cooling	30 min	30 min	3 h
Wrecking	1 ½ h	2 ½ h	1 ½ h
Bottom packing	30 min	30 min	
Former setting	20 min	20 min	30 min
Side packing	30 min	30 min	
Casting			2 ½ h
Former removing		10 min	
First curing	1 h	3 h	2 h
Former removing	10 min		10 min
Second curing	1 h		3 h
Total	5 ½ h	7 ½ h	12 ⅔ h

*Estimated time: the casting method has not been used in Wakayama

Performance of Dry Trough Materials

Installation Time

Table II shows the installation times of the STD method, the ramming method, and the casting method. The STD method allows a shorter time than the ramming method and the casting method by saving time for dismantling and cooling retained material, as well as for drying and curing new material.

Unlimited Repair Capability of the Patching Technique

In the STD and the casting methods, the minimum quantity of the retained material is removed prior to repair, thus rendering unlimited repair by patching possible. Figure 6 shows the changes in the trough material unit consumption by unlimited repair by patching at Wakayama 4 Blast Furnace (BF) skimmer trough. The actual result of the STD method used in the Wakayama 4 BF skimmer showed a large reduction in the trough material unit consumption. From the first full reline (material 25 T used) to the present (December 1983), unlimited repair has been continued.

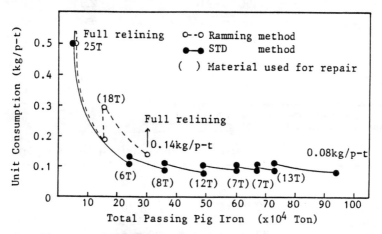

Fig. 6. Changes of the trough material unit consumption in Wakayama 4 BF skimmer trough.

Changes in the Main Trough Unit Consumption

The changes in the main trough unit consumption of Wakayama 2, 3, 4 BFs are given in Fig. 7. The STD method was started as a test in 1977 and introduced on a full scale for the main trough of the second campaign 4 BF in July 1980. Afterward, the scope of application was enlarged to reach 70% of the main trough of 2, 3, 4 BF in December 1983, as shown in Fig. 8. At 3 BF, where the STD method is adopted in most of the main trough, the unit consumption has been reduced to about 60% of that of the ramming method.

Actual Problems and Solutions

Durability Improvement of Tapping Area of Main Trough (Front of Taphole)

Since the main troughs of the Wakayama blast furnace are of the non-pool type, damage of the tapping area of the main trough is serious. The dur-

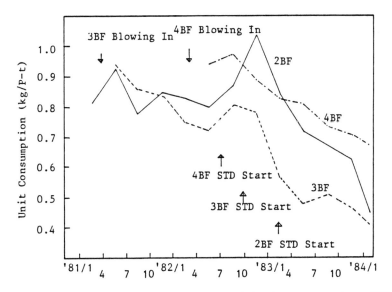

Fig. 7. Changes of the main trough material unit consumption of Wakayama 2, 3, 4 BF.

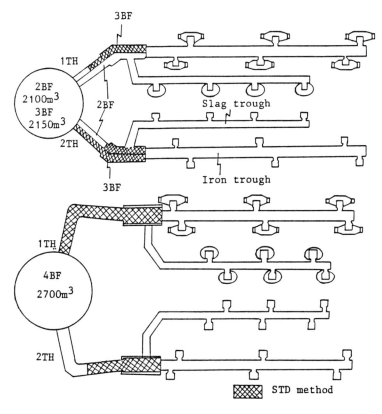

Fig. 8. Scope of STD method applicated to main trough.

Table III. Results of STD Material Tried at Main Trough of 4 BF

Refractory Part	Tapping Area		Middle Area		Skimmer Area	
	STD Material	Ramming Material	STD Material	Ramming Material	STD Material	Ramming Material
Service life (days)	9	8	45	21	98	62
Production pig iron (tons)	22 300	20 300	111 500	47 800	206 600	130 700
Refractory consumption (kg/P-T)	0.62	0.50	0.23	0.40	0.10	0.16
Wearing speed (mm/1000 tons)	18.8 (Bottom)	20.7	3.6 (Bottom)	8.4	1.7 (Side)	2.7

ability improvement of the tapping area (bottom) is very important for further reducing the main trough material unit consumption. We have been engaged in studies and practical furnace tests aimed at strength improvements of the material to increase the resistance of the tapping area STD material to physical damage. Table III shows the durability of STD material in each area. Figure 9 shows the improved cold strength of the STD material.

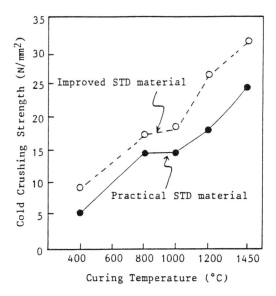

Fig. 9. Comparison of cold crushing strength, practical, and improved STD.

Measures Against Dust

Dust production is a serious problem associated with dry material installation. STD also generates a little dust when pouring the material (no special problem is posed in installation).

As a protection against dust, an addition of 0.2–0.5% surface-active agent liquid (S.A.A.(l)) is effective, and when packing by pressure vibration with Hi-pack, even addition of 0.5% S.A.A.(l) does not greatly lower the packing density, as shown in Fig. 10. The material with 0.5% S.A.A.(l) content is actually used in practice.

Conclusion

This paper reports on the properties, installation techniques, and performance of dry-forming materials in the blast furnaces at the Wakayama Works. The STD method is presently put to practical use in the main trough of Wakayama blast furnaces. High packing densities are obtained by relatively easy installation with backhoe loaders and Hi-pack, and characterized by shorter curing times and less drying energy. Moreover, the trough material unit consumption has been greatly reduced by adoption of unlimited repair by patching techniques. Current developments are intended to promote prac-

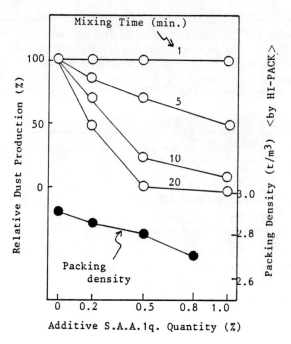

Fig. 10. Relationship between relative dust production, packing density, and additive S.A.A.(*l*).

tical application of measures against dust, improve durability of material in the tapping area of the main trough, and reduce further material unit consumption.

References

[1] Seiji Kajikawa et at., *TETSU-to-HAGANÉ*, **65** [S544] (1979); p. 30.
[2] Genji Nakatani et al., *TETSU-to-HAGANÉ*, **67** [S68] (1981); p. 68.
[3] Shigeru Mizuno et al., *TETSU-to-HAGANÉ*, **68** [S677] (1982); p. 13.
[4] Material for 58th Iron making commission: Dry vibration forming for blast furnace trough, Japan, 1983; No. 33-9.

The Use of Monolithic Refractories in Blast Furnaces

Leonard P. Krietz, Roy Woodhead, Sudhamay Chadhuri, and Akira Egami

Plibrico Company
Chicago, IL 60614

Although currently much attention is being given to the rapid pace of development of monolithics for BF runners and troughs, a number of other areas in the blast furnace have also seen growing use of monolithics. Recent progress on the application of monolithics to these other areas is discussed.

Over the past decade, blast furnace operating practice has been changing worldwide. With the advent of modern, large-capacity furnaces and the current production demands being placed on the older furnaces due to economic considerations, operators are being faced with the problems of optimizing furnace life and reducing maintenance costs. To meet the new operating criteria, refractory producers have been developing new refractories and installation methods to achieve the goals the steel companies demand. Monolithics are playing a large part in current blast furnace operations and maintenance practices.

Much attention is being given to the development of blast furnace runner and trough materials, as well it should. Other refractories and systems are also being developed worldwide to prolong blast furnace life and reduce maintenance costs; they include hot gunning, injection refractories, and techniques for stack maintenance.

Blast Furnace Runners

The development of monolithic refractory materials for blast furnace trough runner application has perhaps been the most active area of monolithic development for blast furnaces, as a brief review clearly shows.

The basic requirements of a blast furnace runner refractory are resistance to chemical attack by hot metal and slag; resistance to erosion, thermal shock, and oxidation; the capability for quick sintering and rapid and easy installation with no air pollution; and low porosity. Various types of monolithics have been developed to meet these requirements: wet ramming mixes, dry vibrating mixes, and low-cement and no-cement castables. The type of monolithic used varies from plant to plant around the world, and the particular choice depends on blast furnace operating parameters and performance history. These materials are generally based on corundum, silicon carbide, and carbon. Typical generic refractory compositions are 55–70% Al_2O_3, 10–25% SiC, and 5–10% C. All materials are formulated to obtain optimum packing density and to meet the basic requirements.

In some cases, runner practice dictates maintenance of the runner or trough by the use of gunning mixes: either conventional mixes where water is added to the nozzle, or plastic gunning mixes that require no additional water. The latter maintenance procedure is currently being used successfully in Germany. Troughs and runners originally rammed with a wet ramming mix are being serviced with a plastic gunning mix of similar composition to the original refractory. Figure 1 shows a worn blast furnace runner which was originally rammed. The left side has been serviced by a plastic gunning repair. Practice has shown that the original rammed refractory requires gunning maintenance after a throughput of approximately 50 000 tons of iron, and plastic gunning maintenance is performed every 40 000 tons throughput thereafter. The repair can be done in a hot or cold condition. This method has been shown to be very economical over the course of a runner campaign.

Fig. 1. Blast furnace trough after plastic gunning maintenance.

Blast Furnace Lining Maintenance

Blast furnace refractory linings naturally deteriorate in use as a result of chemical attack, abrasion, and spalling. In the past, it was impossible to service the stack until the blast furnace was completely shut down. Damage to the lining during a campaign grows progressively worse, resulting in hot spots. To remedy this problem, two methods have been developed to repair a blast furnace stack in the hot condition — hot stack gunning and injection.

Hot Stack Gunning

Hot stack gunning is currently being performed very successfully in Japan and Australia. This method, which has been developed over the past six years to repair the upper stack, allows the lining to be restored to a sufficient thickness while the gunning surface is observed from outside the blast furnace.

The blast furnace burden is lowered below the area to be taken off wind. An opening is provided for the 10–20-m-long gunning nozzle. A standard gunning machine is used. The nozzle is then operated from outside the blast furnace to perform the gunning repairs. Equipment placement is shown in Fig. 2. Typical gunning rates are 3–4 tons/h of refractory. The blast furnace conditions required during hot stack gunning are as follows: (1) The blast furnace gas temperature around the gunning surface area should be below 600°C; (2) the burden level should be at least 1 m below the area to be repaired; and (3) the gas exhaust of the blast furnace is very critical. Gases must be extracted by mechanical means via the exhaust ducts to prevent CO gas leakage and to ensure good visibility. Air quality at the gunning station must be constantly monitored.

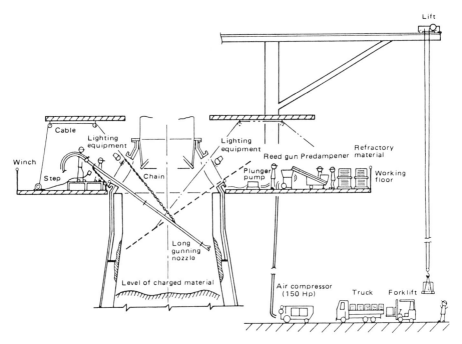

Fig. 2. Equipment setup for blast furnace hot stack gunning repair.

The refractory gunning material is a low-iron, gunning castable with good hot strength and abrasion resistance. Its physical properties are shown in Table I.

The thickness of the applied gunned refractory varies according to the extent of damage to the lining. In general, lining thicknesses are applied between 100 and 600 mm.

The life of a gunned refractory repair for an individual furnace primarily depends on operating conditions. In general, two-thirds of a newly repaired area remains after six months.

Figure 3 illustrates how well the gunned refractory adheres to a firebrick lining. This core-drilled sample was removed from a blast furnace after one

Table I. Refractory Gunning Castable for Hot Gunning Repairs

Properties	Requirements
Maximum temperature (°C)	1300
Weight in service (kg/m³)	1950–2200
Chemical composition (%)	
Al_2O_3	36
SiO_2	48
Hot modulus of rupture (MPa)	
110°C	5.9
1000°C	3.4
1300°C	4.9

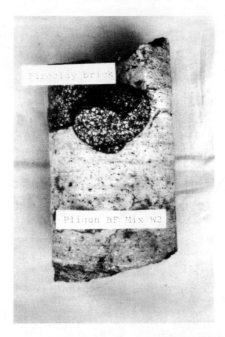

Fig. 3. Adherence of gunned refractory to firebrick.

month of service. Figure 4 shows the adherence of one gunned layer to another. The time between gunnings was one month, and the sample was taken one month later. Physical properties were obtained on core-drilled samples: The density was 2030 kg/m³ and cold crushing strength was 51 MPa.

Blast Furnace Injection Repair

The blast furnace injection process is a method used throughout the world to repair the belly or shaft section of a blast furnace. With this process, it is possible to repair a blast furnace lining while the furnace is off wind by injecting a stiff refractory mix at high pressures through nipples located in areas of hot spots.

Fig. 4. Adherence of gunned refractory to older gunned layer.

There are three main criteria for a successful blast furnace injection: a suitable high-pressure pump, a quality injection refractory, and proper installation technique.

The pump used must be able to deliver a stiff refractory mix at high pressures at a distance of up to 50 m horizontally and 30 m vertically. Injection pressure at the nipple is approximately 30 bars.

The refractory mix used is a heat-setting material with good physical and flow properties at a stiff consistency. Table II lists the physical properties of an 80% alumina injection mix. A key to a successful injection mix is the stiff consistency; a proper mix should have the consistency of toothpaste at the time of injection, which minimizes water and thus optimizes properties.

The injection procedure is as follows: The nipples on the furnace are uncapped and drilled to leave a clear path for the injection refractory through

Table II. 80% Alumina Blast Furnace Injection Mix

Properties	Requirements
Chemical composition (%)	
Al_2O_3	79
SiO_2	11
Water required for injection (%)	12–14
Service limit (°C)	1700
Weight in service (kg/m^3)	2467
Cold modulus of rupture (MPa)	
110°C	4.6
1100°C	19.0
1370°C	22.5
Hot modulus of rupture (MPa)	
815°C	13.5
1100°C	2.9
1370°C	1.5

any remaining refractory into the burden. Normally, nipples are 50 mm in diameter. A valve is then placed on the nipple. A specific amount of injection mix is mixed and introduced to the pump. The amount of refractory used depends on the spacing of the nipples on the blast furnace. This mix is then pumped under high pressure through a 50-mm hose which is connected to the valve on a nipple by a quick release coupling (see Fig. 5). The valve is opened

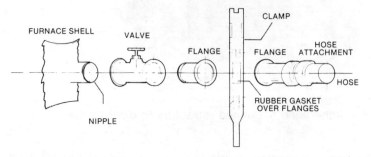

Fig. 5. Nipple, valve, and hose arrangement for blast furnace injection.

and the refractory injection begins (Fig. 6). As the refractory infiltrates the hot burden, it hardens and a scab forms. As injection continues, some movement of the burden can occur, thereby allowing refractory to fill in behind it (Fig. 7). As was previously mentioned, the amount of injection refractory depends on nipple spacing. If too much is injected, it will flow and plug the other nipples. After injection is complete, the valve is closed and the hose

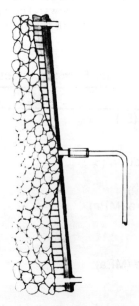

Fig. 6. Injection of refractory and mixing with burden.

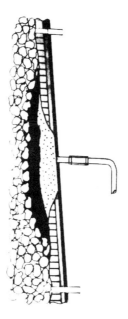

Fig. 7. Scab formation and movement of burden to allow for refractory filling.

Fig. 8. Refractory buildup after an injection repair.

moved to the next nipple. Average injection time per nipple is approximately 10–12 min. The injection process normally proceeds from the bottom of the areas to be repaired, going circumferentially around the repair area, and then moving up to allow a buildup, as shown in Fig. 8. This type of buildup through an injection maintenance program has been confirmed by postmortem analysis of a blast furnace down for rebuild.

Blast furnace injection is not a permanent repair, and wear rates vary based on furnace condition and operating practice. It has been shown to be a viable method for repairing worn shaft linings and yields good results when performed as a regular maintenance procedure.

Miscellaneous Applications

Monolithics are also being used in other areas of the blast furnace. Low-cement castables are being used for various applications, such as skimmer blocks, tilting runners, and trough covers. Specialty silicon carbide containing low-cement castables is also being used to replace standard silicon carbide inserts in copper tuyere tips. As monolithic refractory technology advances and new installation techniques develop, new uses for monolithic refractories in the blast furnace and casthouse will surely develop.

Wear Mechanisms in Alumina–Silicon Carbide–Carbon Blast Furnace Trough Refractories

S. B. BONSALL AND D. K. HENRY

Carborundum Company
Niagara Falls, NY 14302

The wear mechanism of Al_2O_3-SiC-C blast furnace trough refractories was found to vary, depending on location in the trough. At the slag-air-refractory interface, oxidation of free carbon from the refractory, followed by penetration of slag into open porosity, and dissolution of the refractory bond by the slag were seen as the main wear mechanisms. At the slag-molten iron-refractory interface the molten metal participated in the attack on the refractory by dissolving silicon from silicon carbide grains and the refractory bond.

Alumina–silicon carbide–carbon trough lining refractories were originally developed in Japan to meet the performance requirements of large, high top-pressure blast furnaces. Fused or sintered alumina aggregates are generally believed to contribute wear and corrosion resistance, while silicon carbide and carbon contribute volume stability, spalling resistance, and slag resistance. Due to the complexity of the wear mechanisms in these refractories, however, it has not been possible to determine an optimum composition by any theoretical means. The empirically determined optimal composition range for main trough refractories is about 15–25% SiC, 60–70% Al_2O_3, and 1–8% C. A better understanding of the mechanisms through which the components affect trough refractory performance could lead to future improvements in these materials.

Methods of Investigation

Samples of refractories removed from various blast furnace main troughs and runners were collected, and polished sections made of the eroded surface and adjacent refractory. A petrographic microscope was used for primary phase identification and to study the microstructural features. X-ray powder diffraction analyses were used to positively identify crystalline phases. The chemical compositions of phases seen in the polished sections were determined on a scanning electron microscope by semiquantitative energy dispersive X-ray analyses.

Results

Samples of Refractory A (see Table I for bulk composition) were taken 3 to 4 m downstream from the taphole in a main trough sidewall of a 4-taphole blast furnace producing about 8000 tons of hot metal per day.

Microscopic examination of polished sections from the area near the slag-air interface reveals a number of features. From the hot face to a depth of 1 to 2 cm, there is little carbon present. Slag has penetrated through the

Table I. Bulk Compositions of Refractories Studied

	Refractory A (Wt%)	Refractory B (Wt%)	Refractory C (Wt%)
Al_2O_3	59	68	69
SiO_2	5	2	5
SiC	27	25	17
C	6	2	6
% Open porosity (fired to 1450°C reducing conditions)	19	17	14

open pore structure and displays several different zones of reaction. Figure 1 illustrates the slagged zone at the hot face. The mineralogical assemblage is made up of a gehlenite-akermanite member and glass. Semiquantitative analyses by the energy dispersive X-ray technique are listed in Table II and show these phases to be quite similar in composition. Figure 2 shows the character of slag penetration approximately 1 cm from the hot face. This alteration becomes less evident until relatively unaffected refractory is encountered 4 to 5 cm from the hot face (Fig. 3). Figure 4 illustrates the phase relationship in the transition zone. Clusters of bladed crystals are found in a glass matrix. The spot analyses of these phases (Table II) show a relative enrichment in alumina compared with the hot face region. Magnesium spinel is often found bordering fused alumina grains (Fig. 5). These alumina grains are reacting with the slag, but at a much lower rate than the bond matrix. Silicon carbide grains have remained relatively unaffected (Fig. 4).

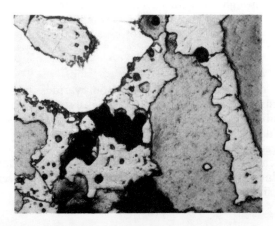

Fig. 1. Refractory A; hot face (×132).

Table II. Analyses of Phases in Slag-Air Reaction Zone of Refractory A

	Outer Zone (Fig. 1)		Intermediate Zone (Fig. 4)		Internal Bond (Fig. 3)		Adjacent Alumina (Fig. 5)		
Amount Present (Wt%)	1 Light Crystals (Gehlenite- Akermanite)	2 Glass around Crystals	3 Bladed Crystals	4 Glass around Crystals	5 Spinel	6 Glass	7 CA$_6$	8 Spinel	9 Glass
MgO	12.70	1.45	5.70	1.97	18.87		2.48	31.50	3.76
Al$_2$O$_3$	11.62	15.69	53.06	41.17	81.13	47.16	85.56	67.83	32.70
SiO$_2$	45.82	49.30	37.11	36.01		41.03	1.98		27.92
SO$_3$				0.92		9.64			
K$_2$O		1.95	0.24	1.25			0.24		
CaO	22.26	23.47	3.89	18.67		0.25	9.39		35.62
TiO$_2$	3.41	2.10					0.42		
MnO$_2$	0.96	2.18						0.67	
Fe$_2$O$_3$	3.23	3.85				1.13			

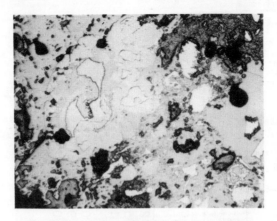

Fig. 2. Refractory A; slag attack, 1 cm from hot face (×132).

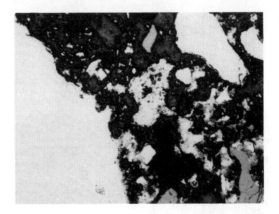

Fig. 3. Refractory A; unaltered refractory 4–5 cm from hot face at slag line (×132).

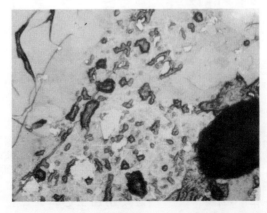

Fig. 4. Silicon carbide grains (white) surrounded by slag (×262).

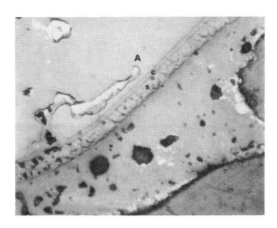

Fig. 5. Refractory A; fused alumina grain, slag attack zone (×262): c = calcium hexaluminate, s = spinel, and A = alumina grain.

The change of composition from silica-rich to alumina-rich phases in the transition zone reflects the depth of penetration of the slag and the degree of assimilation of the bond matrix and alumina aggregrate.

Samples of refractories B and C (see Table I) were taken from slag runners of a two-taphole blast furnace producing about 3000 tons of hot metal per day. The microstructures seen in these samples are similar to those of refractory A in the slag zone. Energy dispersive X-ray analyses of the phases in the slag attack zones are shown in Table III. Note the similarities to the analyses in Table II.

At the slag-metal interface of refractory A, rather different microstructural features are seen. Figure 6 shows the slag-metal zone adjacent the refractory surface. A droplet of metal, spinel crystals, blades of recrystallized corundum, and graphitelike carbon flakes can all be recognized in this photomicrograph. Figure 7 shows the crystalline structure of the slag crust at higher magnification. Anorthite crystals and glass can be seen around the spinel crystals and corundum blades. Figure 8 shows a portion of the reaction zone just inside the refractory. Table IV shows energy dispersive X-ray analyses of the phases in the slag-metal reaction zone of refractory A. In this zone the slag phases are generally more alumina-rich than those in the slag-air reaction zone.

Figure 9 shows silicon carbide grains being attacked at the edges in the slag-metal zone. It is hypothesized that iron metal has dissolved silicon from the grain. The composition of the metal droplet in Fig. 6 (Table IV, analysis 7), supports this hypothesis based on its rather high silicon level.

Discussion

Wu and Yang showed that oxidation of free carbon in blast furnace trough ramming mixes occurred in contact with air above the slag line, at the air-slag-refractory interface, and at the slag-molten iron-refractory interface.[1] The resulting increase in open porosity was said to create passages for slag penetration. As the slag penetrated, it assimilated silica completely and

Table III. Analyses of Phases in Slag Reaction Zones of Refractories B and C

		Amount Present (Wt%)						
		Outer Slag Zone		Inner Slag Zone				Internal Bond
		1	2	3				4
Refractory B	Na₂O		4.33					
	MgO	4.75						70.43
	Al₂O₃	18.86	21.27	50.13				28.47
	SiO₂	38.33	39.59	43.70				
	SO₃	1.25	1.27					
	K₂O	1.28	1.12	0.84				
	CaO	34.64	31.22	2.91				1.10
	TiO₂	0.23	0.31	0.40				
	MnO₂			0.39				
	Fe₂O₃	0.67	0.89	1.63				
		5		6	7			8
Refractory C	Na₂O			1.94	2.67			
	MgO	9.09		2.87	2.55			
	Al₂O₃	14.36		34.81	31.41			31.65
	SiO₂	37.99		45.42	45.60			54.15
	SO₃	3.34						
	K₂O	0.59		4.45	5.75			
	CaO	34.63		10.51	6.83			14.20
	TiO₂			0.55				
	MnO₂							
	Fe₂O₃			4.65				

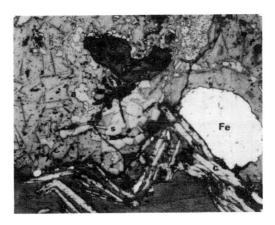

Fig. 6. Refractory A; slag-metal zone, crust (×65): Fe = metal drop; c = carbon flakes; s = spinel; and a = corundum.

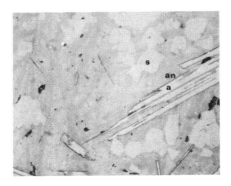

Fig. 7. Refractory A; slag-metal zone, crust (×262): Higher magnification of slag. Phases: s = spinel, an = anorthite, and a = corundum.

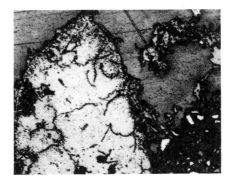

Fig. 8. Refractory A; slag-metal zone inside refractory.

Table IV. Analyses of Phases in Slag-Metal Reaction Zone of Refractory A

	Amount Present (Wt%)										
	Outer Slag/Metal Crust (Fig. 6)				Inner Slag/Metal Crust (Fig. 6)				Internal (Behind Crust) (Fig. 8)		
	1 Spinel	2 Anorthite	3 Glass	4 Spinel	5 Anorthite	6 Glass	7 Metal Drop	8	9 Glasses	10	11 Hematite
Na$_2$O										2.14	
MgO	26.77		5.12	24.74		1.05				2.70	
Al$_2$O$_3$	72.70	39.02	23.50	73.82	39.16	32.71		36.35	31.68	29.69	8.52
SiO$_2$		43.12	39.95		41.56	30.87		43.59	50.56	50.94	
K$_2$O									1.89	3.54	
CaO		17.86	31.43		19.28	35.37		17.77	14.41	8.62	0.82
TiO$_2$				0.87					0.17	0.64	3.29
MnO$_2$	0.54			0.36							
Fe$_2$O$_3$				0.21				2.29	1.29	1.74	87.38
Si							7.73				
Mn							1.06				
Fe							91.22				

Fig. 9. Refractory A; slag-metal zone, silicon carbide grain showing attack.

alumina partially, becoming enriched in both compared with its original composition.

A loss of free carbon near the hot face at the slag-air interface was also noted in the present study. The analyses in Table II confirm the enrichment of the slag in alumina as it assimilates the refractory bond. The observed slow reaction of the slag with large alumina grains and lack of reaction with silicon carbide indicate that most of the wear at the slag line is due to penetration and assimilation of the bond phases.

Refractory B has been observed to wear at a 20-25% slower rate than refractory C under similar field conditions in slag. The difference is difficult to explain in terms of bulk composition. Table III shows that the bond composition of refractory B could be expected to be less reactive with slag than that of refractory C. This seems to confirm the major role of bond assimilation in slag attack.

Wu and Yang also noted that the slag at the slag-molten iron-refractory interface became even more enriched in alumina than at the air-slag-refractory interface because of higher temperatures. Comparison of Tables II and IV shows that the same observation has been made in this study.

In the presence of both molten iron and slag at the slag-molten iron-refractory interface, some reactions may occur in addition to dissolution of the bond by slag. Landfeld discusses possible reactions of constituents in cast irons with refractory oxides.[2] The reaction of carbon with silica to yield metallic silicon and carbon monoxide is quite likely and may occur between carbon in the iron and silica in the slag, between carbon and silica within the refractory, and between carbon in the iron and silica in the refractory. The net result in the refractory would be removal of carbon as carbon monoxide gas and silica as silicon dissolved in the iron. Since both magnesium and aluminum are strong reducing agents, even small quantities of them dissolved in the iron can reduce silica in the slag or refractory to metallic silicon yielding MgO and Al_2O_3 (or, in combination, spinel). The petrographic evidence in this study suggests that silicon carbide also reacts with the iron.

The presence of carbon deposits at the refractory surface (Fig. 6) may be a result of the carbon reaction with silica. The carbon monoxide formed may migrate to the surface and redeposit there as carbon. The alumina-enriched condition of the slag at the slag-molten iron-refractory interface may be partially due to higher temperatures allowing more dissolution of alumina, as suggested by Wu and Yang.[1] Another possible mechanism leading to alumina enrichment of the slag would be the depletion of silicon from the slag and refractory by the reactions of metal constituents (such as C, Al, and Mg) described above.

Summary

From the observations made in this study, it can be concluded that corrosion of blast furnace trough refractories by slag at the slag-air interface is characterized by the following:

(1) Oxidation of free carbon from the refractory adjacent the hot face.

(2) Penetration of slag through open porosity including pores created by the oxidation of free carbon.

(3) Relatively rapid assimilation of the bond phases of the refractory into the slag.

(4) Slow reaction of slag with large fused or sintered alumina aggregates.

(5) Slow or negligible reaction of the slag with silicon carbide.

Based on the importance of the refractory bond phases in the slag corrosion mechanisms described above, it is suggested that developing a more resistant bond is a good approach to improving refractory performance in slag.

The slag at the slag-molten iron interface is more highly alumina-enriched than that at the slag-air interface. The higher temperatures at the slag-molten iron interface may allow greater dissolution of alumina from the refractory, thus enriching the slag. Silica may be depleted from the slag by being reduced to silicon by constituents in the metal, then taken into solution in the metal. The effect of this depletion would be further alumina enrichment of the slag.

The dissolution of silicon from silicon carbide and silica in the refractory may be the major destructive effect of the molten metal, suggesting that refractories for the slag-metal interface could be improved by reducing silicon carbide content as well as developing a less silica-rich bond.

References

[1]Jang-An Wu and Houng-Yi Yang, "Erosion of Blast Furnace Troughs by Molten Iron and Slag," *Am. Ceram. Soc. Bull.*, **62** [7] 793–97, 803 (1983).

[2]C.F. Langfeld, "Thermochemistry of Cast Iron/Refractory Reactions," *Trans. Am. Foundrymen's Soc.*, **88**, 507–14 (1980).

Progress in Casting Trough Materials and Installing Techniques for Large Blast Furnaces

Y. Toritani, T. Yamane, S. Yamasaki, I. Nishijima, T. Kawakami, and Y. Kadota

Kawasaki Refractories Co., Ltd.
Ako, Hyogo, Japan

A castable refractory system has been used for the trough linings of large blast furnaces in Japan. This paper presents the development of casting trough materials, the improvement of installation techniques, the establishment of repairing methods and schedules, the improvement of trough lining structure, and the adaptation of trough design. Longer campaign life and lower refractory consumption of troughs have been obtained under the improved castable system.

The performance of blast furnace trough linings is dependent on the properties of the refractories, trough design, lining and repairing techniques, and overall operational characteristics of the furnace.

The trend toward the larger blast furnace, in conjunction with the increase in hot metal output, has confronted the refractory industry with the task of perpetually improving and developing the effectiveness of blast furnace trough linings.

Many attempts have been made in the past to meet these requirements.[1,2] High-quality casting trough materials have been successfully used at the Chiba No. 6 blast furnace of Kawasaki Steel Corporation. By suitable trough design and effective installation techniques, as well as by the development of suitable refractories, higher hot metal throughput, longer trough life, and lower refractory consumption are today possible.

This paper describes recent advances in the understanding of trough materials, updates the current usage of these materials, and reviews the improved techniques used to install and repair trough materials today. Finally, an update is presented on recent treatments of the blast furnace troughs with casting materials.

Outline of Blast Furnace and Trough Facilities

Table I contains an outline of the facilities of the No. 6 blast furnace and the main trough at the Chiba Works of Kawasaki Steel Corporation. Figure 1 illustrates the layout of trough in the casthouse.

The No. 6 blast furnace is a large blast furnace with 4500 m^3 inner volume, and a daily hot metal output of 9000 tonnes.

The casthouse is divided into east and west sections, each containing two tapholes, and trough and runner systems of identical dimensions.

Table I. Outline of Facilities for Chiba No. 6 Blast Furnace and its Main Trough

Blast Furnace	
Daily production	9000 tonnes/day
Inner volume	4500 m³
Hearth diameter	14.1 m
Tapholes	4
Main Trough	
Construction	Single replaceable pooling type
Slope	0°34'
Length	20000 mm
Width	1100 mm
Depth	960 mm

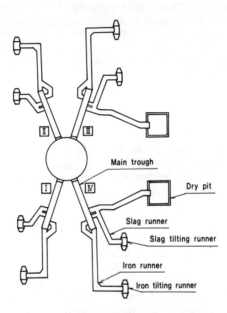

Fig. 1. Layout of trough in the casthouse.

The blast furnace trough is comprised of a main trough, an iron runner, a slag runner, an iron tilting runner, and a slag tilting runner. The main trough is designed as a single replaceable pooling type.

Progress of the Castable Refractory System

Figure 2 indicates the transition of trough refractory consumption and development work.

The refractory consumption changes from 2.20 kg/tonnes pig iron (June–September 1977) to 1.54 (April–September 1978), to 1.16 (April–September 1979), to 0.77 (April–September 1980), and to 0.57 (April–September 1981)—showing a drop of about 1.6 kg/tonnes pig iron over those periods.

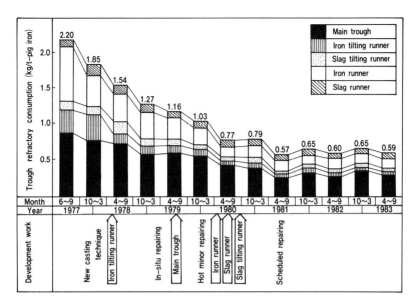

Fig. 2. Transition of trough refractory consumption at Chiba No. 6 blast furnace.

The reduction in consumption was achieved through the following developments.

(1) Introduction of casting trough materials to the iron tilting runner in 1978, to part of the main trough in 1979, to the slag tilting runner in 1980, to the iron runner in 1980, to the slag runner in 1980, and to the main trough in its entirety in 1981.

(2) Partial and whole repair of the main trough in 1979.

(3) Partial hot repairing for the main trough in 1980.

(4) Improvement of the installing technique and execution of scheduled repairs in 1981.

Casting Trough Materials

Material Design

Properties required of casting trough materials include the following: (1) good fluidity with the minimal water content; (2) an appropriate setting time at ordinary air temperature; (3) a high-strength uniform structure; (4) high resistance to explosive spalling; and (5) high resistance to attack by molten metal and slag.

The procedure for trough material design is outlined in Fig. 3.

Chemical and physical changes take place when water is added to the trough material. Hardening is important in its use as a refractory. While hardening occurs at all temperatures, it is sluggish at low temperatures. At high temperatures, the rate increases, and at 20°C and above, the paste fully hardens after several hours. Figure 4 illustrates the temperature dependence of the hardening of trough material.

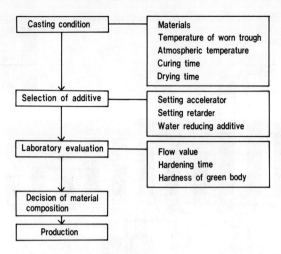

Fig. 3. Procedure for material design.

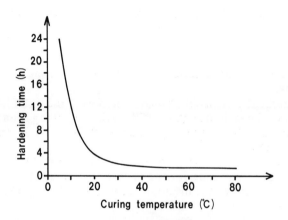

Fig. 4. Temperature dependency of hardening time.

Drying of Casting Trough Material

Casting trough material is basically a mixture that reacts with water at room temperature to form a hard, strong material which may subsequently be heated to high temperatures without cracking or spalling especially without explosive spalling.

To prevent explosive spalling during drying, laboratory evaluation is carried out in two ways.

Visual Observation: Specimens measuring approximately 40 by 40 by 160 mm were cured at 20°C for 24 h. Resistance to explosive spalling was determined by observing crack growth on the surfaces of these specimens which were held at 200°–800°C (at an interval of 100°C) for 10 min.

The explosive spalling apparatus is shown in Fig. 5.[3]

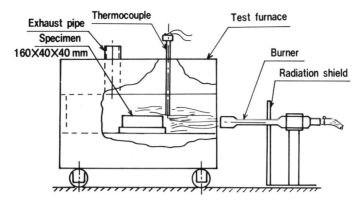

Fig. 5. Schematic side view of the apparatus for explosive spalling.

Figure 6 graphs the relationship of heating temperature, the permeability of green body, and explosive spalling.[4]

It is likely that the increase in permeability plays an important role in the prevention of explosive spalling. The representative trough materials now used at the Chiba Works have high permeability as well as high strength.

Evaluation by Acoustic Emission (AE) Technique: Cylindrical specimens (200 mm in diameter and 150 mm long) were cast and cured at 20°C for 24 h. One side of the specimen was subjected to rapid heating by a furnace. For AE measurement, a transducer was placed in contact with the cold side, opposite the heated side of the specimen. Explosive spalling resistance can be evaluated by AE measurement during heating.

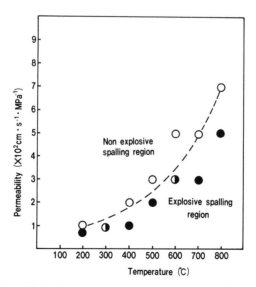

Fig. 6. Relationship between temperature, permeability, and explosive spalling.

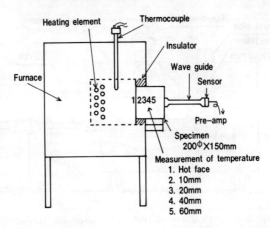

Fig. 7. Schematic illustration of explosive spalling apparatus by acoustic emission measurement.

The AE measurement apparatus is shown in Fig. 7.

AE characteristics of various kinds of trough materials heated at the rate of 100°–300°C/h are shown in Fig. 8. By this method, it has been possible to evaluate explosive spalling resistance, with results that correspond well with results achieved in actual drying conditions.

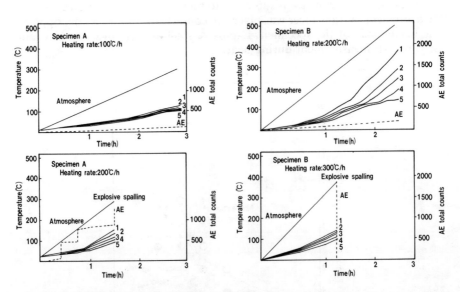

Fig. 8. Relationship between acoustic emission counts and explosive spalling of trough materials heated at 100°–300°C.

Resistance of the Trough Material to Slag and Molten Metal

Trough material failure is often the result of slag and molten metal activity.[5] For example, slag attack may cause failure by both chemical solution and alteration of structure. Alteration of the refractory structure can cause a variety of failure mechanisms, such as sensitivity to thermal shock and excessive shrinkage. The rates of solution and of the alteration processes are affected by the pore structure of the refractory. Therefore, evaluation of a trough material's pore structure should be useful in predicting service performance.

Specimens of ramming and casting trough material were prepared. The distribution of pore size was examined by a method based on mercury penetration. The results are shown in Fig. 9.

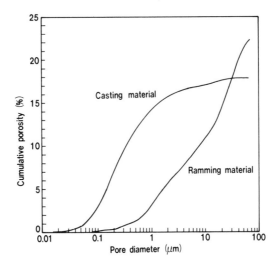

Fig. 9. Pore-size distribution curves for casting material and ramming material (specimens fired at 800 °C).

The casting material had a larger proportion of fine pores than the ramming material, and there was a marked increase in the number of pores below 1 μm, thus causing an abrupt change in the distribution curve.

For trough materials of a given porosity, those having smaller pores should offer better corrosion and penetration resistance.[6] It seems likely, therefore, that the casting trough material will give good service life.

Properties of Casting Trough Materials

The properties of casting trough materials currently used are shown in Table II. Most trough materials now used are Al_2O_3-SiC-C materials. Material A is applied to the main trough from the impact zone to the skimmer zone, both at the slag line and metal line. Materials B, C, and D are used as zone lining. Material B is applied to the wall of the main trough, especially to the slag line, material C to the impact zone of the main trough, and material D to the skimmer zone bottom. Material E is characterized by its

Table II. Properties of Casting Trough Material

Material	A	B	C	D	E	F	G
Application	Main Trough Impact Zone and Skimmer Zone	Main Trough Impact Zone and Skimmer Zone Slag Line	Main Trough Impact Zone Metal Line	Main Trough Skimmer Zone Metal Line	Tilting Runner	Iron Runner	Slag Runner
Chemical composition (%)							
SiO_2	1–3	1–4	1–4	6–9	6–9	48–52	6–9
Al_2O_3	78–82	70–74	68–72	62–68	64–68	30–33	55–59
SiC	11–14	14–17	14–17	16–19	14–17	7–10	18–21
C	1–3	1–3	3–5	2–4	2–4	2–4	4–6
Modulus of rupture (MPa) (cold-crushing strength)							
110°C for 24 h	6.37 (31.7)	2.06 (9.5)	2.45 (11.0)	3.82 (13.3)	2.55 (10.4)	2.16 (7.2)	1.37 (5.5)
1400°C for 2 h	20.10 (90.9)	12.75 (69.1)	10.59 (47.6)	7.26 (25.2)	7.75 (36.4)	5.00 (20.2)	12.94 (51.8)
Bulk density (Mg/m³)							
110°C for 24 h	2.97	2.76	2.71	2.39	2.48	2.15	2.36
1400°C for 2 h	2.92	2.72	2.63	2.34	2.42	1.93	2.32
Hot modulus of rupture (MPa) at 1400°C	6.57	8.83	7.65	3.92	4.61	2.55	1.96

Table III. Performance of Casting Trough Material A at Main Trough

No.	Service period (day)	Throughput (ton)	Wear rate (mm/10³tonnes pig iron)							
			Slag Line				Metal Line			
			Distance from Taphole				Distance from Taphole			
			2 m	4 m	6 m	8 m	2 m	4 m	6 m	8 m
1	17	60 203	1.04	2.24	1.16	1.33	1.45	2.20	1.66	1.25
	18	69 771	0.22	2.22	2.29	1.08	1.86	2.44	1.75	1.43
	18	69 713	1.79	2.87	1.79	1.72	1.58	2.08	1.00	0.14
2	16	64 398	1.78	2.37	1.40	1.40	1.63	1.79	0.97	0.58
	17	72 635	1.31	2.20	1.03	0.96	1.38	1.72	1.65	1.03
	16	65 613	1.68	2.48	1.49	1.18	1.98	2.10	1.30	1.03
3	20	75 815	1.91	2.18	1.45	0.76	2.24	2.37	1.78	1.32
	16	69 471	1.51	2.45	1.48	1.66	2.30	2.34	1.73	1.51
	18	72 165	0.90	1.80	0.83	1.52	2.29	2.22	1.66	1.59

high strength at the early stage of drying and is applied to a tilting runner. Materials F and G are used for the iron runner and slag runner. The performance of casting trough material A is shown in Table III.

Installation Techniques

The casting method aimed at higher installing efficiency, labor savings, and lower refractory consumption. It also offers an installation process that allows partial or entire repair and contributes both technically and economically to the blast furnace operation. This method was adopted to solve such problems as the number of hours necessary for installation and the vast quantity of trough materials used.

The installation process involves the following series of operations: (1) draining and cooling the trough; (2) removal of damaged layer; (3) cleaning the runner; (4) installation of formers; (5) casting; (6) removal of formers; and (7) drying. Layout of installation equipment is shown in Fig. 10. Decrease in the labor necessary and the installation ability are graphed in Fig. 11.

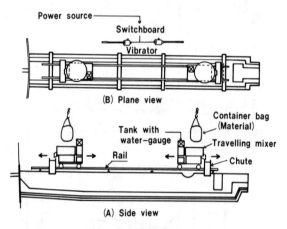

Fig. 10. Layout of installation equipment.

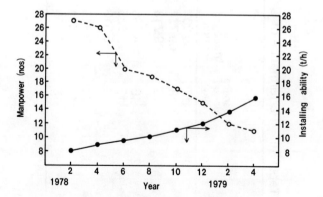

Fig. 11. Transition of labor and installation ability.

The advantages of the casting method are as follows: (1) labor savings (fewer man-hours); (2) cleaner environment; (3) applicable for troughs with wall temperatures of up to 300 °C; (4) short hardening time; and (5) dense and uniform structure.

Repairs Using Casting Materials

Since the high utilization factor of the lining is the primary issue, the residual thickness of the trough lining is under continuous and routine control.

Troughs often exhibit cracks and premature wear in different places, which lower their usability. When worn areas are found in the trough, either interim repair of the worn areas is carried out or the trough lining is broken out for relining.

The materials used have compositions similar to the compositions of those used for casting, with some variations in characteristics.

Under the current technique, the repairs carried out in the trough are fundamentally scheduled repairs. The typical repair pattern is shown in Fig. 12.

The alteration of and improvements in the repair technique for the main trough are shown in Fig. 13.

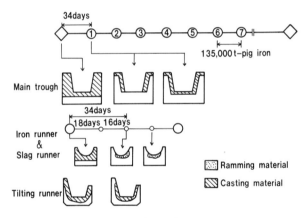

Fig. 12. Repair pattern for trough and runners.

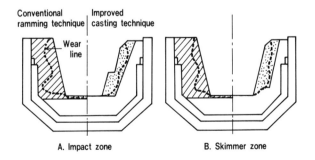

Fig. 13. Improvement of repair technique for main trough.

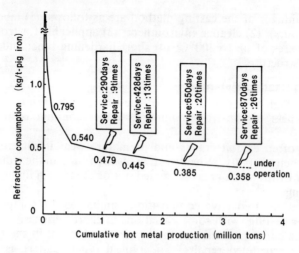

Fig. 14. Service performance of main trough in Chiba No. 6 blast furnace.

Figure 14 shows the refractory material consumption for the main trough of the Chiba No. 6 blast furnace under the ramming technique and the casting techniques, respectively. Refractory consumption for the casting technique is lower than for the ramming technique.

Since 1979, the main trough has been maintaining a record performance of 870 days and a throughput of 3 230 000 tonnes pig iron.

Design of Main Trough

The profile and dimensions of main troughs differ from furnace to furnace. A profile of the main trough of the Chiba No. 6 blast furnace is shown in Fig. 15.

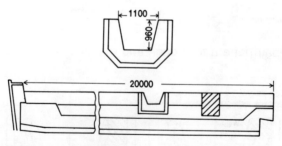

Fig. 15. Profile of main trough.

The main trough has a good hydrodynamic design with optimum geometry. The features are summarized as follows: (1) the single replaceable type (offers simplified repair work and a reduced number of joints); (2) pooling type (offers reduced refractory consumption); (3) back-lined type (offers reduced refractory consumption and prevents leakage trouble); and (4) the natural cooling type (ensures the nondeformation of trough frameworks).

The main trough, which has a gentle slope, is installed on the beam and the trough shell, is reinforced with outer ribs, and has an open construction which facilitates the natural cooling and application of casting trough materials.

The casting trough materials are supported by a safety lining which prevents hot-metal breakthrough.

Control of Wear Zone

The wear of the trough material takes place almost exclusively in the trough walls. In the area separating the slag-atmosphere zone from the slag-pig iron zone, erosion takes place. These wear zones may be controlled and moved by changing the level at the pig iron and slag outlet during service performance. Thus, it can be said that the level control technique gives long service life and is as important as other maintenance techniques. The result of level control is presented in Table IV.

Table IV. Wear Control by Changing the Level at Iron Outlet and Slag Outlet

		1st displacement (conventional)	2nd displacement	3rd displacement
Lining life	(days)	25	30	34
Wear rate (mm/10^3tpig)	Air–slag zone	2.6	2.0	1.8
	Slag–metal zone	2.2	1.6	1.4
Reduction ratio of refractory consumption (%)		—	15	25
Adjustment		no adjust (wrecking line — SL, ML)	lowering of iron outlet level (50–70mm) after 22 days (SL$_1$, SL$_2$, ML$_1$, ML$_2$)	Raising of slag and iron outlet level (ab.100mm) after 10 and 22 days (SL$_3$, SL$_2$, SL$_1$, ML$_3$, ML$_2$, ML$_1$)

Summary

The technology for trough refractories has been outlined; an efficient trough system was obtained by solving the following problems: (1) the development of casting trough materials; (2) the improvement of installation techniques; (3) the establishment of maintenance techniques; and (4) the improvement of the trough design.

The unique system has been established through the closer cooperation between refractory producers and users. It has been successfully applied at the large blast furnace of Kawasaki Steel Corporation and has contributed to the overall economy of ironmaking there.

Acknowledgments

The authors wish especially to acknowledge their indebtedness to T. Haru (General Manager, Ironmaking Department); J. Kurihara (General Manager, Ironmaking Department); and H. Obata (Assistant General Manager, Ironmaking Department) of Kawasaki Steel Corporation.

Helpful discussions have been held with M. Saino (Assistant General Manager); K. Okumura (Manager); and H. Marushima (Manager) of the Ironmaking Department of Kawasaki Steel Corporation's Chiba Works.

References

[1] H. Tanaka, "A Casting Method of the Blast Furnace Trough Materials," *Taikabutsu*, **30** [6] 343-46 (1978).

[2] T. Ochiai et al., "Vibration Forming Method Process for Iron Trough of Blast Furnace," *Taikabutsu*, **28** [3] 108-12 (1976).

[3] T. Yamamoto, T. Yamane, and H. Mitsui, "Drying and Explosive Spalling of Castable Refractories," *Taikabutsu*, **33** [8] 445-49 (1981).

[4] T. Kawakami, T. Yamane, and H. Mitsui, "Gas Permeability and Pore Size Distribution of Unshaped Refractories," *Taikabutsu*, **34** [8] 479-83 (1982).

[5] K. Sugita, "Recent Progress in Refractories for Hot Metal Handling," *Tetsu-to-Hagane*, **69** [15] 235-41 (1983).

[6] T. Ochiai et al., "Development and Application of Special Refractories for Gas Injection," *Taikabutsu*, **32** [4] 179-87 (1980).

Comparison of Monolithic Refractories for Blast Furnace Troughs and Runners

Cecil M. Jones

Norton Company
Worcester, MA 01606

Three types of monolithic refractories (dry-vibration, wet-rams, and low-water castables) for use in the casthouse are compared and discussed based on laboratory tests and statistical analysis. The role of silicon carbide is questioned. The validity of the described slag and iron wear tests is argued.

Every research testing procedure should be reliable and valid. The reliability can be demonstrated in the laboratory by comparing results of repeated tests; such planned repetitions are essential to good research to ensure that test parameters have not changed. The difficult part of testing is to be reasonably certain that R&D tests can be correlated with actual field conditions so that the tests can predict performance in the casthouse.

During the past several years, the evaluation of casthouse monolithic refractories has been relatively easy because of the large difference in quality and metal throughput for dry-vibrated mixes, low-moisture castables, and the new wet rams compared with the old sand-pitch mixes. We are now at a state of development where compositional changes make relatively less improvement in the refractories. It is obvious to anyone connected with the casthouse that small differences in performance are difficult to measure in iron production. It is nearly impossible to get quantitative data on performance of single-block preforms or small patches, and certainly uneconomical to test with 15-ton samples.

Small-scale laboratory tests are needed to compare compositional changes and different monolithic refractories before the best of a lot goes into a main trough or runner. There are as many different slag erosion tests as there are research laboratories conducting tests. The principal forms of slag tests are:

(1) The drip test where molten slag is dripped in a controlled manner onto the hot refractory.

(2) A rotating finger test where a long, thin sample is rotated in a crucible of molten slag.

(3) A rotating drum lined with test samples, e.g., a rotary kiln where slag is melted.

(4) A static test where slag is melted in a refractory cup.

All of these methods ignore the iron-slag interface.

(5) An induction furnace where pig iron is melted with controlled additions of blast furnace slag so that the maximum erosion at the slag-iron interface can be measured.

Field Evaluation of Slag-Iron Interface

Wu and Yang describe the reactions noted when they measured maximum wear rates at the air-slag and slag-iron interfaces in a main trough.[1] Figure 1, from their paper, shows the erosion zones. They suggest that the severe wear at these interfaces is due to (1) mechanical erosion by the fluid slag and iron, and (2) chemical corrosion by oxidation. At the air-slag interface, erosion is affected by oxidation from the air. At the slag-iron interface, they propose oxygen formation according to the reaction:

$$S + CaO \rightarrow CaS + \tfrac{1}{2}O_2$$
Iron Slag

Henry et al. report similar interface notches in a dry-vibrated trough lining.[2] Further, it is well known in the foundry industry that the slag-iron interface is the maximum wear area in an induction furnace.[3]

Possibly any slag test that ignores the slag-iron interface suffers from a lack of rational argument for validity when translating laboratory results to relative predictions of service life in the main trough. Therefore, as a preliminary screening test for comparing compositional changes, Norton has developed a slag-iron test utilizing a 45.3-kg induction furnace.

Description of Test Procedures

A 45.3-kg induction furnace is used with 100 kW, 3000-cycle power. To reduce the testing time (by increasing severity of the procedure) our standard test cycle is 1650°C for 5 h (measured periodically with an immersion thermocouple). Wear rates were calculated by measuring the amount of brick remaining and reporting the rate as $\mu m/°C \cdot h$, where $°C \cdot h$ is the average temperature of a run above the approximate slag melting point of 1250°C, multiplied by time in hours. Based on experience of wear rates in the field, Jones and Planinsek estimate that 2 $\mu m/°C \cdot h$ can be related to a wear rate of 1 mm/1000 tons iron.[4] Pig iron was obtained from a local foundry with an analysis of 4.00% total carbon, 2.25% Si, and 0.9% Mn.

Initially, two slag compositions were used (Table I). BF1 slag was used in all the work reported in this paper. Test samples were prepared by press-

Table I. Oxide Composition of Slags

	BF1	BF2
CaO	40	41
SiO_2	40	35
MgO	10	12
Al_2O_3	8	10
MnO_2	2	
S		2
Slag notch wear rate $\mu m/°C \cdot h$	3.4	3.2
	No statistical difference	

ing, ramming, or casting (as appropriate) to the installed density typically achieved in the field. The specimen shape is shown in Fig. 2 and is designed so that 12 bricks will turn the 17.8-cm ID of the furnace. In all cases, two bricks of each composition were tested to eliminate the edge effect of adjacent specimens. Figure 3 is a schematic of the furnace setup. Test specimens were prefired to 1000°C to correspond to recommended field practice.

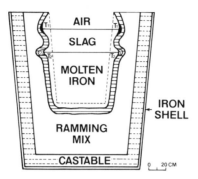

Fig. 1. Main trough cross section.

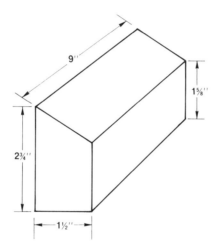

Fig. 2. Melt test brick.

Test Results

There are many reasons that a specific product type and/or brand name of monolithic refractory is used in a given casthouse:
  (1) Resistance to slag and iron.
  (2) Cost per ton of iron produced.
  (3) Reliability of the specific product and manufacturer.
  (4) Ease of installation, drying, and presintering.
  (5) Impact on the casthouse environment.

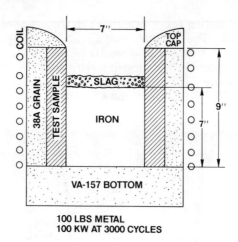

Fig. 3. Test furnace.

(6) Hot strength.
(7) Thermal and mechanical stability.

The test results in this discussion relate to only one of these criteria, the Norton screening test results of resistance to slag and iron. The data are from 16 melt tests. All the °C·h are between 2000 and 2100.

Table II is a tabulation of the wear rate performance of the three types of monolithics used in this analysis.

Table II. Wear Rate Performance

		Wear Rate (μm/°C·h)			
		Slag		Iron	
Monolithic Type	Number of Samples	Mean	Standard Deviation	Mean	Standard Deviation
Dry-vibration	17	2.9	0.8	1.7	0.4
Wet-ram	6	3.3	0.4	0.1	0.1
Low-water castable					
with SiC	8	2.6	0.8	1.4	0.3
without SiC	6	2.1	0.5	0.4	0.3

For statistical analysis, t values were calculated from:

$$t = \frac{\overline{X}_1 - \overline{X}_2}{\left[\frac{N_1 S_1^2 + N_2 S_2^2}{N_1 + N_2 - 2} \cdot \frac{N_1 + N_2}{N_1 N_2}\right]^{1/2}}$$

and are tabulated in Table III.

Table III. t Values from Table II

Comparisons	DF	t Value for Slag	t Value for Iron	Critical t Value at 0.05 Confidence Level
Monolithic types				
Dry-vib vs wet-ram	21	1.12	9.26*	2.08
Dry-vib vs castable	23	0.83	1.81	2.07
Castable vs wet-ram	12	1.82	9.44*	2.18
Castable types				
With SiC vs without SiC	12	1.24	5.71*	2.18

*Figures show significant differences.

Based on this analysis, we can make the following suggestion: These three types of monolithics, i.e., dry-vibratable, castable, and wet-ram, have statistically equivalent performance with the one exception that the wet ram has appreciably less wear in iron. Therefore the three types of casthouse monolithic refractories may perform equally well in slag and iron in the field, and the choice of one monolithic type over the others will depend on factors other than wear rates. The closeness of field data, as presented by Henry et al.,[2] supports this contention.

Each of the monolithic types has a characteristic wear profile after being subjected to the slag-iron tests. Figure 4 shows cross sections of typical specimens after testing. Because of the darkness of these specimens, good photographs are difficult to obtain. Therefore sketches showing the essential features of each type are shown in Fig. 5.

Fig. 4. Cross section of several specimens after slag-iron tests.

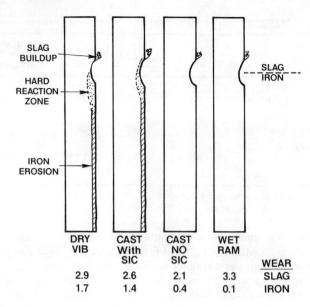

Fig. 5. Sketches of specimen cross sections highlighting structural changes and wear rates (from Fig. 4).

There is controversy about the role of silicon carbide in casthouse refractories. Some refractory researchers feel that silicon carbide is essential to reduce iron wear, but that it does not have much effect on slag wear. Other researchers hold that silicon carbide gives superior slag resistance, but is not as good as tabular alumina in resisting iron wear. Thus, the value of SiC has yet to be resolved.

In the development of Norton's low-water castable, compositions were made with and without silicon carbide. Nishi et al. found equal performance in the iron runner if the alumina content was above 80%.[5] Nishi compared performance with and without silicon carbide using an N-3 blast furnace at Fukuyama, Japan. These results are compared with Norton laboratory results in Table IV. Figure 6 shows the wear faces of cast compositions with and without SiC after exposure to iron in our tests.

Summary

The testing procedure if the refractory is exposed to a slag-iron interface is validated with reference to actual tests in the main trough.

Table IV. Wear Performance of High-Alumina Castables in Iron

Castable Type	Nishi (Ref. 5)		This Paper	
	Al_2O_3 (%)	Wear (mm/kilotons iron)	Al_2O_3 (%)	Wear ($\mu m/°C \cdot h$)
With SiC	68	2.8	74	1.4
Without SiC	84	1.8	95	0.4

Fig. 6. Wear faces of cast specimens after exposure to iron in tests.

So far as resistance to molten iron and slag is concerned, there does not appear to be any reason that equal performance cannot be achieved with dry-vibration mixes, low-water castables, and/or wet rams. Significant differences found in some instances probably are due to material composition differences and not to the method of installation (vibration, ramming, or casting). Criteria other than resistance to slag and iron will be the principal selection parameters in the future.

The value of silicon carbide in the trough mixes has yet to be resolved. However, the results of slag and/or metal attack tests indicate that very-high-alumina products have a superior resistance compared with high-alumina/silicon carbide mixes, in agreement with practical Japanese studies.

In the initial phases of this revolution in casthouse refractories it was easy to differentiate between the old sand-pitch mixes and the new, high-performance products, even with the crudest of laboratory tests. In the present state of the art, we are now trying to differentiate between high-performance products and both steel and refractories researchers must review their laboratory tests and adapt them for the more rigorous job of evaluating the high-performance monolithic refractories.

References

[1] Yang-An Wu and Houng-Yi Yang, "Erosion of Blast Furnace Troughs by Molten Iron and Slag," *Am. Ceram. Soc. Bull.*, **62** [7] 793-97 (1983).

[2] F.W. Henry, G.E.D. Snyder, and J.P. Willi, "Materials for Blast Furnace Cast House Application;" for abstract see *Am. Ceram. Soc. Bull.*, **62** [8] 885 (1983).

[3] R.A. Stark, Norton Company; private communication.
[4] C.M. Jones and B.L. Planinsek, "Characterization of Dry Vibrated Cements from Blast Furnace Troughs," *Industrial Heating*, **3** 18-20, (1983).
[5] Masaaki Nishi et al., "Application of Al_2O_3-SiO_2 Casting Materials for Blast Furnace Iron Troughs," translated from *Taikabutsu*, **34** [289] 26-29 (1982) The Refractories Institute, TRI Translation No. 78.

Section VI
Miscellaneous New Applications for Monolithics

Reactions of Alumina-Rich Ramming Mixes with Lignite Ashes in Reducing Atmospheres at High Temperatures 365
    P. Dietrichs and W. Kronert

The Properties and Applications of Dolomite Ramming Mixes 388
    J. W. Stendera

SiC Monolithics for Waste Incinerators: Experiences, Problems, and Possible Improvements 395
    G. S. Dhupia, W. Kronert, and E. Goerenz

Development of Spinel-Based Specialties: Mortars to Monoliths 411
    A. Cisar, W. W. Henslee, and G. W. Strother

Reactions of Alumina-Rich Ramming Mixes with Lignite Ashes in Reducing Atmospheres at High Temperatures

P. DIETRICHS AND W. KRÖNERT

Institut für Gesteinshüttenkunde der RWTH Aachen
Mauerstrasse 5, D-5100
Aachen, Federal Republic of Germany

Samples of three alumina-rich ramming mixes containing 60, 75, and 90% Al_2O_3 were examined. The ramming mixes, installed in a rotary hearth furnace for lignite coke production, were severely attacked by basic lignite fly ashes under moderate reducing conditions and H_2O vapor. The complex corrosion reactions were studied by optical microscopy, scanning electron microscopy, X-ray microanalysis, X-ray diffraction, and chemical methods. The slag cover formed at the hot face mainly consists of melilite, spinel, and residual eutectic melts; in the reacted layers, mainly anorthite and corundum were found. Residual eutectic melts again occurred with widely varying compositions. Cavity and crack formation due to these reactions cause increasing structural corrosion, especially in the mullite-based mix.

The technique of producing low-ash, lignite coke in a rotary hearth furnace is relatively new. Therefore, neither sufficient data concerning the effect of different refractory materials nor selection criteria are available. The refractory lining on the ceiling of the rotary hearth, as well as the refractory lining in the waste-gas flue, are severely attacked by basic lignite fly-ash, which causes considerable corrosive and erosive wear at temperatures of 1300°-1400°C as a maximum in reducing atmospheres with vapor contents up to 25%.

Only a few experiences with refractories being used under similar conditions exist for coal gasification plants. The rotary hearth kiln and the waste-gas flue were lined with different ramming mixes with alumina contents of 60, 75, and 90%. These materials reached lifetimes of 0.5-1 year.

Service Conditions

The product of the rotary hearth process is a coke, which is used increasingly for electrochemical and metallurgical applications because of its purity, electrical properties, and reactivity. The main part of the rotary hearth kiln is a hearth plate 23 m in diameter.

The furnace ceiling, lined with ramming mixes, is suspended above the hearth bottom; together with the side linings and the hearth plate, they form the heating chamber. A water seal is the joint between the sides and the hearth plate. The lignite with a humidity of 14-18 wt% is fed through two down-pipes to the periphery of the hearth plate. Water-cooled rabbles cause the turnover of the lignite and the transportation toward the discharge opening at the center of the hearth plate. The temperature of the coke is

850°–900°C, the temperature of the ceiling about 1300°C. The waste gases are passed off through a square waste-gas flue (3 by 3 m²) which runs into the center of the furnace ceiling, rises vertically about 5 m, then runs horizontally about 15 m, and leads to the waste-heat vessel. This waste-gas flue is lined with three ramming mixes: the front horizontal part with mix C (90% Al_2O_3), the intermediate part with mix B (75% Al_2O_3), and the last third with mix A (60% Al_2O_3). The different panels are mounted with ceramic anchor bricks. After a short service period, the refractory lining is coated with ash deposits up to 10-cm thick at the ceiling. There are areas with partial spallings of the deposit and of refractory material evident. Samples were taken from different parts of the ceiling to investigate the wear mechanism.

The following stresses act on this part of the ceiling under service conditions:

(1) Temperature: 1300°, maximum 1400°, minimum 1000°C.

(2) Atmosphere: reducing (moderate), containing 8% CO; 6% H_2; 0.5% CH_4; 25% H_2O (steam); the remainders, N_2, CO_2, SO_2, and SO_3, oxidize during heatup and cooldown.

(3) Corrosion by attack of adhering ash: (changing in composition and ash content) CaO, 51–55%; MgO, 12–15%; SiO_2, 6–15%; FeO, and/or Fe_2O_3, 6–8%; and Al_2O_3, 2–12%.

(4) Erosion by ash particles in the waste gas: 6 g fly ash per m³ waste gas.

(5) Thermal shock: moderate.

Chemical, Mineralogical, and Technological Data of the Ramming Mixes under Investigation (Table I)

The mixes are characterized as follows: Ramming mix A contains 60% Al_2O_3 and is bound ceramically by clay and fine-grained corundum. The aggregates are mullite, cristobalite, and glassy phase. Ramming mix B contains 75% Al_2O_3 and consists mainly of mullite and minor amounts of corundum within the aggregate, with fine-grained corundum in the binding phase. The binding is achieved with aluminum phosphates, as in the case of ramming mix C (90% Al_2O_3). Here again, the binding phase contains fine-grained corundum.

Experimental Techniques and Results

The following test methods were used to investigate the wear mechanism of the refractory lining of waste-gas flue ceilings of post-mortem samples: optical microscopy, scanning electron microscopy, X-ray microanalysis, X-ray diffraction, and chemical methods.

Chemical and Mineralogical Composition of the Fly Ash

Fly ash samples adhering to the slag deposit of the refractory lining were analyzed. The sample positions are listed in Table II. Lignite ash consists mainly of CaO (51–55%), MgO (12–15%), SiO_2 (6–16%), Al_2O_3 (2–11%), and FeO and/or Fe_2O_3 (6–8%). Depending on the SiO_2 content, the fly ash consists of C_2S, periclase, portlandite, and calcite, as well as of such iron-containing minerals as C_4AF and C_2F.

According to the different chemical composition of the fly-ash samples, the melting behavior of two samples was measured under reducing conditions in the hot-stage microscope. The hemisphere points of the ashes containing 15.6% and 8.6% SiO_2 were determined to 1425° and 1530°C, respectively,

Table I. Chemical, Mineralogical, and Technological Data of the Ramming Mixes Employed in the Rotary Hearth Kiln

Chemical Composition	Ramming Mix A	Ramming Mix B	Ramming Mix C
Oxide (Wt%)			
SiO_2	34.00	16.94	4.56
Al_2O_3	59.57	74.66	88.98
TiO_2	1.56	1.52	0.13
Fe_2O_3	0.99	1.03	0.39
Cr_2O_3	0.05	0.05	0.05
MnO	0.05	0.05	0.05
CaO	0.11	0.20	0.19
MgO	0.10	0.10	0.10
Na_2O	0.07	0.18	0.29
K_2O	0.18	0.23	0.02
ZrO_2	0.12	0.11	0.13
P_2O_5	0.10	3.23	3.49
SO_3	0.22	0.03	0.03
LOI	3.11	2.32	2.40
Phase Assemblage (X-ray diffraction)			
Corundum	+ +	+ + +	P.C.
β-Al_2O_3			+
Mullite	P.C.*	P.C.	
Rutile		+	
Cristobalite	+ +	+	
Quartz	+	+	
Clay Minerals (kaolinite, illite)	+ +		
$AlPO_4$		+	+
Glassy Phase	+	+	+

	Ceramic bond	Inorganic-chemical bond	Inorganic-chemical bond
HMOR (N/mm²) 1400°C	5.3	25.3	10.5
1650°C	6.6		
1700°C		27.7	13.1
Linear shrinkage (%) 1000°C	−0.55	−0.15	−0.34
1650°C	−0.04		
1700°C		−0.4	+0.5

*P.C.: Principal Constituent

and the melting point to 1440° and 1535°C, respectively. To a great extent, the differences in chemical and mineralogical composition of the fly ash can be explained by the specific conditions under service.

Ramming Mix A after Service in the Waste-Gas Flue

Parts of the suspended roof in the last third of the waste-gas flue were installed with ramming mix A. Several samples were taken.

Table II. Chemical and Mineralogical Compositions of Lignite Fly Ash Adhering to the Roof of Waste-Gas Flue

Chemical Composition	Transition Rotary Hearth Kiln-Waste-Gas Flue (Vertical part)	Waste-Gas Flue (horizontal part)		
		Roof (near vertical part)	Roof (middle part)	Side wall (middle part)
Oxide (Wt%)				
CaO	51.44	55.48	54.21	54.33
MgO	12.41	14.68	13.75	12.32
SiO_2	15.60	8.61	6.14	15.33
Fe_2O_3	6.61	6.97	7.20	8.50
Al_2O_3	11.49	3.76	2.73	1.98
TiO_2	0.41	0.37	0.39	0.36
Cr_2O_3	0.05	0.05	0.05	0.05
MnO	0.15	0.17	0.15	0.18
K_2O	0.02	0.02	0.02	0.02
Na_2O	0.07	0.07	0.07	0.05
ZrO_2	0.14	0.16	0.05	0.03
P_2O_5	0.24	0.06	0.05	0.02
SO_3	0.04	0.01	0.05	0.04
LOI	1.83	10.22	14.60	6.07
C	0.84	1.89	1.82	0.76
H_2O (Chemical bonded)	1.50	8.0	8.6	4.1
F	0.02	0.02	0.02	0.02
Mineralogical composition	C_2S (s) Periclase (w) C_4AF (w) $Ca(OH)_2$ (vw) (Portlandite)	$Ca(OH)_2$ (m) MgO (m) C_2S (w) C_4AF (vw)	$Ca(OH)_2$ (m) MgO (w) C_2S (w) C_4AF (vw) CaO (t)	C_2S (m) $Ca(OH)_2$ (w) MgO (w) $CaCO_3$ (w) C_4AF (vw) C_2F (t)

Relative X-ray diffraction intensities: vs, very strong; s, strong; m, medium; w, weak; vw, very weak; t, trace.

Macroscopic Description of the Scorified Samples

Four areas with widely different structures can be distinguished macroscopically (Fig. 1): A dark gray slag cover (zone 1) 5–10 mm thick with inclusions of large bubbles; a gray and brownish penetrated zone (zones 2 and 3) 15–25 and 5–10 mm thick, respectively, with relics of white coarse grains of the refractory material; and a bright transition zone to the nearly unaffected refractory material (zone 4). Numerous pores and cavities have formed between the slag and the penetration zone. More cracks and cavities can be seen in zones 2 and 3, causing partial spalling that extends over all four zones.

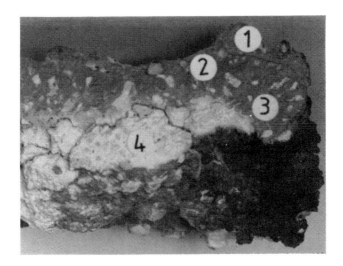

Fig. 1. Cross-sectional area of slag ramming mix A; sample from roof of the waste-gas flue.

Chemical and Mineralogical Composition of the Scorified Layers (Table III): In zone 1, a slag cover has been formed through the reaction of the fly ash with the refractory material; it is predominantly composed of the main constituents of the fly ash, but also contains compounds of the original refractory material. The phase assemblage varies according to the position of the sample and is characterized by the paragenetic, simultaneous appearance of melilite and spinel. In zone 2, melilite is changed to anorthite with a lower content of Ca; in addition, mullite, cristobalite, and corundum from the original refractory material are still present. The amount of fly-ash constituents decreases strongly the greater the distance from the slag cover. Therefore, the reaction products, anorthite and perowskite, can be detected only in minor amounts (zones 3 and 4).

Microscopic, Scanning Electron Microscopy, and X-ray Microanalysis Investigations of the Scorified Samples: The slag cover (zone 1) can be divided into three subzones according to the variances in mineral composition and structure. At the fly ash-slag cover interface, big, mostly idiomorphic, spinel crystals have precipitated within the melilite areas (Fig. 2). This subzone is very porous, with the pores reaching remarkable dimensions. Within eutectic areas, the melilite is interlaced with a phase rich in Ca and Si and containing Al, Fe, and Ti. The adjacent subzone largely consists of melilite. Often, eutectic structures can be detected between the melilite areas (Fig. 3). Near zone 2, the structure consists of folded, ridge-shaped melilite and anorthite crystals. Small amounts of crystallized phases of the residual melt occur at the wedges.

In zone 2, the structure of the refractory material becomes distinguishable, but there are still marked differences between the coarse-grained and the fine-grained areas. As in zones 3 and 4, the bulk of the coarse grains, being only slightly densified, consists of mullite, cristobalite, and a molten

Table III. Chemical and Mineralogical Composition of the Various Slag Zones for Ramming Mix A (Roof of Waste-Gas Flue)

	Phase Assemblage (X-ray diffraction)				Chemical Composition		
	Sample 1		Sample 2		Sample 1		
					(wt%)		
Zone 1 (5–10 mm) Reacted layer (mainly slag) dark gray	Melilite Spinel (Mg·Fe)O·(Al·Fe)$_2$O$_3$ Anorthite CA C$_2$S	(vs) (m-w) (vw) (vw) (t)	Melilite Spinel C$_2$S	(vs) (m) (w)	SiO$_2$ Al$_2$O$_3$ TiO$_2$ Fe$_2$O$_3$ CaO MgO Na$_2$O	22.31 32.85 0.97 6.17 30.47 5.16 0.28	K$_2$O < 0.02 P$_2$O$_5$ 0.06 SO$_3$ 0.01 LOI 0.09 Sample 2
Zone 2 (15–25 mm) Penetrated area gray	Anorthite Cristobalite Corundum Mullite	(vs) (vw) (vw) (t)	Anorthite Mullite Corundum Quartz	(vs) (vw) (t) (t)	SiO$_2$ Al$_2$O$_3$ TiO$_2$ Fe$_2$O$_3$ CaO MgO Na$_2$O	33.86 40.38 1.49 3.87 17.92 1.46 0.51	34.04 42.69 2.22 3.80 13.76 1.52 0.62
Zone 3 (5–10 mm) Penetrated-infiltrated area brown	Mullite Cristobalite Anorthite Perovskite CaO·TiO$_2$ Corundum	(s) (w) (w) (w) (vw)	Anorthite Mullite Cristobalite Quartz	(vs-s) (w) (vw) (t)	SiO$_2$ Al$_2$O$_3$ TiO$_2$ Fe$_2$O$_3$ CaO MgO Na$_2$O	38.43 52.63 1.50 0.71 6.09 < 0.10 0.12	39.52 53.50 1.54 0.71 5.80 < 0.10 0.08

Zone 4	Mullite	(s)
	Cristobalite	(w)
	Corundum	(m)
	Anorthite	(w)
Transition zone to unaffected area	Mullite	(s)
	Cristobalite	(w)
	Anorthite	(m-w)
	Corundum	(w)
	Hematite	(t)

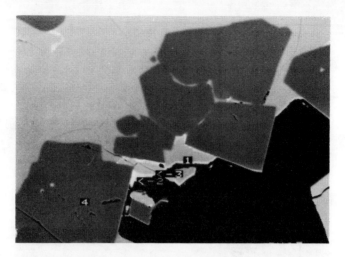

Fig. 2. Ramming mix A: slag cover (zone 1) near adhering fly ash; SEM micrograph, polished section; (1) melilite, (2,3) Ca-, Si-rich melt containing Al, Ti, Fe, (4) spinel $(Mg,Fe)O \cdot (Al,Fe)_2O_3$.

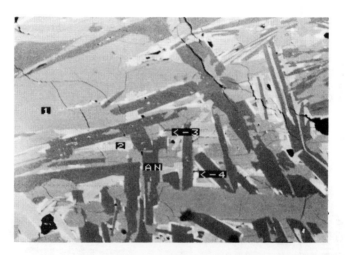

Fig. 3. Ramming mix A: slag cover (zone 1) near reacted zone 2, SEM micrograph, polished section; (1) melilite, (2) aluminosilicate melt rich in Ca, Fe, Mg, (3) Ba-Ca silicate, (4) Ca silicate with Na, K, Al, Fe; An: anorthite.

phase. Reaction seams of anorthite have been formed by the infiltration of Ca into the surface area. The coarse grains are decomposed only in zone 1, which contains higher amounts of Ca, Fe, and Ti than the compacted seams (Fig. 4). In zone 2, the coarse grain (center, dark gray area) is surrounded by two reaction seams: the lighter gray area, (B), and by the fine-grained area

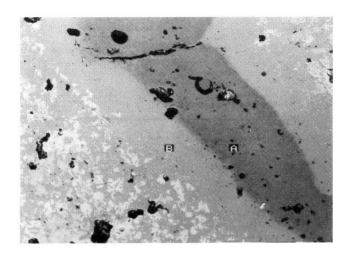

Fig. 4. Ramming mix A: reacted zone (2), coarse grain with reaction seam (center) and fine-grained area (lower, left), SEM micrograph, polished section; (A) mullite; (B) anorthite.

(bottom left) which consists of ridge-shaped anorthite and corundum. Also, some rounded, eutectoid modes of anorthite can be detected in the molten phase (Fig. 5) which mainly consists of Al, Si, Ca, Fe, and the minor components, Mg and Ti. Furthermore, several minor compounds are often found in the fine-grained area that is composed of varying amounts of Al, Si, Ca, Ti, and Fe.

Fig. 5. Ramming mix A: reacted zone (2), fine-grained area, SEM micrograph, polished section; (1) anorthite, (2) glassy phase: Al, Si, Ca, Fe, traces of Mg, Ti, (3) corundum.

Within the transition area of zone 2 and the slag cover, the reaction seams are dissolved by constituents of the slag (Fig. 6). Deep cracks that run parallel to the ceiling are formed during the cooling period. The fine-grained area of zone 2 is more densified than that in zone 3, and the big pores are often spherical. Some areas consist mainly of ridge-shaped, densely intergrown anorthites. At the wedges, phases are found which presumably have formed an aluminosilicate melt rich in Ca and Fe at operating temperature. Nonetheless, in the fine-grained region often there are areas where fine, platelike crystals of corundum are found in direct contact with zone 1. In these areas, the corundum crystals are interspersed in a ground mass of feldspar and melilite.

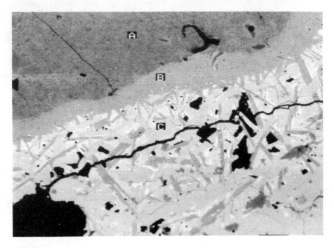

Fig. 6. Ramming mix A: transition area from slag cover (lower part) to reacted zone (2), coarse grain: SEM micrograph, polished section; (A) mullite, (B) anorthite, (C) ≈ B, but higher amounts of Ca, Fe, Ti.

In zone 3, the porosity decreases in both the coarse-grained and fine-grained areas. As in zone 4, the coarse grains consist of mullite, cristobalite, and molten phase. The tetragonal, needlelike crystals of mullite are interconnected by a matrix of molten phase and cristobalite. Reaction seams of anorthite have been formed in the surface area of the coarse grains, and residues of mullite can also be detected (Fig. 7). The fine-grained area consists mainly of corundum and anorthite. The corundum often appears in the form of idiomorphic hexagonal plates of partly irregular shapes and is embedded in a matrix of anorthite and molten phase. Anorthite mainly crystallizes in a ridgelike form, but often the ridges are strongly intergrown, and therefore only few idiomorphic facets can be detected. Also, the coarse grains in zone 4 show that reaction seams increase toward the reaction layer, but the individual crystallites (mullite) are submicroscopically small. The very porous, fine-grained area consists of corundum, mullite, cristobalite, and silicate melt, as well as of anorthite. The amount of anorthite and the structure compaction

Fig. 7. Ramming mix A: penetrated zone (3), coarse grain with reaction seam (upper part) and fine-grained areas (lower part): SEM micrograph, polished section.

increase strongly toward the hot face, especially in the areas rich in molten phase.

Ramming Mix B after Service in the Waste-Gas Flue

As mentioned before, the first third of the roof of the waste-gas flue was lined with ramming mix B (75% Al_2O_3). Due to the slag zones' irregular appearances and their variations in thickness, numerous specimens were required.

Macroscopic Characterization of the Slag-Attacked Samples

Five zones are clearly macroscopically distinguishable: a dark gray slag cover (zone 1), a light gray reacted zone (zone 2), a reddish-brown, reacted infiltrated zone (zone 3), and a relatively unaffected zone (zone 4). Many samples show dark gray veins (zone 5) passing through zones 2 and 3 (Figs. 8 and 9). In addition to the ordered sequence of layers, specimen regions were found with a disordered sequence of layers. Thus in sample 1, zone 2 directly follows zone 4 (right half of Fig. 8). The transition from zone 2 to zone 3 is sharply marked here.

In some specimens, large pores (up to 100 mm long) occur between zones 3 and 4 parallel to the roof, some of them being infiltrated by slag. The slag reacts with the fine-grained areas, which results in an irregular network of veins (zone 5). By progressive reaction, a structure is formed that consists of isolated coarse grains distributed in the matrix of reaction products between slag and fine-grained areas (Fig. 9, center).

Chemical and Mineralogical Composition of the Slag Zones: The results of chemical and X-ray diffraction analyses (Table IV) show considerable amounts of the fly-ash major components (CaO, MgO) in the slag cover (zone 1). Iron oxide increased from approximately 7% to 9% in the slag

Fig. 8. Cross-sectional area of slag ramming mix B.

Fig. 9. Cross-sectional area of slag ramming mix B; disturbed sequence of reacted layers and veins (5).

cover. Depending on the regions from which the samples were taken, the scorified zone 1 consists of the fly-ash components (C_2S, C_4AF, C_2F, and periclase) and newly formed phases of melilite, spinel, and anorthite.

Increased distance from the hot face (zone 2) exhibits decreased amounts of foreign components CaO, MgO, and FeO and/or Fe_2O_3, so no minerals containing Fe and Mg can be detected there by X-ray diffraction. Due to the higher amounts of SiO_2 and Al_2O_3, anorthite is formed as the main component. In addition, small amounts of the original refractory phases of mullite, corundum, and $AlPO_4$ were detected.

Zone 3, with about 5% CaO, contains little anorthite or phases of the original refractory material; this phase decreases considerably in zone 4.

Microscopic, Scanning Electron Microscopy, and X-Ray Microanalysis Investigations of the Slag Samples: As already discussed, macroscopic observation reveals that different sequences of layers can occur besides the ordered layer sequence of zone 1 to zone 4. These phenomena can be explained by spalling at the slag-refractory interface, which leads to direct contact of the fly ash with secondary layers and results in higher wear by structural corrosion. Accelerated wear also results from slag penetration along cracks and interconnected greater pores. The structural corrosion thus causes different wear rates as follows: regular slag attack with ordered zonal sequence; slag attack with higher wear rate and disordered zonal sequence; slag attack with higher wear rate resulting from spalling at the slag-refractory material interface; and slag attack with higher wear rate resulting from slag penetration along veins.

Slag attack with ordered zonal sequence: The phase assemblage is similar in all of the ramming mixes, yet, with the help of petrographic studies, subzones of different appearance were found (Fig. 10). Zone 2 is mainly characterized by the paragenesis of the corundum and anorthite phases. There is a great variation in their habit and crystalline form, and the size and occurrence of pores often change. Zone 2 contains areas with widely differing microstructures, so this zone can again be subdivided into four subzones. A fine-grained anorthite matrix with large idiomorphic corundum crystals can be seen in some areas which are highly densified. In other areas, large, acicular anorthite crystals occur, with evidence of interspace filling of secondary and highly porous corundum (Fig. 11). The residual melt is deposited next to the pores on grain boundaries and in wedges. That most of the anorthite and corundum must have been crystalline in service can be concluded from their size and form, and also from the appearance of structural relics with still visible coarse grains, although the mullite has decomposed to anorthite and corundum. On the other hand, a number of different phases which melted at the service temperature were identified. They are found in wedges, grain boundaries, and next to pores. X-ray microanalysis shows that their Mg, Al, Si, P, Ca, Ti, and Fe content varies widely in amount and combination.

In zone 3, the coarse grains consist mainly of cryptocrystalline mullite and melt, and the fine-grained areas are highly densified. X-ray area maps show that the formation of glassy phase results from penetration of Ca. Pores with particularly large diameters have been formed as a result of the densification of structure in fine- and coarse-grained areas in the direction of zone 2 (Fig. 12). The left half of Fig. 12 shows the fine-grained region with

Table IV. Chemical and Mineralogical Composition of the Various Slag Zones for Ramming Mix B (Roof of Waste-Gas Flue)

	Phase Assemblage (X-ray diffraction)				Chemical Composition Sample 1			
	Sample 1		Sample 2		(wt%)			(wt%)
Zone 1 (5–50 mm) Reacted layer (mainly slag) dark gray	Melilite C_2S Anorthite Periclase Spinel $(Mg \cdot Fe)O \cdot (Al \cdot Fe)_2O_3$	(m) (w) (w) (w) (w)	C_2S Spinel C_4AF (Brownmillerite) C_2F	(s) (w) (vw) (t)	SiO_2 Al_2O_3 TiO_2 Fe_2O_3 CaO MgO	20.48 31.08 1.15 8.78 29.65 5.68	Na_2O K_2O P_2O_5 SO_3 LOI	0.43 0.07 1.65 0.02 0.14
Zone 2 (5–10 mm) Reacted area light gray	Anorthite Corundum Mullite $AlPO_4$ (Quartz Type)	(vs) (w) (vw) (t)	Anorthite Corundum Mullite $AlPO_4$ (Quartz Type)	(vs) (w) (vw) (vw)	SiO_2 Al_2O_3 TiO_2 Fe_2O_3 CaO MgO	26.07 49.13 1.60 4.80 12.76 1.97	Na_2O K_2O P_2O_5 SO_3 LOI	0.79 0.06 2.14 0.01 0.01
Zone 3 (2–30 mm) unfiltrated area reddish-brown	Mullite Corundum Anorthite $AlPO_4$ (Quartz type)	(vs) (m) (w) (t)			SiO_2 Al_2O_3 TiO_2 Fe_2O_3 CaO MgO	24.80 59.39 2.04 1.78 5.16 1.06	Na_2O K_2O P_2O_5 SO_3 LOI	0.59 0.06 3.80 0.01 0.11

Zone 4 Transition zone to unaffected area white	Mullite Corundum Anorthite AlPO₄ (Quartz type)	(vs) (m) (vw) (vw)		Mullite Corundum Anorthite AlPO₄ (Cristobalite and tridymite type)	(vs) (w-m) (w) (vw)
Zone 5 (dark gray vein)				Melilite C₄AF	(vs) (w)

Fig. 10. Ramming mix B: transition from densified areas of reacted zone 2 (upper part) to slag cover with the melilite-rich subzone (middle) and the spinel-rich subzone (lower part): SEM micrograph, polished section.

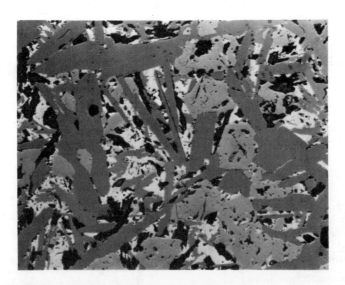

Fig. 11. Ramming mix B: reacted zone (2), polished section: anorthite (gray), corundum (light gray), eutectic phase (bright).

finely plated corundum crystals in a matrix of anorthite and glassy phase. Some of the corundum grains have been formed as a result of the decomposition of mullite after reaction with CaO. Large acicular corundum crystals formed preferentially on the boundaries of the coarse grains. Pores with respectively large diameters can be found next to them. Near these corundum

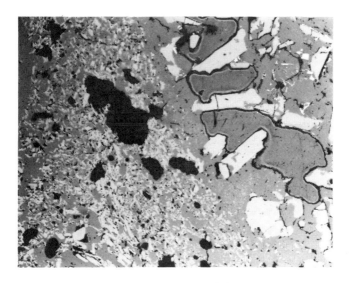

Fig. 12. Ramming mix B: infiltrated zone (3) near reacted zone (2), fine-grained area (left) and coarse grain (right); polished section: anorthite and glassy phase (gray), corundum (bright).

grains (Fig. 12, right half) bright, reflecting titanium phases, mainly rutile, and relatively large, recrystallized mullite grains are discernible in an anorthite and glassy phase matrix.

The relatively unattacked zone 4 consists mainly of coarse mullite grains with small amounts of melt. The bonding matrix consists of intergrown, acicular mullite and corundum plates in which the coarse grains are embedded. In the fine-grained areas, some isolated crystals of anorthite and $AlPO_4$ in various modifications appear. Toward the hot face, the amount of Ca-containing phases increases. In addition to anorthite, X-ray microanalysis reveals a phase of CA_6, a phase containing mainly Ca with medium amounts of Si and small amounts of Al, and a phase containing predominantly Al with minor amounts of Mg, Si, Ca, and Fe.

Acceleration of slag attack through infiltration of slag via veins: In some areas, slag penetrates into the pores and the crack system. In the finer veins, the slag has a phase assemblage (mainly melilite and Fe-rich spinel) as in zone 1. In the larger veins with diameters up to 10 mm, the penetrated slag consists of melilite and brown millerite, which is generally a main component of fly ash. C_2S and spinel could be detected on a microscopic level only.

The vein-shaped slag is Ca- and Fe-rich, especially in the center; however, it is poorer in Al than the slag in zone 1. In transition to the refractory material (Fig. 13), there is a high concentration of melilite. Its conversion to anorthite is sharply marked. Corundum grains belonging to the refractory often project into the melilite regions and are partially dissolved by the slag. In the refractory itself, Ca and P phase concentrations are enriched near the large veins. The fine-grained regions mainly have crack and pore formation followed by slag penetration. Since the reaction rate of coarse grains is lower, a microstructure of coarse grains is formed, although a greater amount

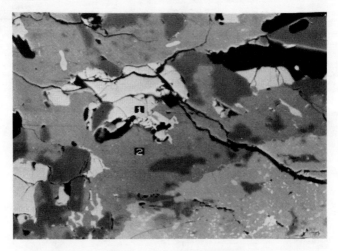

Fig. 13. Ramming mix B: vein (zone 5, lower right): (1) Ca, P, (2) anorthite; corundum (dark gray), melilite (light gray).

reacted out to anorthite and corundum. The fine-grained region consists of anorthite, melilite, and spinel.

Ramming Mix C after Service in the Waste-Gas Flue

This material has been installed in the forward part of the waste-gas flue where the gas enters the horizontal part.

Macroscopical Description of the Scorified Samples

At the flue ash-slag interface, the lining is partly fused through reactions with the fly ash and blown off. Large amounts of fly ash, however, often stick to the refractory lining. Four zones are macroscopically distinguishable: slag cover (zone 1, black, up to 25 mm); a reacted zone (zone 2, dark brown, 2–5 mm); a penetrated zone (zone 3, light gray, about 10 mm); and an unaffected refractory material (zone 4). Often, pores parallel to the roof, 15 mm long and 5 mm in diameter, have been formed between the reacted and the penetrated zones.

Mineralogical Composition of the Slag Layer: A mineralogical composition similar to that of samples A and B is detected within the slag layer—melilite, spinel, and C_4AF. Within zone 2, the mineral phases of zone 1 occur in addition to the main phases of corundum and anorthite; within penetration zone 3, the structure consists of corundum and small amounts of anorthite and melilite.

Microscopic, Scanning Electron Microscopy, and X-Ray Microanalysis Investigations of the Scorified Samples: The slag cover (zone 1) consists mainly of melilite, with eutectic residual melt precipitated at the grain boundaries and at the wedges. X-ray microanalysis reveals small amounts of this eutectic phase with Mg, Fe, and Ca, as well as with Ca, Al, Si, and Ti. At the fly ash-slag cover interface, strong porous areas of spinel, brown millerite, and calcium-aluminates are found.

Table V. Mineralogical Composition of the Various Slag Zones for Ramming Mix C (Roof of Waste-Gas Flue)

	Phase Assemblage (X-ray diffraction)	
Zone 1		
Reacted layer	Melilite	(m-s)
(mainly slag)	Spinel	(w)
dark gray	$(\underline{Mg} \cdot Fe)O \cdot (\underline{Al} \cdot Fe)_2O_3$	
to 25 mm	C_4AF	(vw)
Zone 2		
Reacted area	Corundum	(m-s)
dark brown	Anorthite	(w)
2–5 mm	Spinel	(vw)
	$(\underline{Mg} \cdot Fe)O \cdot (\underline{Al} \cdot Fe)_2O_3$	
	Melilite	(vw)
	C_4AF	(t)
Zone 3		
Infiltrated area	Corundum	(vs)
light gray	Anorthite	(w)
	Melilite	(t)

In zone 2, near zone 1, the shape of the corundum crystals changes to platelike structures which are fascicularly intergrown (Fig. 14). Molten phases rich in phosphorous containing Al, Si, Ca, Ti, and Fe are associated with the fascicular corundum crystals.

Fig. 14. Ramming mix C: reacted zone (2) near slag cover; SEM micrograph, polished section: corundum (dark gray), anorthite (gray), eutectic phases (light gray).

Near zone 3, a structure of coarse corundum crystals embedded in a ground mass of anorthite and eutectic phases is formed. In that ground mass, Al-Si-Fe titanates and Al-Si-Ca phosphorous compounds are precipitated into an aluminosilicate melt containing Na, Mg, P, K, Ca, Ti, and Fe (Fig. 15). In zone 3, within the adjacent subzone, much more corundum in the shape of smaller plates and ridges occurs embedded in the ground mass

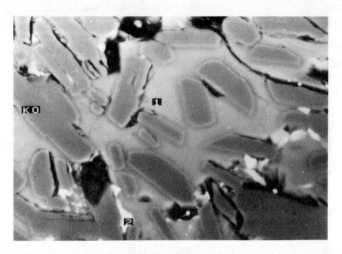

Fig. 15. Ramming mix C: reacted zone (2) near zone 3, SEM micrograph, polished section: corundum (dark gray), anorthite (gray), eutectic phases (light gray).

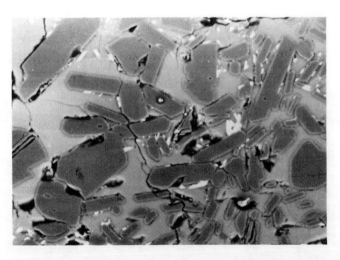

Fig. 16. Ramming mix C: infiltration zone (3) near reacted zone (2), SEM micrograph, polished section: (1) anorthite, (2) Al-Fe titanate; Ko: corundum.

that consists of anorthite in which Al-Fe titanates also are detected (Fig. 16). Much larger pores are identified within the transition to zone 2 (Fig. 17).

In the area toward the original refractory material, big, ridge-shaped corundum crystals occur embedded in a ground mass consisting of Al, Si, and traces of Ca, Ti, and Fe. Several submicroscopic phases can be detected (Fig. 18). Reaction seams have been formed around the corundum crystals containing much more Si than the bulk of the ground mass. Big, layer-shaped pores have been formed wthin this zone.

Fig. 17. Ramming mix C: cracks and cavities in infiltrated zone (3) and reacted zone (2).

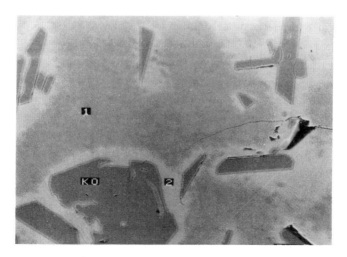

Fig. 18. Ramming mix C: infiltrated zone (3) near zone (4); (1) Al, Si, traces of Ca, Ti, Fe; (2) Al, Si; Ti; Fe (trace).

Discussion

The corrosion reactions in the ramming mixes investigated are complex because of the great number of reacting oxides (CaO, MgO, FeO and/or Fe_2O_3) under moderate reducing atmosphere and H_2O vapor. The transport mechanisms of the fly-ash and volatile components from the hearth kiln atmosphere into the refractory material are influenced by the reducing conditions and the H_2O vapor, as are the melting behavior, the viscosity of the ashes, and the slag structure zones. In the different zones formed, the deposition of the volatile components was indicated analytically. Depending on the SiO_2 content, the ash deposit is composed of C_2S, periclase, and CaO, which reacted with components of the atmosphere to form portlandite and calcite, as well as such Fe-containing minerals as C_4AF and C_2F. Under service temperatures of 1300°–1400°C maximum, there is hardly any melt formation to be seen in the ash deposits, although the hemisphere point for 15.6% SiO_2 fly ash in a reducing atmosphere is at 1425°C. In zone 1 (deposit of slags) the hemisphere point falls to 1280°C due to the higher SiO_2-Al_2O_3 content. This zone was formed by the reaction of the slag refractory material and the fly ash. In the slag cover, the mineral paragenesis is far from equilibrium. However, reactions occurring in the fly-ash deposits can be described in terms of the system CaO-MgO-SiO_2-Al_2O_3.[1-8] The presence of iron oxide makes the actual situation more complex.

The reactions occurring in the reacted layer (zone 2) can be described in terms of the system CaO-Al_2O_3-SiO_2.[9] According to that description, the already small changes in the ratios of SiO_2/Al_2O_3 and Al_2O_3/CaO theoretically lead to the expected formation of corundum and anorthite, and of melilite and anorthite, respectively. Ramming mixes B and C show no relics of the original refractory material; the reactions with CaO have completely transformed the microstructure. In case of an "accelerated" slagging, however, mix B shows structure relics of the coarse grains. In this case, ramming mixes B and C consist of anorthite, corundum, and residual eutectic melt. The strongly different appearance of the corundum crystals gives a hint of the complex reaction kinetics involved. Under service conditions, all three ramming mixes undergo wear, implying the renewal of single panels from time to time. The wear is due not only to the chemical corrosion by the compounds of the fly ash, but also considerably to the structural corrosion by the crack and pore formation due to the densification caused by the reactions of the refractory components with CaO, MgO, and FeO and/or Fe_2O_3, which results in a complete transformation of the microstructure. Within ramming mix A, which contains the least amount of Al_2O_3, silicate melts cause densified areas, particularly in the coarse grains. Ramming mix B, based on mullite, however, undergoes this strong densification in a more narrow region. Because zone 2 consists only of reaction products and the slag deposit is much thicker, thermal stresses can be released only by crack formation in this narrow area. In combination with growing porosity caused by the transformation of the structure, the slag can penetrate into the structure to fill veins and accelerate the corrosion (structural corrosion). Ramming mixes A and C, like ramming mix B, are also corroded structurally, but not in so pronounced a fashion. The position of the original composition of the three ramming mixes within the phase diagrams lets one assume that a decrease in chemical corrosion corresponds with an increase in Al_2O_3 content. Mixes A and C corrode more slowly than B because of the superimposed structural corrosion. Taking into consideration possible temperature raises, ramming mix C is supposed to have a longer lifetime.

Acknowledgment

The authors thank the Deutsche Forschungsgemeinschaft for providing the scanning electron microscope with accessories used during this investigation.

References

[1]A.T. Prince, "Liquidus Relationships on 10% MgO Plane of the System Lime-Magnesia-Alumina-Silica," *J. Am. Ceram. Soc.*, **37** [9] 402 (1954).
[2]E.F. Osborn, R.C. De Vries, K.H. Gee, and H.M. Kraner, "Liquidus Diagrams for the System CaO-MgO-SiO$_2$ with Addition of Al$_2$O$_3$ between 5 and 35 wt%," *Trans. AIME*, **200**, 38–39 (1954).
[3]A. Muan, "Phase Equilibra at Liquidus Temperatures in the System Iron Oxide-Al$_2$O$_3$-SiO$_2$ in Air Atmosphere," *J. Am. Ceram. Soc.*, **40** [4] 121 (1957).
[4]L.M. Atlas and W.K. Sumida, "Solidus, Subsolidus, and Subdissociation Phase Equilibra in the System Fe-Al-O," *J. Am. Ceram. Soc.*, **41** [5] 150 (1958).
[5]A. Muan, "Reactions Between Iron Oxides and Alumina-Silica Refractories," *J. Am. Ceram. Soc.*, **41** [8] 275 (1958).
[6]F.P. Sorrentino and F.P. Glasser, "Phase Relations in the System CaO-Al$_2$O$_3$-SiO$_2$-Fe-O: Part II, The System CaO-Al$_2$O$_3$-Fe$_2$O$_3$-SiO$_2$," *Trans. J.Br. Ceram. Soc.*, **75**, 95 (1976).
[7]F.P. Sorrentino and F.P. Glasser, "The System CaO-Fe$_2$O$_3$-Al$_2$O$_3$-SiO$_2$: I, The Pseudoternary Section Ca$_2$Fe$_2$O$_5$-Ca$_2$Al$_2$O$_5$-SiO$_2$," *Trans. J. Br. Ceram. Soc.*, **74**, 253 (1975).
[8]W.T. Bakker and G.E.D. Snyder, "Permeable Alumina Refractories," *Am. Ceram. Soc. Bull.*, **49** [7] 664 (1970).
[9]A.L. Gentile and W.R. Forster, "System CaO-Al$_2$O$_3$-SiO$_2$ in the Primary Field of CaO-6Al$_2$O$_3$," *J. Am. Ceram. Soc.*, **46** [2] 76 (1963).

The Properties and Applications of Dolomite Ramming Mixes

J. W. STENDERA

J.E. Baker Company
York, PA 17405

Ready-to-use refractory monolithics such as plastics and rams are widely used throughout the steel industry. Most of these materials use either an alumina or aluminosilicate aggregate as the primary constituent. Materials using such basic aggregates as dolomite and magnesite have suffered from a number of problems, including poor shelf life, poor bonding at low to intermediate temperatures, and poor hot strength. In the last 10 years, there has been a dramatic increase in the use of basic refractories in the steel industry, particularly evident with ladle refractories. The increased hold times associated with continuous casting and increased attack due to the advent of ladle metallurgy have placed severe requirements on ladle refractories. In particular, dolomite refractories have found application in ladles, due to their low cost, good wear properties, compatibility with desulfurization processes, and other metallurgical benefits. This growth in the use of dolomite refractories required the development of a full range of products using dolomite aggregates. This, of course, included a ready-to-use plastic ramming mix. This paper will describe the development of this material, its properties, and its major use applications.

Survey of Literature

Magnesite and dolomite monolithics have been used for many years in the form of gunning mixes or vibratables, where no water is used, or is mixed only just prior to use. Archibald and Bigge described the procedure for lining an electric furnace bottom using homemade mixtures of magnesite, dolomite, and sodium silicate binders.[1,2] This procedure has been modified substantially over the years, and many excellent commercial mixtures are available. All of these materials, however, must be mixed with water on-site to be used. Mixing is inconvenient when only a small quantity is required and, in any case, would be unsuitable for use with dolomite bricks due to the possibility of hydration. For this situation, dolomite and magnesite mixes have been produced using pitch and/or oil as a plasticizing and binding agent. Pirogov described a magnesite, plastic ramming mixture that used a binder which consisted of 70% coal tar pitch and 30% anthracene oil.[3] This was similar to the pitch and creosote oil mixture described in an ISI special report.[4] In this report, it was stated that binder level was critical, and a difference of 4.7–5.7% binder made the difference between a material that slumped and one that did not. The usual practice of using mill scale as a flux to obtain intermediate-temperature strength would, of course, not work in these types of mixtures, and dependence for strength was placed on the carbon remaining after burnout. It should be mentioned in passing that there is at least one sodium silicate-bonded, chrome magnesite, basic plastic on the market.

Although it has a reasonable shelf life, it suffers from low hot strength at intermediate temperatures. This problem is most likely due to silicate liquid formation in the 800°–1200°C range. In addition, one of the major uses of a dolomite ram would be in conjunction with other dolomitic products, and a water-based material such as this would be unacceptable from that stanpoint.

Material Development

A major function of refractory monolithics is to easily fill areas in vessel linings which would be difficult or too time-consuming to accomplish with bricks. Since a product such as this would often be used in conjunction with dolomite bricks, two criteria had to be met which are unique to the application: (1) The material must be virtually water-free, and cannot release water on heatup; and (2) the material must be compatible with dolomite bricks at the service temperature, and form no low-temperature liquid phases in conjunction with dolomite or any process slag or clinker likely to be encountered. These two criteria are severe restrictions, and virtually limit the system to one which uses an organic vehicle with either a dolomite or MgO aggregate. Although they were not absolute requirements, several other criteria were desirable: (1) The material should exhibit good strength at all temperature ranges; (2) the bond should be stable under oxidizing as well as reducing conditions; (3) the material should be nonhazardous in the field environment, containing no flammable or toxic solvents; and (4) the appearance and application technique should be similar to existing materials in its class. With these criteria in mind, a high-purity, dolomite aggregate was chosen.

The chemistry of the ram, with the exception of the loss on ignition, is virtually identical to the aggregate used.* The loss on ignition reflects the organic liquid used in the bonding system. This liquid plays a similar role to the water contained in conventional systems, in that it acts as the vehicle that carries the binder and gives the material the proper consistency. The bonding system of the dolomite ram is activated by the removal of the organic liquid during burning and is not reversible by reapplication of liquid. The bond formed is refractory at steelmaking temperatures, unlike bonds based on fluxing liquids, such as sodium silicate. The bond is also good under oxidizing conditions, unlike preparations based on pitch or carbon-yielding foundry binders. In addition, neither phosphorous nor sulfur is used in any compound or form, so there is no danger of phosphorous or sulfur contamination of the steel. In short, as far as has been determined, the bond formed is very close to the direct bond that would be achieved from sintering the material at high temperature.

Because an organic liquid is used instead of water, several points should be mentioned to contrast this system to water-based systems. Water, of course, is an evaporable solvent. In an organic system, however, an evaporable solvent cannot be used for the obvious reason of fire hazard. Only materials which have very low vapor pressure at ambient temperatures can be considered. The practical results of this constraint entail the use of a material which will not air-dry, and is heat-setting only. This development constraint also causes problems with rheological stability at high temperatures. In a

---

*Chemical composition of ram: Al_2O_3, 0.8%; CaO, 57.0%; MgO, 42.0%; SiO_2, 0.9%; Fe_2O_3, 1.1%; Na_2O, Tr; P_2O_5, Tr; LOI, 4%.

water-based system, water starts to evaporate even at room temperature, which stiffens the material. This means that a water-based material is softest right out of the container, and that once it is exposed to air, any application of heat or passage of time makes the consistency stiffer. With the dolomite ram, the organic liquid does not start to come off until 230°–260°C (450°–500°F), and during the heatup, the organic liquid becomes less viscous, allowing the material to become softer than it was initially. This fact limits the use of dolomite ram to applications where it does not have to be suspended in preheat. The practical considerations of this limitation will be covered under use. Experimentation has shown that the effects of this problem can be limited by using a "gap-type" particle sizing rather than a graded batch. Table I presents the approximate particle-size distribution used. It should be noted that at least one advantage is gained by using an organic liquid over water; preheat is much less critical, and preheating schedules can be much faster without the danger of steam buildup and explosive spalling.

Table I. Screen Analysis

Particle-Size Distribution	Wt% Retained
1/2 by 6	40
6 by 20	10
20 by down	50

Properties

Figure 1 shows the modulus of rupture (MOR) strength of dolomite ram compared with a typical, 85% alumina phosphate-bonded plastic. For this test 5 by 5 by 23-cm bars were pressed at 4.2×10^5 Pa (60 psi) to simulate a rammed density. The bars were then dried to remove water or organics as required. Heat treatment was for 2 h at the temperature indicated on the graph. Heatup and cool-down were accomplished at 100°C/h, and the kiln atmosphere was oxidizing at all times. Also included in this graph is the high-strength, low-shrinkage dolomite ram described below.

To test the effect of ramming vs pressing, small bars have been tested that were actually formed by ramming. Results, as expected, were much more variable than for pressed bars, with an average strength 10–20% below pressed bars. Apparently ramming does not change the intrinsic properties of the material; it merely adds an additional detrimental variable to the test.

Shrinkage has rarely been a problem with most applications of dolomite ram. However, where shrinkage has been a problem, two methods of controlling it have been used. The first method consisted of the addition of carbon, which was very effective but suffered from two problems. First, in an oxidizing environment, the carbon burns out, allowing shrinkage to occur, and second, carbon was found to be a diluent to the system, which interferred with the bonding. To get around this difficulty, the carbon was added in a coarse particle size and at levels no higher than 5%. The second method of controlling shrinkage is with an inorganic sintering inhibitor. It has the advantage over carbon of being effective even under oxidizing conditions, and the additive actually increases the strength of bond. However, it is not

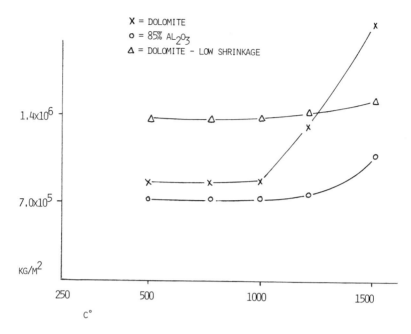

Fig. 1. Room-temperature MOR vs heat treatment.

quite as effective as carbon. In a test of permanent linear change to 1600°C for 2 h, the ram without additives has 2-3% shrinkage. The carbon-containing material has virtually no shrinkage, while the material containing a shrinkage inhibitor has 1-1½% shrinkage.

Hot modulus of rupture has been determined on the dolomite ram using the apparatus described by Stendera et al.[5] Results are shown in Fig. 2, along with the results from a direct-bonded, fired, dolomite brick as would be used in a steel ladle application. As can be seen, the behavior is very similar except for the difference in initial strength. It should be noted that there is no dip in strength seen at the intermediate temperatures. At high temperatures, lower strength is seen in both the ram and brick. This has been postulated by Landini to be due to a "creep deformation" type of mechanism in the lime phase.[6] Regardless, this evidence supports the earlier conclusion that the bond developed has the characteristics of a true, direct bond.

Use

With any material of this class, the uses are very diverse. In this section, the more important uses will be described. As would be expected, the major use found to date for the ram is in conjunction with dolomite bricks. A cross-sectional view of a typical, large, steel ladle such as would be used in a continuous casting shop is shown in Fig. 3. In such a ladle, anywhere from 1-2 tons of ram are used in the areas indicated. The right-hand side of the diagram shows the sidewall-over-bottom construction method, where the ram is used on the lip to fill between the well block and bricks, and the brick and safety lining. This type of construction is used in shops where the sidewalls tend to wear faster than the bottom. On the left side, the "plug bottom" type

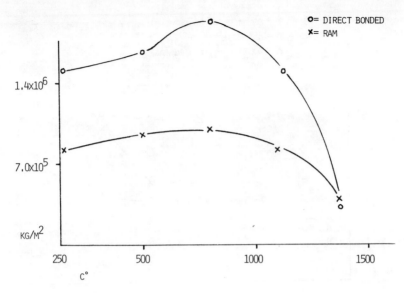

Fig. 2. Hot MOR.

of construction is shown, where the sidewalls rest on the safety lining, and the bottom is installed last. Ram is used with this type of construction between the bottom and working sidewalls as well as on the lip and around the well block. This type of construction facilitates easy bottom repairs in a shop where the bottom wears out before the sidewalls.

In small, steel ladles of 10 tons or less, the ram has found use as the primary refractory. Figure 4 shows a teapot foundry ladle being rammed using a dolomite ram. Because such a small vessel is inconvenient to brick, many steel foundry operators have been using alumina or zircon lining materials which are incompatible with their slag practice. This has been especially true with the high-calcium slags required for desulfurization. A previously published paper described this application for dolomite ram in detail.[7]

Other uses for the ram have included soaking pit burner ports, AOD hoods, walking-beam furnace hearths, electric furnace runners, as a closure material in cement rotary kilns, and in blast furnace metal runners.

One interesting aspect of the dolomite ram is that is presents the opportunity to fabricate large, custom-preformed shapes out of dolomite. Figure 5 shows a blast furnace runner and two well blocks made with dolomite ram. Aside from the obvious limitations in mold design that using a ram presents, almost any size or shape can be made. The runner sections shown weighed about 318 kg, and the well blocks were 113 kg each. Blocks such as these are made using two techniques. The first technique heats the formed shape to 260° to 315°C for 6–10 h, depending on the size. The organic liquid used in the ram will take on a mild thermoset; the bond formed is not as good as that formed with resins designed specifically for the purpose, but the strength, 3.4×10^6–4.8×10^7 Pa, is generally sufficient for most applications. The second method is to actually burn the organics off, and form the ceramic bond described earlier. This procedure is the more difficult of the two and is rarely required.

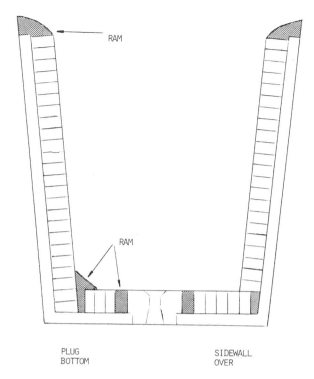

Fig. 3. Large concasting ladle.

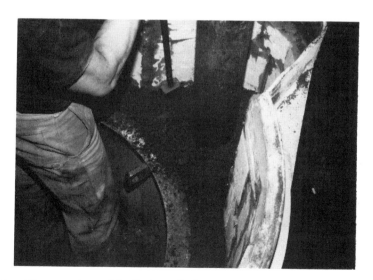

Fig. 4. Five-ton teapot ladle installation with dolomite ram.

Fig. 5. Well blocks and blast furnace runner made with dolomite ram.

Summary

A dolomite ramming mix has been developed using an organic liquid as a vehicle. A low-temperature bonding system is used, which has the characteristics of a direct bond. Properties of the ram were discussed, and major use applications described.

References

[1] Archibald, "Recent Developments in Refractories in the British Steel Industry," *Metallurgia*, p. 175 (1950).
[2] H.C. Bigge, "Bottom Building and Maintenance," *Electric Furnace Steel Proceedings*, pp. 49-67 (1950).
[3] A.A. Pirogov, "Properties of Magnesite Ramming Mixes," *Ogneuport;* pp. 92-98 (1960).
[4] ISI Special Report No. 87, "Steelmaking in the Basic Arc Furnace," pp. 146-52.
[5] J. Stendera, G. Bartkowski, and J. Firestone, "A Simple Apparatus for Testing Hot Strength of Bulk Refractory Samples"; for abstract see *Am. Ceram. Soc. Bull.*, **62** [3] 431 (1983).
[6] D.V. Landini, "Fracture of Dolomite Refractories"; Ph. D. Thesis. The Pennsylvania State University, University Park, 1982.
[7] J.W. Stendera, "The Use of Dolomite Ram in Small Steel Ladles," *AIME Electric Furnace Proceedings*, **40**, Kansas City, 1982.

SiC Monolithics for Waste Incinerators: Experiences, Problems, and Possible Improvements

G.S. Dhupia and W. Krönert

Institut für Gesteinshuettenkunde der RWTH Aachen,
Mauerstrasse 5, D-5100
Aachen, Federal Republic of Germany

E. Goerenz

DYKO Industriekeramik GmbH
Wiesenstrasse 61, D-4000
Düsseldorf, Federal Republic of Germany

The heat produced in waste incinerators is utilized for steam generation through a system of water-bearing tubes. SiC mixes of various compositions are used to protect these tubes against corrosion, ensuring at the same time high thermal conductivity. Properties required of such mixes, as well as problems encountered during installation, are briefly reviewed. Results of experiments conducted with silicate, fluid phosphate, and dry phosphate-bonded mixes are presented and compared. In addition to technological properties, distributions of the binder components are studied using scanning electron microscopy and X-ray microanalysis.

In the Federal Republic of Germany, about 25% of the 200–400-kg waste produced per inhabitant every year is burned in over 40 incinerators nationwide.[1] Due to the changing compositions of household and industrial waste, such incinerators are subject to widely varying thermal and environmental stresses during operation. In the past, silicon carbide materials have found widespread use in waste incinerators. Silicon carbide brick is employed more for its resistance to chemical attack, whereas with SiC monolithics the high thermal conductivity of the material is of primary importance.[2,3]

Refractories for Waste Incinerators

The main requirements of refractory linings for waste incinerators are refractoriness up to 1400°C, good thermal conductivity, high strength, high resistance to corrosion, and economical refractory life. These requirements are satisfied to the largest extent by silicon carbide materials. For SiC monolithics, however, good thermal conductivity, high strength, and economical refractory life are most important.

In state-of-the-art waste incinerators, the waste heat generated is used to produce steam. This is achieved with the help of water-bearing tubes projected into the burning zone of the boilers. Figure 1 shows the cross section of a typical waste incinerator. These tubes are protected against corrosion by a refractory lining. Due to present technology, only SiC monolithics are used

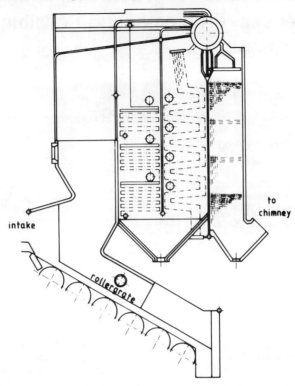

Fig. 1. Cross section of a typical waste incineration plant.

for this purpose. The tubes are covered with up to 800 steel studs per m^2 (preferably heat-resistant steel) to both support the monolithic lining and increase the heat conductivity.[4] Generally, the studs are about 10 mm in diameter and about 14 mm high. Figure 2 illustrates the general appearance of such tubes, and Fig. 3 shows the cross section with the surrounding protec-

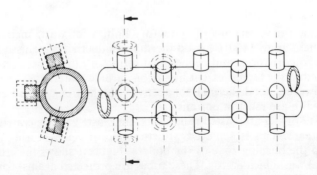

Fig. 2. Arrangement of studs for individual pipes.

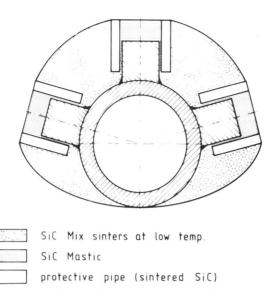

Fig. 3. Arrangement of studs (pipe cross section).

tive ceramic mix. The studs are protected by refractory collars (sintered SiC) glued to the stud using an SiC mastic.[5] Other systems envision ceramic caps instead of collars for studs.[6] The room between the studs is filled with a silicon carbide monolithic.

Fig. 4. Section of pipe with collars.

Fig. 5. View of pipes with studs.

SiC Monolithics Presently in Use

Depending on the manufacturer, the SiC mixes are supplied in the following different states: (1) dry mix including all additives (ready to use); (2) dry mix with the bonding agent supplied separately, which must be mixed before use; (3) premoistened mix including all additives (ready to use); and (4) premoistened with the bonding agent supplied separately, which must be mixed before use.

The spectrum of the bonding agents in use includes silicates (bentonites and clays), fluid phosphates, dry phosphates, and boric acid, vanadium compounds, and vitrifiers, e.g., powdered glass. These bonding agents are employed in various combinations. To achieve satisfactory strengths and densities, the grain-size distribution includes, according to experience, about 30% coarse grains and about 30% fine meal.

Installation Techniques

The most prevalent technique employed is gunning. Smaller areas are also troweled manually. In terms of strength and density, however, ramming would be the most effective method, although economic considerations generally preclude its use.

As a result, SiC mixes have to be such that they can be gunned, rammed, or troweled, as the need arises. To achieve this at the site, the mixes must satisfy the following requirements: constant, adjustable moisture content; homogeneity of the grain-size distribution; and the rebound should be completely reusable.

Figure 6 shows an example of the procedure recommended for such linings. The initial heat-up period must be slow enough to enable the moisture to escape. Deviations from recommended procedure can lead to moisture-induced cracks and spalling during heatup, thus having a detrimental effect on lining life, which later is often mistakenly attributed to other causes.

Mistakes often made during installation of such mixes can be summarized as: gunning of the mix with insufficient pressure; gunning with false moisture content; using a shorter than recommended drying period after

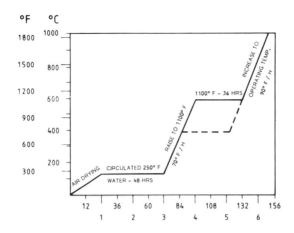

Fig. 6. Drying and firing procedure for silicon carbide gunning, ramming, and troweling mixes.

installation; and operating without preheating the mix to sintering temperature.

In the past, a number of publications have dealt with the tendency of SiC to oxidize in the 900°–1200°C temperature range. This cannot be corroborated for SiC mixes in waste incinerators. Premature failure of the mix due to oxidation is not known, which can be explained by the protective layer of fly ash that adheres to the hot face during operation. Spalling of the mix, due to the difference in the thermal coefficients of expansion of the steel tube and refractory mix, has also not been observed. According to experience, premature failure of the SiC mix is most often caused by faulty installation techniques and nonadherence to the recommended drying procedure.

Experiments with Various Binders

To study the properties of the SiC mixes, a number of experiments with a variety of different binders were undertaken. Specimens were prepared with fluid and dry phosphate binders, as well as with silicates. Specimen A contains a boron additive, B and C were prepared with fluid phosphates, and specimens D, E, and F contain dry phosphate binders.

Experimental Procedure

The various mixes were pressed into cylinders (50 by 50 mm) and blocks (114 by 62 by 62 mm). From the different classes of binders investigated, the following were chosen for discussion: one specimen with a silicate bond and a boron-based additive (A), two with fluid phosphate binders of differing compositions (B and C), and three with dry phosphate binders (D, E, and F) also differing in composition. The specimens were fired at 100°–1600°C at 100°C intervals for 6 h each. The macroscopic appearance of the specimens after firing at 500°, 700°, 1000° and 1200°C is presented in Fig. 7.

The specimens with fluid binders (B and C) are distinctly better sintered at the peripheries than in the middle, especially below 1000°C. Similar behavior is evident with specimen D (dry phosphate binder); in addition, a

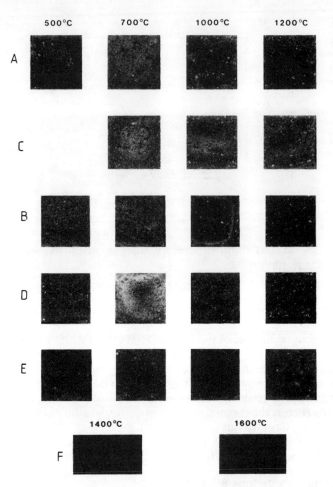

Fig. 7. Macroscopic appearance of the specimens after firing at various temperatures.

bloating and blistering of the specimen cylinder occurs at 1200 °C. Specimens A, E, and F show a more homogeneous structure, and specimen A appears to be less dense. For specimen F, only the samples fired at 1400° and 1600 °C are shown, since this was the only mix that showed no macroscopic destruction above 1200 °C. For all other specimens, the samples fired above 1200 °C had an appearance similar to that of specimen D at 1200 °C. An example of this destruction is shown in Fig. 8.

Experimental Results

The technological properties of the various compositions investigated are presented in Table I. It is evident that the silicate-bonded mix has the lowest bulk density, and the highest is exhibited by the dry phosphate-bonded samples. The bulk densities of the fluid phosphate-bonded mixes are also distinctly higher than that of the silicate-bonded mix. The silicate-bonded mix

Fig. 8. Macroscopic appearance of specimen IV after firing to 1200°C.

has the highest apparent porosity; the differences between the phosphate-bonded mixes fluctuate within a narrow region, although, within this region, a clear correlation between apparent porosity and bulk density cannot be determined. The differences in grain-size distribution and in the ceramic bond are probably responsible for these minor deviations in bulk density.

After firing at various temperatures, the cold crushing strengths of the samples were determined. The results are also tabulated in Table I. The silicate-bonded sample A shows the lowest strength values, particularly in the temperature range below 500°C. Below 300°C, sample A has extremely low strengths, which suggests a very weak bond between the grains. Only above 700°C is the strengthening effect of the boron-based additives evident.

The fluid phosphate-bonded specimens B and C have distinctly higher strengths at low temperatures than the silicate-bonded specimen A. In contrast to specimen B, specimen C contains a vanadium-based sintering aid. However, they have the lowest strengths at high temperatures compared with the other categories. All specimens with dry phosphate bonds already have at 100°C strengths higher than both the silicate-bonded sample at 700°C and the fluid phosphate-bonded samples at any temperatures. Specimen D shows a steep increase in strength between 500° and 1000°C, due to the combined effect of the higher fine-meal content and the sodium phosphate binder used. However, as already stated, temperatures above 1000°C lead to destruction of this mix. Specimen F was the only composition that withstood temperatures up to 1600°C without visible destruction.

The microstructures of the specimens were also investigated with the help of scanning electron microscopy and X-ray microanalysis to study the temperature-dependent behavior of the binders. The microstructures of the

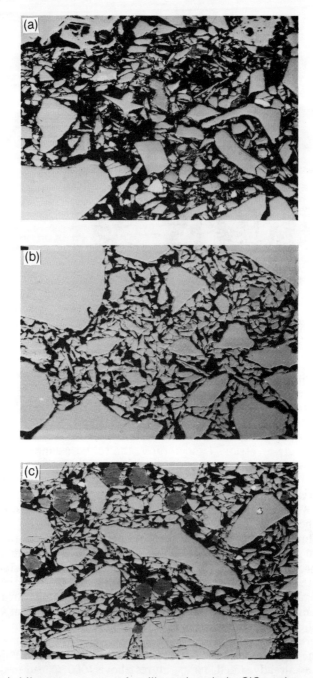

Fig. 9 (*a*) Microstructure of silicate-bonded SiC mix at 500°C.
(*b*) Microstructure of fluid phosphate-bonded SiC mix at 500°C.
(*c*) Microstructure of dry phosphate-bonded SiC mix at 500°C.

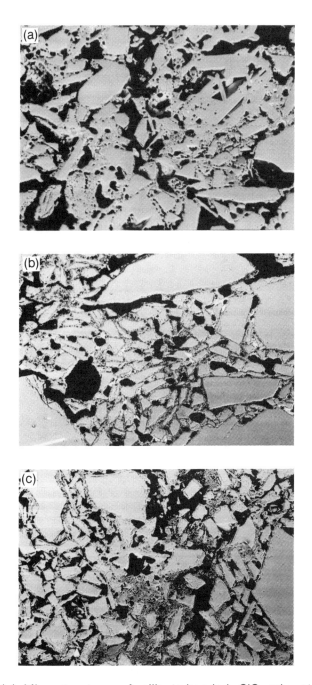

Fig. 10 (a) Microstructure of silicate-bonded SiC mix at 1000°C.
(b) Microstructure of fluid phosphate-bonded SiC mix at 1000°C.
(c) Microstructure of dry phosphate-bonded SiC mix at 1000°C.

silicate-bonded fluid phosphate-bonded, and dry phosphate-bonded specimens at 500°C are compared in Fig. 9. The silicate-bonded mix exhibits the loosest microstructure. The dry phosphate and alumina additives are evident in Fig. 9(c) as dark gray islands interspersed among the fine grains. A similar comparison at 1000°C (Fig. 10) shows a well-developed bond in all three specimens. Although the silicate-bonded specimen (Fig. 10(a)) exhibits more pores, the bond seems more cohesive than for the fluid-bonded specimen (Fig. 10(b)). The rather isolated islands evident in the dry phosphate specimen (Fig. 9(c)) at 500°C have dispersed at 1000°C (Fig. 10(c)) to surround

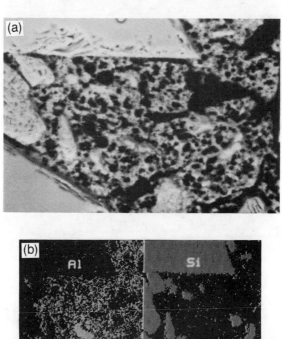

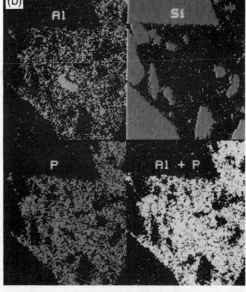

Fig. 11 (a) Fluid phosphate-bonded SiC mix: specimen edge (scanning electron micrograph, BSE image). (b) Element distribution maps (left half) and color-coded video image (right half) corresponding to Fig. 11(a).

and bridge the SiC grains. A comparison of the elemental distributions for the fluid phosphate-bonded specimen (700°C) between the specimen edge and center are presented in Figs. 11(a) and 11(b), and in Figs. 12(a) and 12(b). The left half of Figs. 11(b) and 12(b) show distributions of the elements Al, Si, and P, as well as a composite of the elements Al + P (lower right); in the right half, a color-coded, nearest-neighbor, averaged video image is shown, corresponding to the backscattered electron (BSE) images in Figs. 11(a) and 12(a). A comparison of the P distributions in both micrographs show high concentrations of this element at the outer edge. This phe-

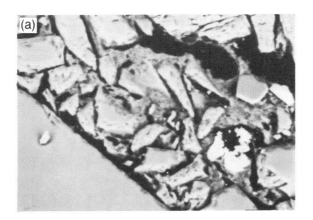

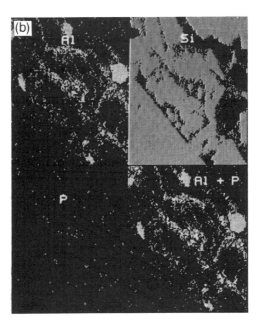

Fig. 12 (a) Fluid phosphate-bonded SiC mix: specimen center (scanning electron micrograph, BSE image). (b) Element distribution maps corresponding to Fig. 12(a).

nomenon of fluid phosphate-bonded mixes is well documented in practice; the fluid phosphate wanders toward the hot outer face leading to high densification near the hot face, with a correspondingly unsatisfactory bond behind this sintered layer. As a result, spalling at this layer can lead to quick destruction of the remaining powdery mix.

A similar comparison of different regions of a dry phosphate-bonded specimen (700 °C) is presented in Figs. 13(*a*)–(*d*) (outer edge) and 14(*a*)–(*d*) (center). The BSE images in Figs. 13(*a*) and 14(*a*) reveal that bonds between the SiC grains are just as good at the center as at the outer edge. A comparison of the P distributions for both micrographs (Figs. 13(*d*) and 14(*d*)) shows very little difference in concentration. At the center (Figs. 13(*a*) and 13(*d*)), P is well distributed between the SiC grains, thus leading to the high strengths documented in Table I.

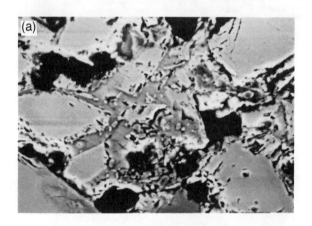

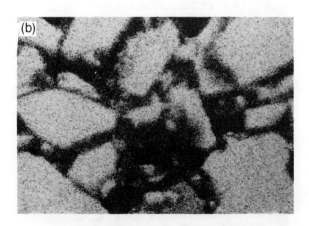

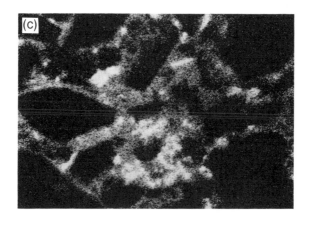

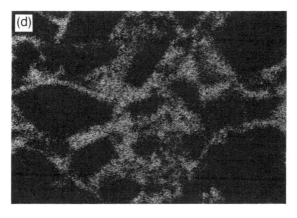

Fig. 13. Dry phosphate-bonded mix: micrograph of specimen edge (a) with corresponding elemental distributions, Si (b), Al (c), and P (d).

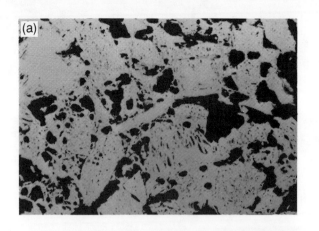

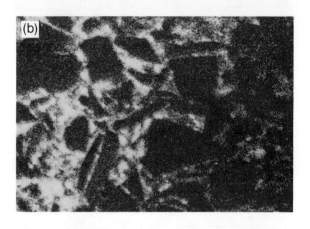

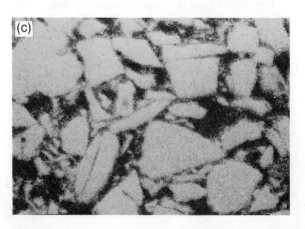

Fig. 14. Dry phosphate-bonded mix: micrograph of specimen center (a) with corresponding elemental distributions, Si (b), Al (c), and P (d).

Table I. Properties of the SiC Ramming Mixes Investigated

Properties	A	B	C	D	E	F
Cold crushing strength in N/mm^2 after firing (°C)						
100	5	24	15	60	50	50
300	14	24	20	65	56	55
500	38	24	25	85	60	60
700	46	38	30	165	72	75
1000	98	39	44	180	116	100
1200	71	42	45		135	110
1400						110
1600						114
Bulk density in g·cm^2 after firing at 1000°C	2.27	2.51	2.54	2.60	2.62	2.58
Apparent porosity (vol%)	24.2	17.2	18.5	17.2	17.0	16.4

Conclusions

The investigations show that strengths of SiC mixes can be sharply increased through the addition of phosphate binders. However, there are large differences between fluid and dry phosphates. At 100°C, dry phosphate-bonded mixes had strengths equivalent to those achieved by fluid phosphate-bonded mixes at 1200°C. For waste incinerators in particular, SiC mixes should have sufficient strength at low temperatures since, due to the quick dissipation of heat through the mix, it is quite likely that the cold face does not achieve temperatures high enough to form a ceramic bond. The boron-based mix, which was compared in this test series with the phosphate-bonded mixes, shows an unsatisfactory strength development in the lower temperature ranges, so that it seems unable to match the requirements for use in waste incinerators (although in practice these types of mixes are used). They show, however, different physical properties due to adapted grain sizes and binder contents. At high temperatures, however, they achieved strengths that were distinctly higher than fluid phosphate-bonded mixes, due largely to the mullitization of the clays.

Practical experience with SiC mixes for waste incinerators has shown that it is essential to have sufficient strength at 100°C (drying temperature) to achieve satisfactory lining lives. The strength should hold over the whole temperature range and also be capable of absorbing short-term overheating. These demands can be met only with the help of a multiple-stage, temperature-dependent bonding system, for example, an organic bond, followed by a phosphate, and finally by a ceramic bond at high temperatures. The silicate bond is effective only at higher temperatures, whereas the phosphate bond begins at low temperatures and is supported by a ceramic bond at high temperatures. Our investigations have shown that dry phosphate-bonded, SiC mixes exhibit a higher level of strength over a relatively wide temperature range and are thus better suited to fulfill the demands made on such materials by the service conditions encounered in waste incinerators.

Regardless of the composition and the binders of SiC mixes it is imperative that the mixes allow easy and economical processing of the material at the site. In other words, the method of introduction should be changeable solely through manipulation of the moisture content. Economic considerations also dictate the almost complete reuse of the rebound. This is easier with dry, ready-to-use mixes than with premoistened ones.

Acknowledgment

We thank the Deutsche Forschungsgemeinschaft for providing the scanning electron microscope with accessories used during this investigation.

References

[1] L. Barniske, "Waste Incinerators in the Federal Republic of Germany (German)," *Muell und Abfall,* 4 111–15 (1975).

[2] E.H.P. Wecht, Refractory Silicon Carbide (German). Springer-Verlag, Vienna, 1977.

[3] E.H.P. Wecht, "The Role of Silicon Carbide for Refractory Linings in Waste Incinerators (German)," XXI International Refractories Colloquium, Aachen, Federal Republic of Germany, (1978).

[4] W.E. Fuchs, "Silicon Carbide Mixes—Their Use in Steam Generators, Waste Incinerators and Blast Furnaces (German)," *Der Maschinenschaden,* 44 205–15 (1971).

[5] F. Pings and H. Lange, "Improved Ramming Mixes for Tube Linings in Waste Incinerators (German)," *VGB Kraftwerkstechnik,* 61 [5] 403–05 (1981).

[6] H. Stein, G. Gelsdorf, and M. Schwalb, "Ceramic Linings of Studded Tubes in Waste Incinerators (German)," *VGB Kraftwerkstechnik,* 4 332–35 (1979).

Development of Spinel-Based Specialties: Mortars to Monoliths

A. Cisar, W. W. Henslee, and G. W. Strother

Dow Chemical U.S.A.
Freeport, TX 77541

Chemically synthesized magnesium aluminate spinel was used to formulate a range of refractory specialties. Controlled powder stoichiometry and surface area offered advantages from variable set times and densities to compatibility with molten nonferrous alloys. Examples include high-strength mortars, coatings, and castables.

A family of magnesium aluminum spinel-based refractory specialties has been developed at The Dow Chemical Co. These acid-bonded systems include high-strength mortars for use with refractory shapes, castables for monolithic applications, and coatings to protect less-resistant materials in corrosive, high-temperature environments. How and why these materials were developed, and the advantages and properties of spinel-based refractories such as superior resistance to attack by molten, nonferrous metals and a neutral salt chemistry that minimizes chemical attack are discussed.

The key to this work is the use of a spinel produced by coprecipitation that can be tailored to the optimum reactivity level for each of a wide variety of uses, only some of which will be covered here. Since our process has been described in detail elsewhere,[1,2] it will only be briefly touched on here. Figure 1 gives a flow chart for this process. All feedstocks are clear, easily handled solutions prepared from industrially available sources. By controlling reactor conditions, a variety of product compositions and consistencies can be produced. Stoichiometry can be precisely controlled to make aluminum oxide- or magnesium oxide-rich materials over a wide range of compositions (e.g.,

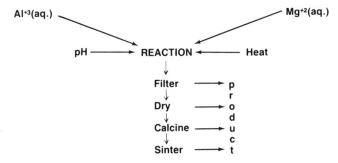

Fig. 1. Synthesis of magnesium aluminum spinel by Dow precipitation process.

Al/Mg from 0.7–3.6) as well as a precisely stoichiometric spinel (Al/Mg = 2.00 ±0.03). As will be shown later, this precise control of product stoichiometry is a valuable tool for controlling refractory properties.

All of these compositions are reproducible, and the precipitation product is homogeneous down to the individual crystallite scale, about 100 nm (1000 Å). Calcination of this homogeneous, fine-grained intermediate yields a single-phase product when the stoichiometry is 2.0 Al/Mg. Most spinel produced by oxide fusion or hydroxide sintering exhibits some residue of one or more unreacted oxides. Even when the aluminum/magnesium ratio is deliberately adjusted away from 2.0, the chemically produced material contains only two phases—never the three sometimes observed in other materials. Optional postprecipitation treatments of the hydroxide precursor material and special filtration conditions offer further controls on the spinel's properties.

Drying and firing of the wet precursor product offer the final control over the product's properties. Fusion or sintering processes produce a coarse aggregate that must be crushed and/or ground to the desired size. In either case, the clinker has a very low surface area which is not altered by grinding. Our process allows the direct production of a fine-grained material of various surface areas. As Fig. 2 shows, by varying time and temperature, the surface area can be changed from over 200 m^2/g to less than 0.5 m^2/g. These surface areas are maintained independently of the state of aggregation of the spinel.

Since most oxide surfaces are hydroxide-terminated, especially those that have not been specially treated or heated to near the oxide melting point,

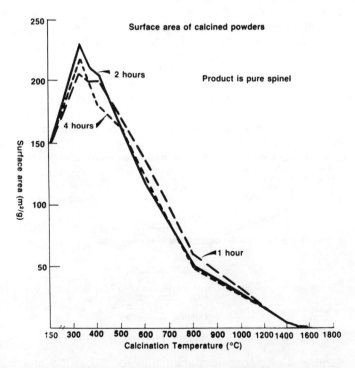

Fig. 2. Surface area of calcined powders (SPP-65; 2:1::Al:Mg).

it is a safe assumption that this spinel has a hydroxide-terminated surface also. Because surface hydroxide groups are far more active than a simple oxide surface, the chemistry of these hydroxide groups will dominate the chemistry of the powder. Although heating to higher temperature will reduce hydroxide concentration on a surface, this same treatment also significantly reduces the surface area, which is the simplest monitor of activity.

Spinel produced by oxide fusion will, by virtue of having been melted, have a very inactive surface (very few hydroxyl groups). Coupled with a low specific surface area and large particle size, these properties make for a material poorly suited for use in the binder phase of a refractory specialty. Efforts to reduce the particle size to something more suitable generally lead to contamination from the grinding equipment and exposes impurities at grain boundaries. While a sintered product will have a higher surface area, it still suffers the same impurity problems on grinding as the fused grain.

Properties and Advantages of Spinel-Based Specialties

A spinel mortar has been developed for use in Dow molten salt electrolysis cells used for producing magnesium. Magnesium electrowinning cells are a severe test for any refractory. In this environment, refractories are alternately exposed to molten salts, molten magnesium, steam, HCl, and thermal cycling. Most commercially available mortars are unable to withstand the attack of molten magnesium and, as a result, mortar selection for these cells is based on longest survival, with corrosive failure of the mortar being a weak link in the system.

The thermodynamics of the failure for both silicate and high-alumina systems are shown below. In both cases, the reason for the failure is the extremely high reducing power of magnesium. Silicate-bonded or silicate-containing systems are the most rapidly attacked, with the silicon produced dissolving in the molten metal. The reaction starts immediately on contact of the silicate material with molten magnesium and continues until all of the mortar is consumed. As would be expected, the rate does fall off as the depth into a relatively narrow mortar joint increases and as exchange with the bulk melt is impeded by distance and the buildup of debris.

$$2Mg + SiO_2 \rightarrow 2MgO + Si$$
$$\Delta H = -68.12 \text{ kcal/mol}$$

$$3Mg + 4Al_2O_3 \rightarrow 3MgAl_2O_4 + 2Al$$
$$\Delta H = -56.7 \text{ kcal/mol}$$
$$\Delta V = +15.8 \text{ cm}^3/\text{mol}$$

High-alumina materials fail similarly. The reaction products of alumina and magnesium are aluminum metal and spinel. Since the latter is resistant to molten magnesium, a passivation effect of the surface being converted to a layer of resistant spinel might be expected, but this is not the case. Conversion of the alumina to spinel so damages the joint that the converted layer is degraded and lost. This reaction is general to all alumina products, although high-density products are destroyed more slowly than alumina-based mortars and castables; it will occur independent of the electrochemistry of the cell as well.[3]

Phosphate-bonded spinel mortars fare far better in this application. In tests with electrolytic cells constructed with phosphate-bonded spinel mortars, the mortar joints last much longer and have better physical integrity at every stage than cells constructed with aluminosilicate products.

In many aluminum applications, such alloying metals as Mg, Zn, Li, and Cu are the most active agents of attack.[4] Besides resisting magnesium attack, preliminary work has shown that spinel components are more stable to molten aluminum alloys, molten brass, and zinc vapor than alumina refractories.

Another major advantage of mortars based on chemically produced spinel is bond strength. By controlling the surface properties of the spinel, the activity of spinel can be adjusted to react with the acid phosphate solution and bond to the phosphate matrix over a large fraction of its surface. At the same time, the activity is not so great that a detrimental amount of water is required to produce a manageable mortar. With this control comes the ability to make mortar joints with more strength than typically found for 85% alumina shapes (see Fig. 3).

Fig. 3. Mortar exceeding the strength of 85% Al_2O_3 brick.

Although a mortar made from pure phosphate-bonded spinel has the most chemical resistance, it is also expensive to produce. To reduce the cost for this mortar, we have found several ways to fill and extend the mortar with inert grain materials, some of which also act to improve its handling characteristics. The primary filler materials investigated have been two forms of alumina. Graded electric furnace-fused alumina produces a grainy texture in the final mortar, but is too expensive for use where its properties are not specifically needed. Tabular alumina has given the best results. In comparison with all spinel formulations, tabular alumina-filled formulations typically have most of the chemical resistance, about the same strength, and improved rheology. A tabular alumina-filled formulation is currently a developmental product. Other fillers have been tested, such as zircon, silicon carbide, mullite, and graphite. The results have been good, especially with zircon, but this work is still at an early stage.

Coatings that are closely related to mortars in composition have also been developed. We currently have formulations that permit the application of a thin, monolithic spinel layer to various materials. The successfully coated materials include alumina and aluminosilicate bricks, high-alumina cast or rammed refractories, refractory fiber boards and vacuum-formed pieces, and steel parts.

Methods of application vary with coatings applied by brush, spray, or trowel, whichever is most appropriate to the specific situation. Results have been quite impressive. An alumina-silica fiber board, which would normally be rapidly attacked and destroyed by molten magnesium, was coated with a layer of phosphate-bonded spinel. After curing, the coating withstood a steady flow of molten magnesium for hours, with no apparent effect.

An excellent example is the protection of a rammed crucible from molten magnesium attack. Troweling a thin layer of phosphate-bonded spinel onto an uncured, phosphate-bonded, high-alumina monolith and then curing the entire unit produces a spinel coating which is integral to the piece. This 0.6-cm coating greatly extends the life of crucibles for holding nonferrous metals. At lower viscosities, phosphate-bonded spinel coatings can be sprayed onto irregular surfaces with excellent results.

Another specialty area where we have had excellent results using chemically produced spinel has been monolithic castables. With our spinel and various acid phosphates as binder phases, virtually any grain material can be used to produce a castable or rammable refractory. A good example is the production of a 95% spinel castable. Any of a number of spinel sources may be used for the bulk aggregate, ranging from the currently commercially available fused or sintered spinel products, which will not bind well on their own, to high-density, irregular aggregate or sintered beads produced from our active spinel. This aggregate is combined with our active fine spinel in a ratio appropriate to the grain used. These ratios vary considerably and are generally dictated by the packing efficiency of the pure aggregate and the desired fluidity in the final product (greater for castables than rammables).

Depending on the coarse aggregate and the final properties desired, the chemical composition of the fine spinel is also varied. By using a magnesium-rich material for part of the fine phase, the setting reactions can be accelerated in a controlled way, ensuring prompt setting and permitting rapid demolding. Because of the coprecipitation process, the exact amount of excess magnesium in the mixture can be accurately controlled, thus furnishing a method for reliably compensating for variations in the composition of fused or sintered materials. An acid phosphate, for example, monoaluminum phosphate, is added to the dry components, and the shape is formed. Our experience has shown that for best results these spinel phosphate-bonded castables should be dried at 50°C for up to 24 h before firing to avoid steam damage. Rapid curing is certainly feasible, at some sacrifice in strength. When fired, strength is partially a function of firing temperature, with higher temperatures giving higher strengths. To date, we have obtained reproducible modulus of rupture (MOR) values to 33.1 MPa (4800 psi) and compressive strengths to 64 MPa (9300 psi) for specimens with densities to 2.88 g/cm^3 (180 lb/ft^3). Maximum service temperature is in excess of 1700 °C.

Another example involves silicon carbide as an aggregate. Using active spinel and monoaluminum phosphate as a binder system, we have produced rammable refractory mixtures with up to 85% silicon carbide. The setting time can be controlled from ½ h at room temperature to days at room temperature (the latter is actually a heat-set material). Strength is good, with MOR values to 10 MPa (1500 psi) and compressive strengths to 36 MPa (5300 psi). If the silicon carbide level is cut to about 70%, the latter rises to 57 MPa (8300 psi). Since silicon carbide is less dense than spinel (3.22 vs 3.58 g/cm^3), density of the cast piece is reduced to about 2.26 g/cm^3 (141 lb/ft^3).

The common theme through all of this work is the binder system produced by active spinel and acid phosphates. Monoaluminum phosphate converts by loss of water to aluminum metaphosphate (Al(PO$_3$)$_3$) on heating, and this metaphosphate has a glassy, polymeric structure which could serve as a binder on its own. One might therefore question whether the spinel is even involved. Figure 4 shows the thermogravimetric curves for both pure monoaluminum phosphate air-dried to a 74% Al(H$_2$PO$_4$)$_3$ paste and a mixture of spinel and monoaluminum phosphate. Two differences are readily apparent. Both transitions occur about 35 °C lower when spinel is present. The final weight loss observed in the pure material is absent when spinel is present, with all losses essentially over by 500 °C. These differences are clear evidence that the spinel is interacting with the aluminum phosphate.

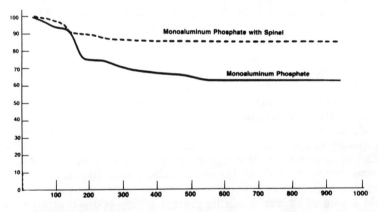

Fig. 4. Gravimetric analysis.

These data are further reinforced by X-ray diffraction data collected on both pure monoaluminum phosphate and monoaluminum phosphate-spinel mixtures. In the case of pure monoaluminum phosphate, a clear step-by-step dehydration takes place through two intermediates to Al(PO$_3$)$_3$. This reaction is complete below 700 °C. Above 1000 °C, phosphorous pentoxide is lost and aluminum orthophosphate (AlPO$_4$, low cristobalite form) is left behind. Continued heating gradually drives off even more phosphorous pentoxide, to leave corundum as an end product. This last reaction is slow, since a small sample was only about one-third converted after 8 h at 1700°C, and an additional 4 h between 1600° and 1700°C.

$$Al(PO_3)_3 \xrightarrow{>1000°C} AlPO_4 + P_2O_5$$

The chemistry is very different when active spinel is present. No crystalline phosphate is detectable unless the firing temperature reaches 500°C. At this temperature, a phase appears with an X-ray pattern that most closely matches that of the tridymite form of aluminum orthophosphate (referred to as pseudotridymite), but the match is not good. By 700°C, a small amount of aluminum metaphosphate appears, but the psuedotridymite still dominates. This situation continues until around 1200°C, when the metaphosphate begins to disappear, and a small amount of the compound designated as MgAlPO$_5$ by Holland and Segnit[5] appears. If heating is carried to 1700°C, all of these phosphates are altered, wth some loss of phosphorous pentoxide to form the cristobalite form of AlPO$_4$, corundum, and a small amount of some as yet unidentified compound.

Table I. Variation of the Lattice Parameter of Spinel in the Binding Phase

Specimen Firing Temperature (°C)	a_0 for Spinel Å (at 20°C)
325	8.084(1)
500	8.083(2)
700	8.074(4)
1000	8.014(4)
1200	8.019(3)
1700	8.056(6)

Further evidence of a unique role for active spinel comes from the variation in the size of the spinel crystal lattice for the spinel participating in the bond phase. Table I shows how the lattice contracts by nearly 1% (19 σ) for specimens fired at 1000° compared with specimens fired at or below 500°C. The latter show no change from either the value obtained for the starting spinel (8.0834(15) Å) or the NBS value (8.0831 Å).[6] At higher firing temperatures, the lattice shrinks less, but even at 1700°C is still significantly smaller. Shrinkage like this indicates a partial removal of magnesium from the spinel.[7]

It is clear that chemically produced, active spinel is an interesting material that can be tailored to a wide variety of applications. When used with acid phosphates, it produces a unique bond phase that can be used to produce a wide variety of specialty products, including mortars, castables, ramming mixes, and coatings.[8]

References

[1]W. W. Henslee, J. S. Lindsey, S. J. Morrow, J. N. Periard, and C. R. Whitworth, "Magnesium Aluminum Spinels," U. S. Pat. No. 4 400 431, August 23, 1983.

[2]R. Smyth, W. W. Henslee, and T. S. Witkowski, "Preparation and Evaluation of Chemically Co-precipitated Spinel Ceramics"; for abstract see *Am. Ceram. Soc. Bull.,* **62** [3] 434 (1983).

[3]W. W. Henslee, S. J. Morrow, J. S. Lindsey, C. P. Christenson, and H. H. Schwantje, "Spinel Surfaced Objects," U.S. Pat. No. 4 382 997, May 10, 1983.

[4]E. M. DeLiso and V. L. Hammersmith, "Testing Refractories for Molten Aluminum Contact," *Am. Ceram. Soc. Bull.,* **62** [7] 804–08 (1983).

[5]A. E. Holland and E. R. Segnit, "Subsolidus Relationships in the Ternary System $MgO\text{-}Al_2O_3\text{-}P_2O_5$," *J. Aus. Ceram. Sec.,* **16** [2] 17–20 (1980).

[6]Nat. Bur. Stand. (U.S.) Monogr. No. 25, Sec. 9, 1971.

[7]W. T. Donlon, T. E. Mitchell, and A. H. Heuer, "Precipitation in Non-Stoichiometric Spinel," *J. Mater. Sci.,* **17** [5] 1389–97 (1982).

[8]W. W. Henslee and G. W. Strother, "Phosphate Bonding of Reactive Spinels for Use as Refractory Materials," U.S. Pat. No. 4 459 156, July 10, 1984.

Author Index

Alder, W. R. Compressive Stress/Strain Measurement of Monolithic Refractories at Elevated Temperatures, 97
Banerjee, S. Low-Moisture Castables: Properties and Applications, 257
Bentsen, S. See Monsen, B.
Bonsall, S. B. Wear Mechanisms in Alumina-Silicon Carbide-Carbon Blast Furnace Trough Refractories, 331
Bradt, R. C. See Homeny, J.
Bray, D. J. Creep of Refractories: Mathematical Modeling, 69
Chaille, C. E. See Richmond, C.
Caprio, J. A. See Kleeb, T. R.
Chadhuri, S. See Krietz, L.
Cisar, A. Development of Spinel-Based Specialties: Mortars to Monoliths, 411
Clavaud, B. A New Generation of Low-Cement Castables, 274
Dannemiller, T. A. See Howe, R. A.
Dietrichs, P. Reactions of Alumina-Rich Ramming Mixes with Lignite Ashes in Reducing Atmospheres at High Temperatures, 365
Dhupia, G. S. SiC Monolithics for Waste Incinerators: Experiences, Problems, and Possible Improvements, 395
Egami, A. See Krietz, L.
Engel, A. J. See Weaver, E. P.
Fentiman, C. H. The Heat Evolution Test for Setting Time of Cements and Castables, 131
Fujimoto, S. See Naruse, Y.
George, C. M. See Fentiman, C. H.
Goerenz, E. See Dhupia, G. S.
Henry, D. K. See Bonsall, S. B.
Henslee, W. W. See Cisar, A.
Hofmann, D. See Krietz, L. P.
Homeny, J. Aggregate Distribution Effects on the Mechanical Properties and Thermal Shock Behavior of Model Monolithic Refractory Systems, 110
Howe, R. A. Designing a Casthouse for Preformed Shapes, 305
Jones, C. M. Comparison of Monolithic Refractories for Blast Furnace Troughs and Runners, 356
Kadota, Y. See Toritani, Y.
Kawakami, T. See Toritani, Y.
Kelley, J. W. See Howe, R. A.
Kiehl, J. P. See Clavaud, B.
Kilgore, R. V. See Banerjee, S.
Kiwaki, S. See Naruse, Y.
Kleeb, T. R. Properties and Service Experience of Organic Fiber-Containing Monoliths, 149
Knowlton, D. A. See Banerjee, S.
Kondo, A. See Nishizawa, S.
Kopanda, J. E. See MacZura, G.
Krietz, L. The Use of Monolithic Refractories in Blast Furnaces, 323
Krietz, L. P. A Review of International Experiences in Plastic Gunning, 165
Krönert, W. Recent Progress in the Use of Monolithic Refractories in Europe, 21
Krönert, W. See Dietrichs, P.
Krönert, W. See Dhupia, G. S.
Kunkel, H.-J. See Stieling, R.
Lankard, D. R. Evolution of Monolithic Refractory Technology in the United States, 46
MacZura, G. Calcium Aluminate Cements for Emerging Castable Technology, 285
Martin, U. See Stieling, R.
Masaryk, J. S. See Alder, W. R.
Matsuo, A. See Morimoto, T.
Mishima, M. See Naruse, Y.
Miyagawa, S. See Morimoto, T.
Monsen, B. Effect of Microsilica on Physical Properties and Mineralogical Composition of Refractory Concretes, 201
Montgomery, R. G. J. See Fentiman, C. H.
Morimoto, T. Introduction of Automatic Gunning Machines for Tundish Linings, 139
Myers, D. M. See Turner, J. L., Jr.
Naruse, Y. Progress of Additives in Monolithic Refractories, 245
Nishijima, I. See Toritani, Y.
Nishizawa, S. Application of Dry-Forming Method to Blast Furnace Troughs, 313
Ogasahara, K. See Morimoto, T.
Padgett, G. C. Test Methods for Monolithic Materials, 81
Palin, F. T. See Padgett, G. C.
Radal, J. F. See Clavaud, B.
Richmond, C. High-Performance Castables for Severe Applications, 230
Rohr, F. J. See MacZura, G.
Sandberg, B. See Monsen, B.
Seltveit, A. See Monsen, B.
Severin, N. W. Dryouts and Heatups of Refractory Monoliths, 192
Shinohara, Y. Recent Progress in Monolithic Refractories Usage in the Japanese Steel Industry, 1
Siegl, W. Viscosity and Gunning of Basic Specialties, 175
Stendera, J. W. The Properties and Applications of Dolomite Ramming Mixes, 388
Stieling, R. Vibrated Castables with a Thixotropic Behavior, 211
Strother, W. See Cisar, A.
Sugita, K. See Shinohara, Y.
Talley, R. W. See Weaver, E. P.
Toritani, Y. Progress in Casting Trough Materials and Installing Techniques for Large Blast Furnaces, 341
Tsukino, M. See Krietz, L. P.
Turner, J. L., Jr., The Development of Dry Refractory Technology in the United States, 161
Weaver, E. P. High-Technology Castables, 219
Wilson, G. See Krietz, L. P.
Woodhead, R. See Krietz, L.
Yamane, T. See Toritani, Y.
Yamasaki, S. See Toritani, Y.
Yaoi, H. See Shinohara, Y.

Subject Index

Abrasion, 222
Aggregates, 110
  chamotte, 249
  fireclay, 201
  fused alumina, 331
  graded refractory, 230
  hard, 276
  sintered alumina, 331
  tabular alumina, 58
    high-purity, 111
  sillimanite, 52
Akermanite, 332
Alumina, 13, 25, 52, 96, 162, 257, 331, 404
  fused, 230
  graded, electric furnace-fused, 414
  high-purity, 110
  tabular, 55, 212, 230, 287, 414
Aluminum oxide, 212
  phosphate, 366
American Refractories Institute, 58
Analogs, mechanical, 72
Anelasticity, 72
Anhydrous aluminate phase, 57
Anorthite, 220, 245, 298, 369
ASTM, 81
Atmosphere, firing, 92
Ball in hand test, 82
Bauxite, 52, 110, 230, 278
Behavior, stress/strain, 97
  thermomechanical, 97
Bentonite, 398
Binder, calcium aluminate cement, 267
  low-temperature ceramic, 164
Blocks, preformed, 13
Boards, insulating, 137
Bond, boric acid, 162
  clay-air, 168
  hydraulic, 161
  mullite, 280
  phosphate, 168
  pseudozeolithic, 28
  wet, green-strength, 161
Bonding, aluminum phosphate, 58
  phosphate, 55, 58
Bosch electric tamper, 162
Boundaries, types, 70
Brick, aluminosilicate, dry-pressed, refractory, 75
  basic, 107
  clay-alumina, 105
  dolomite, 389
    direct-bonded, fired, 391
  fired refractory, 232
  roseki, 11
British Standard Institution, 81
Brucite, 190
Calcination, 412
Calciner, rotary coke, 102
Calcium aluminate, 52
  phosphate, tribasic, 253
Calorimetry, adiabatic, 132
Carbon, 26, 331
  black, 277
Castables, cement-free, 26, 230
  chemically bonded, 214, 219
  chrome ore, 148
  clay-bonded, 219, 254
  conventional, 257
    cement-bonded, low-cement, 219
  dense, 81
  fiber-containing, 156
  fiber-free, 156
  fireclay, 148
  fused-alumina-based, low cement, 278
  generic tabular, 287
  high-alumina, 76, 148, 232, 260
  high-strength, dense, 248
  high-technology, 220
  hydraulically bonded, 131
  insulating, 54, 84
  low-cement, 26, 211, 230, 274, 285, 330
  low-iron, gunning, 325
  low moisture, 257
  magnesia, 163
  prepackaged, proprietary, 60
  proprietary, 147
  reduced-cement, 230
  sintered alumina-based, 153
  thermal shock-resistant, volume-stable, clay-bonded, 311
  ultrafine-bonded, 219
  ultralow-cement, 280
  vibrated, cement-free, 212
    low-cement, 105
Casting, 1, 112, 357
  continuous, 127
  vibration, 211
Catalytic crackers, 243
Caterpillar body, 10
Cement, 110
  alumina, 211
  aluminous, 131, 245
    hydraulic, 230
  calcium aluminate, high-purity, 58, 111
  calcium silicate, 132
  high-alumina, 201
  high-iron, 82
  high-purity, 156
  hydraulic, 47
  intermediate purity, 57
  low-iron, 82
  portland, 52, 131
  Refcon, 57
Chemical compatibility, 223
Chrome ore/periclase, 58
Ciment Fondu, 57
Classification, bulk density, 93
  chemistry, 96
  cold strength, 96
  hot strength, 96
  of unshaped refractories, 93
Clay, 249
  alumina, 107
  calcined, 47, 55, 110
  expanded, 55
  plastic, 47, 55
Coatings, 411
Coke, 365
  low-ash, lignite, 365
  ovens, 272
Compaction, vibration, 84
Compactness, 276
Compressive stress, 98
Concrete, continuously graded, 127
  conventional, 28
  dense refractory, 26
  fireclay, 33
  generic, 127
  high-cement, 105
  refractory, 4, 52
  steel-fiber reinforced, 32
Continuous casting-direct rolling method, 17
Copper, 414
Corrosion, 222
  structural, 386
Corundum, 29, 52, 204, 335, 366, 416
  acicular, 380
  facicular, 383
Crack minimization, 159
  propagation, 126
Creep, 169
  definition of, 69
  deformation, 263, 391
  high-temperature, 69
  metals, 69

421

primary stage, 69, 108
pure oxides, 69
secondary stage, 69, 108
steady-state, 70
tertiary region, 69
Cristobalite, 366
Curing, 112, 157, 415
  field, 305
  oven, 305
Dashpot, 72
Deflocculant, 28, 254
Dehydration, 110, 268
Densification, 406
Density, 195, 223, 257, 285, 321
  bulk, 93, 112, 216, 259, 400
  packing, 285, 314
Design, casthouse, 305
Diaspore, Missouri, 54
Diatomaceous earth, 55
Dicalcium ferrite, 184
  silicate, 184
Dolomite, 388
Drip test, 355
Dry forming, 313
Drying, 112
  microwave, 19
Dry mixing, 112
Dryout, 194
Dust, 321
Elastic modulus, 108, 113
Element, Kelvin, 73
  Maxwell, 73
Energy, fracture, 33
  kinetic, 178
Erosion, 222, 353, 356
European Community, 21
European Federation for Refractories Producers, 81
Exfoliation, 317
Extensometers, 109
Feldspar, 374
Fiber board, alumina-silica, 415
  copolymer, 148
  insulation, 193
  organic copolymer, 147
  polyethylene, 147
  polypropylene, 147
  properties, 158
  stainless steel, 102
Finite element analysis, 97
Firebrick, 6, 19, 55
  insulating, 170, 308
  plastic, 49
Fireclay, 18, 29, 201, 278
  calcined, 202
  zircon, 12
Firing, 112
Flocculation, 247
Fluidity, 343
Fluxes, 26, 296
Fly ash, lignite, 365
Fracture stress, 118
Fuel, gaseous, 6
  liquid, 6
  solid, 6
Furnace, aluminum double-inductor, 42
  induction, 159
  blast oven, 6, 269, 323
  basic oxygen, 175
  bogie-hearth, 32
  channel induction, 160, 162
  control, dynamic, 6
  electric, 58
    coreless, 161
  end-pusher, 39
  ferrosilicon, 243
  hearths, reheat, 240
  induction, 355
  metallurgical, 175
  muffle, 153

open hearth, 58
reheat, 172, 272, 283
resistance-heated, 164
reverberatory, 164, 244
ribbon, 267
rotary hearth, 365
  ring, 42
single-taphole blast, 164
slab reheating, 1
two-taphole blast, 335
walking-beam, 32
  reheat, 43
Gases, waste, 366
Gehlenite, 220, 246, 298, 332
German Refractory Association, 21
Gradients, chemical, 111
  phase, 111
  property, 111
Grain size distribution, 161
Graphite, 415
Grog, 26
  alumina-rich, 26
  crushed-brick, 47
Gunning, dry, 143
  hot, stack, 324
  machine, automatic, 137
  plastic, 37
  pneumatic, 165
  repair, 6
    dry, 6
    flame, 6
    wet, 6
  wet, 143
Guns, rotary, 166
Hardening, 343
Hearths, rammed-chrome, ore-based, 52
  vitrified clay, 46
Heat stability, 224
Heatup, 194
Hex mesh, 37
High-temperature material test system, 108
Hooke's law, 72, 99
Hot metal transfer, 4
Hot wire test, 88
Hydration, 190, 206, 247
Hydroxides, 413
Impact, elastic, 178
  plastic, 178
Incinerators, waste, 395
Interface, air-slag, 356
  slag-iron, 356
Iron-constantan, 132
  oxide, 92, 375
Ironmaking, 2
Insulation board, 14, 15
Kaolin, calcined, 230
Kiln cars, 243
  cement, 242
  direct reduction, 241
  pelletizing, 241
  rotary hearth, 32
Kyanite, 54
Ladle, 1, 7, 163, 272
  aluminum, 244
  hot metal, 4
  steel, 282
  torpedo, 5, 240
Ladle relining, rammming method, 9
  slinger method, 8
  stamping method (LSM), 9
Laminations, 166
Lifter blocks, 243
Limestone, 53
Linings, blast furnace, 324
  trough, 341
Lithium, 414
Loss, rebound, 143
Lumnite, 53, 57
Magnesia, 16, 162

carbon brick, 7
iron-rich, 184
Magnesioferrite, 184
Magnesite, 43, 388
  sodium silicate-bonded, chrome, 389
Magnesium, 413, 414
  chloride (sorel bond), 31
  oxide, 32
  spinel, 332
Materials, natural, fibrous, 147
  straw-shaped, 147
  synthetic, fibrous, 147
  straw-shaped, 147
Mechanisms, boundary, 70
  lattice, 70
Melilite, 369
Microsilica, 201
Millerite, brown, 381
Mixes, moldable, 84
  ramming, 84
Modulus of elasticity, 97,
  rupture, 85, 97, 210, 213, 390, 416
  cold, 259
  hot, 214, 235, 278, 290
Monoaluminum phosphate, 415
Monoliths, cement-bonded, 147
  vibration-cast 105
Montmorillonite, 251
Mortar, ceramic bond-forming, 25
  chemically bonded, 25
  high-strength, 411
  phosphate-bonded spinel, 414
  siliceous sand, 133
Mullite, 71, 162, 268, 366, 415
  cryptocrystalline, 377
Multioperator trials, 82
N-CAST Process, 4
Needle test, Vicat, 287
Notched-beam fracture surface energy, 114, 118
Parallel-wire tests, 88
Particle-size distribution, 298
Particles, ultrafine, 276
Patching mix, phosphate-bonded, 107
Penetration tests, Gilmore, 131
  Vicat, 131
Periclase, 184, 189, 377
Perlite, 54
  expanded, 55
Permanent linear change, 88
Permeability, 147, 195, 276
Perovskite, 369
Phase, eutectic, 384
Phosphates, dry, 409
  fluid, 409
Phosphoric acid, 31
Pig Iron, 245, 356
Pilny-Martienssen cone test, 82
Planetary cooler blends, 243
Plastic, alumina phosphate-bonded, 390
  graphitic, 58
  high-alumina, 58
  wet graphitic, 164
Poisson's ratio, 97, 108, 113
Polymers, inorganic, 31
Porosity, 11, 28, 30, 223, 226, 257, 285, 401
Powder technology, 2
Pressing, 356
Pressure, internal, 147
Pseudotridymite, 417
Quartzite, high-purity, 162
Rabbles, water-cooled, 365
Ramming, 1, 357
  devices, pneumatic, 52
Ratio, continuous casting, 8
  secondary refining, 8
Refractories, air-set-type, 16
  clay-bond castable, 16
  collars, 397
  dry-vibration, 161

fireclay-based, 76
high-alumina, 71, 137
  low-cement, castable, 5
high-corrosion-resistant, 19
lightweight, 193
low-cement castable, 16
magnesia, 137
monolithic, dry-vibration, 355
  low-water, 355
wet rams, 355
plastic, 16, 49, 76, 165, 168, 388
  extruded, 168
  granular, 168
  superduty, 169
Refractoriness, 224, 245, 254, 286
  sag deformation, 298
Reheat, 257
Relining, intermediate, 5
Reynolds number, 178
Resistance, abrasion, 30, 237
  corrosion, 20
  explosion, 140
  slag, 142
  spalling, 51
  thermal shock, 30, 33, 267
Resonance frequency, 98
Retaining rings, 243
Rheology, 2, 176, 181, 251, 298, 415
Rotating drum, 355
  finger test, 355
Rotational viscosimeter, 183
Round Robin tests, 82
Runners, 164
  blast furnace, 282
Rutile, 381
Sand, basic, 9
  Belgian natural, 9
  high-silica, 9
  silica, 55
  -slinger method, 4
  synthetic, 9
  zircon, 9
Scorification, 368
Secant modulus, 99
Sensors, thermal, 131
Shale, 54
  expanded, 55
Shape, particle, 297
  prefabricated, 25
  preformed, 305
Shear modulus, 113
Shinagawa vibration process, 2
Shrinkage, 33, 36, 92, 390
Silica, high-purity, 162
  volatilized, 201
Silicon carbide, 331, 360, 395, 415
  clay-bonded, 26
  nitride, 26
Simulative testing, 91
Sintering, 161
  liquid-phase, 57
Slag, 331, 335
  blast furnace, 55
  control, 6
Sliding, grain-boundary, 71
Soaking pits, 1, 16, 32, 165
Sodium hexametaphosphate, 247
  pyrophosphate, 251
  ultrapolyphosphate, 247
Spalling, 222, 377
  explosive, 254, 267, 343
  mechanical, 226
  steam, 147, 149
  thermal, 226
Specialties, 274
  magnesium aluminum spinel-based, 411
  refractory, 56
Specific heat, 97
Spinel, 369

423

phosphate-bonded, 415
stoichiometric, 412
Spray nozzle, 137
Static test, 355
Steelmaking, 6, 245, 272
Stiffening, premature, 134
Strength, 224, 225 310
  bond, 414
  cold compressive, 259
    crushing, 84, 208, 214, 232, 326, 401
  crushing, 33, 96
  hot, 30, 86, 257, 299
  mechanical, 246, 275
  shape-holding, 316
  tensile, 32
  thermomechanical, 280
  ultimate, 108
  yield, 99, 108
Structural integrity, 310
Support configuration, 309
Temperature, BOF tapping, 8
  gradient, 91
Testing, thermomechanical, 93
Texture, 232
Thermal conductivity, 33, 35, 88, 97, 227
  expansion, 92, 97, 112, 226
  gradient, 111
Thermodynamics, 413
Thixotropy, 2, 180
Titanium, 382

Traprock, 55
Troughs, blast furnace, 313
  iron, 1, 2
  lining, dry-vibrated, 356
  removable, 164
  single-taphole, main, 164
  stationary, 3
  wall, air-cooled, 308
Tundishes, 1, 137, 175, 283
  concast, 240
Turbulence, 310
Vanadium, 398
Vermiculite, 54
  expanded, 55
Vibration, 1
  compaction, 29
  forming, 2
Vibrators, external pneumatic, 163
Viscosity, 176ff.
Visqueen, 193
Welded wire tests, 88
Wet mixing, 112
Work of fracture, 114, 118
Yield value, 180
Young's elastic modulus, 113
Zeta potential, 246
Zinc, 414
Zircon, 10ff., 162, 415
  -roseki, 10ff.

HORACE BARKS
REFERENCE LIBRARY

STOKE-ON-TRENT